International Terrorism and Threats to Security

NEW HORIZONS IN MANAGEMENT

Series Editor: Cary L. Cooper, CBE, *Pro Vice Chancellor (External Relations) and Distinguished Professor of Organizational Psychology and Health, Lancaster University, UK.*

This important series makes a significant contribution to the development of management thought. This field has expanded dramatically in recent years and the series provides an invaluable forum for the publication of high quality work in management science, human resource management, organizational behaviour, marketing, management information systems, operations management, business ethics, strategic management and international management.

The main emphasis of the series is on the development and application of new original ideas. International in its approach, it will include some of the best theoretical and empirical work from both well-established researchers and the new generation of scholars.

Titles in the series include:

Employment of Women in Chinese Cultures
Half the Sky
Cherlyn Granrose

Competing Values Leadership
Creating Value in Organizations
Kim S. Cameron, Robert E. Quinn, Jeff DeGraff and Anjan V. Thakor

Research Companion to Working Time and Work Addiction
Edited by Ronald J. Burke

Happy-Performing Managers
The Impact of Affective Wellbeing and Intrinsic Job Satisfaction in the Workplace
Peter J. Hosie, Peter P. Sevastos and Cary L. Cooper

Women in Leadership and Management
Edited by Duncan McTavish and Karen Miller

Appreciative Inquiry and Knowledge Management
A Social Constructionist Perspective
Tojo Thatchenkery and Dilpreet Chowdhry

Research Companion to the Dysfunctional Workplace
Management Challenges and Symptoms
Edited by Janice Langan-Fox, Cary L. Cooper and Richard J. Klimoski

Research Companion to Emotion in Organizations
Edited by Neal M. Ashkanasy and Cary L. Cooper

International Terrorism and Threats to Security
Managerial and Organizational Challenges
Edited by Ronald J. Burke and Cary L. Cooper

Women on Corporate Boards of Directors
International Research and Practice
Edited by Susan Vinnicombe, Val Singh, Ronald J. Burke, Diana Bilimoria and Morten Huse

Handbook of Managerial Behavior and Occupational Health
Edited by Alexander-Stamatios G. Antoniou, Cary L. Cooper, George P. Chrousos, Charles D. Spielberger and Michael William Eysenck

International Terrorism and Threats to Security

Managerial and Organizational Challenges

Edited by

Ronald J. Burke

Professor of Organizational Behavior, Schulich School of Business, York University, Canada

and

Cary L. Cooper, CBE

Pro Vice Chancellor (External Relations) and Distinguished Professor of Organizational Psychology and Health, Lancaster University, UK

NEW HORIZONS IN MANAGEMENT

Edward Elgar

Cheltenham, UK • Northampton, MA, USA

Published by
Edward Elgar Publishing Limited
The Lypiatts
15 Lansdown Road
Cheltenham
Glos GL50 2JA
UK

Edward Elgar Publishing, Inc.
William Pratt House
9 Dewey Court
Northampton
Massachusetts 01060
USA

A catalogue record for this book
is available from the British Library

Library of Congress Control Number: 2008932866

ISBN 978 1 84720 150 8

Printed and bound in Great Britain by MPG Books Ltd, Bodmin, Cornwall

Contents

PART III BUILDING INDIVIDUAL AND ORGANIZATIONAL RESILIENCE

Contributors

Elizabeth Arwood Pamplin School of Business Administration, University of Portland, OR, USA.

Briana Barocas Center on Violence and Recovery, New York University, USA.

Shlomo Berliner Faculty of Medicine, Tel Aviv University, Israel.

Ronald J. Burke Schulich School of Business, York University, Canada.

Kim Cameron Ross School of Business, University of Michigan, USA.

Antonio Cano-Vindel Department of Psychology, Complutense University of Madrid, Spain.

Cary L. Cooper School of Management, Lancaster University, UK.

Stephanie A. Dopson Terrorism Preparedness and Emergency Response Centers for Disease Control and Prevention, Atlanta, GA, USA.

Brian W. Flynn Department of Psychiatry, Uniformed Services University of the Health Sciences, Bethesda, MD, USA.

Robyn R.M. Gershon Mailman School of Public Health, Columbia University, NY, USA.

Jody Hoffer Gittell Heller School for Social Policy and Management, Brandeis University, MA, USA.

Hector González-Ordi Department of Psychology, Complutense University of Madrid, Spain.

Brian J. Hall Department of Psychology, Kent State University, OH, USA.

Stevan E. Hobfoll Department of Psychology, Kent State University, OH, USA.

Iciar Iruarrizaga Department of Psychology, Complutense University of Madrid, Spain.

Jack N. Kondrasuk Pamplin School of Business Administration, University of Portland, OR, USA.

C. Frederick Lane DBL and Associates, Manhattan Beach, CA, USA.

Sandy Lim School of Economics and Social Sciences, Singapore University, Singapore.

Samuel Melamed Faculty of Medicine, Tel Aviv University, Israel.

Juan Jose Miguel-Tobal Department of Psychology, Complutense University of Madrid, Spain.

Ross H. Pastel Medical and Clinical Psychology, Uniformed Services University of the Health Sciences, Bethesda, MD, USA.

Julie Pearson Mailman School of Public Health, Columbia University, NY, USA.

Kristine A. Qureshi School of Nursing and Dental Hygiene, University of Hawaii at Manoa, Hawaii.

Victor Rivas Budget Department, Massachusetts Bay Transportation Authority, USA.

Itzhak Shapira Faculty of Medicine, Tel Aviv University, Israel.

Arie Shirom Faculty of Management, Tel Aviv University, Israel.

Anat Shoshani Department of Psychology, Tel Aviv University, Israel.

Parbudyal Singh School of Administrative Studies, York University, Canada.

Michelle Slone Department of Psychology, Tel Aviv University, Israel.

John L. Taylor JT Associates, Salisbury, UK.

Sharon Toker Faculty of Management, Tel Aviv University, Israel.

Kristen H. Walter Department of Psychology, Kent State University, OH, USA.

Acknowledgements

I have always been interested in the application of organizational psychology's research findings to improve the lives of individuals inside and outside of the workplace. Although my PhD from the University of Michigan was in organizational psychology and obtained in the Department of Psychology, I have only worked in schools of management since then, believing that these units were more strongly interested in building healthy organizations and healthy people. The attacks of 9/11 in the US raised some new challenges for individuals and their employing organizations, hence this collection. Other incidents since then that have occurred in several other countries have reinforced the importance of this topic for organizational leaders, government policy makers and academic writers and researchers. The world has become a more dangerous place.

I want to acknowledge my association with Cary Cooper. Cary and I have previously worked together in producing other volumes; I continue to value his friendship, energy and talents.

Both Cary and I would like to thank our international contributors for sharing their latest thinking. Gerry Wood, at Lancaster University, was outstanding in working with our authors in coordinating the manuscripts and linking with Edward Elgar. Pulling this off would have been significantly more difficult without her. As always, the staff at Edward Elgar was first rate. The preparation of this collection was supported in part by York University, Canada, and Lancaster University, UK.

Finally, I would like to thank family, friends and colleagues who have supported me during the journey.

Ronald J. Burke
Toronto, Canada

Preface Terrorism, security and the management of organizations: impact, preparation and response

Ronald J. Burke and Cary L. Cooper[1]

On September 11, 2001, at 8:46 a.m., terrorists flew an American Airlines Boeing 767 passenger plane into Tower 1 of the World Trade Center (WTC). The plane hit the north side of the building between the 94th and 98th floors. The plane was traveling at 470 mph and carried 10,000 gallons of fuel. At 9:03 a.m., a second plane flown by terrorists, a United Airlines Boeing 767, crashed into the south side of Tower 2 of the WTC between the 78th and 84th floors. This plane was traveling at 590 mph and also carried 10,000 gallons of fuel. The collapse of Tower 2 occurred 57 minutes after impact, followed by the collapse of Tower 1, 102 minutes following impact (Gershon et al. 2007). Almost 2800 lives were lost and this was the worst terrorist attack in US history; the Pearl Harbor death toll was 2403. Although terrorist attacks had occurred previously in the US and in many other countries, the events of 9/11 have changed many parts of the world forever. The events of 9/11 were unique. The number of deaths was unprecedented, and include those of 343 firefighters who lost their lives responding to the attacks. The terrorists did not need weapons of mass destruction to cause mass casualties and more than $90 billion in losses. The airline, insurance and tourism industries were particularly hard hit. It highlighted the extent of the brutality that fundamentalist terrorists would undertake, and has since caused many countries to spend considerable amounts of money on homeland security. The attacks highlighted the vulnerabilities that existed at the time and still exist. It raised the awareness of the general public of the threat posed by present-day terrorists. And it raised the bar for future attacks as terrorists try to attract media attention.

Furthermore, 9/11 is still killing people. Some individuals have become sick as a result of toxins in the air (asbestos, PCBs (polychlorinated biphenyls), dioxins) immediately following the WTC attacks while others have more recently become sick after an earlier exposure. Two firefighters died in August 2007 battling a blaze in the Deutsche Bank building damaged during 9/11 but only just being demolished. A blaze broke out as a result of careless smoking and the firefighters had little access to water as this was shut off in the building following the terrorist attacks.

Indeed, 9/11 highlighted how risky life in organizations has become. Parcels left unattended now frighten people, close down buildings and bring out the police. White powder on envelopes leads to the evacuation of buildings. Increases in the level of threat announced by the government, coupled with vague threats of possible attacks, increase levels of anxiety.

The time, inconvenience, and sometimes absurdity of airport security checks has increased the stress of traveling, as airlines, many facing insolvency, simultaneously reduce the level and quality of their services. The financial costs of this increased security have been borne mainly by government agencies, often requiring them (or others) to reduce expenditures in other important areas such as education and healthcare. The clean-up after 9/11 alone cost several hundred million dollars.

The 9/11 terrorists may also be winning on the economic front, one of their stated objectives. The powerful US Department of Homeland Security has made it much harder to move goods, services and people across the Canada–US border since then. Canadian productivity is likely to suffer with the potential of Canadian job losses. New passport requirements have increased paperwork and delayed approvals, and computer foul-ups have increased the time needed to cross the border. Pre-clearance initiatives designed to reduce such delays have also been undermined, delayed or stopped by the US Department of Homeland Security. The US has invested millions of dollars since 9/11 on computer programs that search through financial, communication, travel and other personal records of people in the US and around the world for connections to potential terror plots (data mining).

The 2008 US presidential election race is being influenced by how 'tough' the candidates are on national security, a code word for protection against terrorists and whether one is the 'law and order' candidate. Data from Israel (Berribi and Klor, 2006) have shown that support for the right-wing party increased following periods of high terrorism and decreased following periods of relative calm, and terrorism was higher when the left-wing party was in power than when the right-wing party held power.

The attacks had a short-term negative effect on the stock market and economy in the US. Other wars (for example, between the Israelis and the Palestinians, and in Yugoslavia and Iraq) have also had negative effects on stock market performance, at least in the short run. Since 9/11, national governments in some countries have lowered interest rates to spur their economies. This has worked: the economies in many countries have boomed since then. Events since 2002 indicative of this boom include increases in merger activities and in the stock market, short-term increases in the housing market (since fallen in the US), and a shortage of skilled workers.

The events of 9/11 took place almost seven years ago. International terrorism continues to be a threat in many countries. Some experts have argued that the responses to 9/11 reflected in the 'war on terror' have in fact made the world less safe. The threat of terrorism, particularly against the US, will be present for decades. The accumulating evidence shows that terrorist attacks have immediate and negative effects on those directly or indirectly affected by them, and on those who repeatedly watched the events on television. Furthermore, 9/11 has increased the costs of doing business for both organizations and consumers. The immediate negative effects of 9/11 (on the airline industry and on tourism; increased anxiety and depression), however, were short term. Many economies have since rebounded; most individuals and some organizations seem to have coped successfully with the events of 9/11, particularly those having resilience. Some governments, those supporting the war in Iraq in particular, have lost the support of their citizens (Spain, the US, the UK, Australia). In addition, attempts to reduce terrorism through the use of force do not seem to have worked too well. Indeed, these efforts often seem to create more terrorists, and it is clear that citizens in many countries do not feel safer as a result of efforts by their govenments to reduce terrorism; an increasing number expect to be attacked again by terrorists. On the sixth anniversary of 9/11, the *Toronto Star* conducted an on-line poll of its readers on the question of whether global anti-terror measures made them feel safer. Almost 80 percent said 'no'.

There are a considerable number of incidents that contribute to these feelings of not being any safer. Consider the following:

- German police arrested three people in an attempted bomb plot against Americans (September 6, 2007).
- A Serbian man was arrested after attempting to enter the American Embassy in Vienna (October 1, 2007).
- Authorities in Denmark arrested several people in Copenhagen on September 4, 2007, charging them with plotting terrorist attacks.
- The Pakistani military attacked the Red Mosque in Islamabad where there were some Islamic extremists; dozens of people died.
- British police arrested five people in connection with plots to attack airports in London and Scotland. Two cars packed with explosives were found in London. Two men rammed a flaming jeep into the main entrance of Glasgow airport.

- American authorities averted an attack on the Army base in Fort Dix, New Jersey (May, 2007).
- American authorities stopped the potential detonation of explosive devices at JFK International Airport in New York City (June, 2007).
- Global terrorism is present in more than 50 nations (Korten, 2006).
- Terrorism is a widespread problem throughout the world, occurring in 54 countries from May 2003 to May 2004 and taking place on all but one continent (Antarctica).
- US diplomats abroad have been subjected to unprecedented security restrictions, with several being confined to fortress-like compounds incorporating both travel restrictions and lockdowns.

CHANGES SINCE 9/11

The *Globe and Mail* (2006) chronicled some effects of 9/11 five years after the attacks. Some of their findings are as follows:

- 2 number of days that mail addressed to the US government is delayed so it can be irradiated for biohazards.
- 39 number of organizations, including their aliases, that Canada considers to the terrorist groups. Before 9/11, Canada had no such list of its own.
- 11 number of countries added to the list whose citizens now need passports to visit Canada.
- 2 the length of waiting time to receive a passport in Canada has doubled.
- 75 percent percentage increase in Canadian defense spending from 2001 to 2010.
- 10 percent the drop in business travel from Canada to the US.
- 66 percent percentage of respondents in a US poll indicating that people who look 'Middle Eastern' should be singled out for security screening at airports and train stations.
- 12 percent percentage of Turks having a favorable impression of the US, down from 53 percent in 2000.
- 10 percent the drop in earnings of Muslim and ethnically Arab men working in the US since 9/11. Some Arab and Muslim Americans have changed their name because of stereotypes, discrimination and prejudice. Thus Isam Abu Zaid became Sam Paul St. Germain.
- 391:1 ratio of pre-terrorism websites compared with 1998.

- 15 percent of the 48 governments that originally joined the 'coalition of the willing' in Iraq, seven have since been defeated at the polls (Dominican Republic, Hungary, Italy, Norway, Slovakia, Spain and the Ukraine). In addition, the popularity of Tony Blair and George Bush reached almost record lows. Blair has since left office and Bush's Republican party lost control of both the House and Senate.
- 300 percent in the five years before 9/11, 8309 deaths were attributed to terrorism; after 9/11, the figures for five years rose to 26,620.
- 50 percent security hassles in crossing the US–Canada border have resulted in half as many US visitors entering Canada for one-day visits.

WHAT IS TERRORISM?

Terrorism is an intentional act designed to intimidate or coerce a civilian population, usually accomplished through an action that evokes fear and stress in others than those directly involved in the terrorist event. Terrorism is a psycho-social attack on people, and as a political tool it has a long history.

Interestingly there is more terrorism in the more democratic countries than in non-democratic countries. Democracies are more vulnerable to terrorism both from inside and outside the country (WTC, bombings in Madrid and London, the Oklahoma City bombing). This results from terrorists' dislike of the democratic countries and what they are seen to stand for, and under-reporting of terrorist attacks in the non-democratic countries. In addition, the terrorists' seeking of publicity and the freedom of the press in more democratic countries makes such countries valued targets.

Violence and terror have come to be central products in the media; terrorists and the media have come to need each other. The general public also seems to prefer media coverage of violence and terrorism. The events of 9/11 seemed to be repeated on all major US networks non-stop for days following the attack. The news media have learned that focusing on terrorism and terrorist attacks attracts viewers, so they are now continuously bombarded with stories and pictures of death and destruction.

A terrorism database in the US, as of June 2006, contained data on over 28,445 terrorist events in more than 5000 cities across 187 countries worldwide since 1968. Some 10,878 of them resulted in at least one person being killed or injured. The frequency of major terrorist events had not changed over this period, but the frequency of less severe attacks had increased (Clauset et al., 2007). Severe terrorist attacks, in contrast to smaller attacks, get significantly more media attention, and terrorists seek publicity. More severe attacks tend to occur in industrialized countries, while more frequent terrorist events occur in non-industrialized countries. More severe attacks also have a greater psychological, financial and economic impact. Furthermore, they also suggest that the terrorists have more resources, skill and persistence. In addition, governments of countries receiving severe attacks can also increase the magnitude of their responses. Governments can use the 'stick' – tough counter-terrorism policies and measures – or the 'carrot' – more accommodating policies and actions – to reduce the terrorists' base of support. There is some suggestion that heavy-handed government actions may increase the levels of terrorism, however.

WHAT ARE THE CAUSES OF TERRORISM?

A few different causes of terrorism have been proposed. One states that terrorists hate the US for their use of military and economic power to 'oppress and humiliate other nations and peoples' (Korten, 2006, p. 245). This allows the targets of terrorism to shift the focus away from the real causes of this hatred, which may include economic injustices and poverty. Economic conditions such as poverty and income inequality influence terrorism by increasing feelings of deprivation, injustice and political tension. Thus promoting economic growth and reducing inequality through economic aid and the development of welfare programs may reduce these seeds of discontent. Countries with more generous social welfare policies and practices have had fewer terrorist incidents on their soil as well as fewer terrorist acts perpetrated by their citizens. These findings suggest that strengthening of social policies at home and abroad may reduce terrorist violence.

Another view states that terrorism results from a 'clash of civilizations'. Terrorists come from countries that have an ideology that runs counter to the capitalist and materialistic values of the West. Most of the 19 hijackers involved in 9/11 came from middle-class families and were reasonably well educated; they were not economically deprived. There are some terrorists, primarily Islamic radicals, who hate the 'infidels' and what the West represents and are actively pursuing Jihad. These individuals, hopefully few in number, will not be influenced by Western efforts to reduce levels of economic hardship.

International cooperation is needed to bring terrorists to justice and prevent future attacks. Unfortunately several countries, while paying lip-service to the need for international cooperation and information sharing, perform half-heartedly (for example, Saudi Arabia, Syria and Pakistan). Launching unilateral pre-emptive strikes against weak nations fails because it increases the number of terrorists, uses military resources in no-win conflicts, undercuts international trust and cooperation, and reduces the moral authority of the invading country.

TERRORISM AND ORGANIZATIONS

Organizations and their workforces are attractive targets for terrorists (Inness and Barling, 2005). Many people are present at particular times of the day, attacking organizations and their employees receives immediate media attention, and attacking organizations represents an attack on an ideology (for example, progress, capitalism and materialism) that is different from that of the terrorists. Examples of workplaces that have been attacked include the WTC, government embassies, government buildings, military buildings such as the Pentagon, and an army barracks in Saudi Arabia. Attacks have also occurred on transportation infrastructure where people congregate, such as airplanes, trains, buses and subways. The targets of terrorist attacks are not random; most require considerable long-term planning and resources. Individuals apprehended before they were able to carry out a potential terrorist attack often had taken pictures and videotapes of potential targets. Some of the individual hijackers on 9/11 had spent months in the US taking flying lessons.

WHY ARE POTENTIAL OR ACTUAL TERRORIST ATTACKS STRESSFUL?

The vast literature on occupational stressors and their effects offers some suggestions as to why terrorism has such a negative impact on individuals (ibid.). The terrorist event or attack cannot be controlled, has an external cause, is more detrimental to psychological health because it is done deliberately to cause harm, is undertaken by other people who are evil, is difficult to predict when the event has ended, and is associated with an unknown future. Actual terrorist attacks causing death, destruction and loss, impact directly on victims, those in close proximity to the attacks, those who have personally witnessed the attacks, and those losing friends and family in the attacks. Furthermore, the size of the disaster is likely to have an influence on the magnitude of its effects on individuals.

EFFECTS OF TERRORISM ON INDIVIDUALS

Terrorist attacks have been found to have the following effects on individuals: increases in psychological health symptoms such as anxiety, depression and fear; post-traumatic stress disorder (PTSD); inability to concentrate; sleep disturbances; increased alcohol consumption; feelings of helplessness; and negative attitudes towards 'outsiders'. We shall review only a sample of these research findings, as some of the chapters that follow consider them in more detail.

A 2004 survey of New York City (NYC) residents showed that:

- 56 percent feared they would be the victims of a terrorist attack;
- 35 percent felt less safe than two years ago;
- 23 percent seriously considered relocating from NYC;
- 48 percent experienced flashbacks of 9/11 during the August 2003 blackout; and
- 23 percent were now more vigilant, careful, and alert.

A survey of NYC commuters revealed the following:

- 55 percent feared an attack on a subway or train;
- 52 percent feared an attack on a ferry; and
- 35 percent feared an attack on a bus.

Galea et al. (2003) reported in a study of NYC residents after 9/11 that:

- 35 percent thought about 9/11 almost every day;
- 35 percent reported anxiety, depression or mental stress from 9/11; and
- 20 percent of those living below Canal Street (closer to WTC) reported PTSD symptoms while 7.5 percent living between 110th and Canal Street (further from WTC) reported PTSD symptoms.

Stein et al. (2004) studied psychological and behavioral reactions following 9/11 in a nationally representative sample of 560 adults, resurveying them about two months later (n = 395). Sixteen percent had persistent distress at both measurement periods. Adults with persistent distress reported accomplishing less at work (65 percent), avoiding public gathering places (24 percent), using alcohol or other medications or drugs to relax, sleep, or feel better about terrorism (38 percent). Seventy-five percent talked with family or friends but 43 percent reported being unable to share their terrorism-related thoughts and feelings with others as it made others uncomfortable. Few reported receiving counseling or information about psychological distress from general medical providers (11 percent).

Felton (2004) reviewed what we have learned since 9/11 about the impact on mental health of the attacks, and how best to respond to victims and the general public. Rates of PTSD and depression were heightened immediately after 9/11 but had subsided since then. Factors associated with elevated PTSD and depression were close proximity to the attacks, low social support, more life stressors prior to 9/11, major loss due to the attacks (for example, death of a friend, loss of a job or possessions), watching the attacks on television with high frequency, and experiencing a panic attack during or immediately after the attack. Individuals far away from the attack could also be affected by it. Felton reports on the experiences of Project Liberty, a mental health service provided to NYC citizens. Over 400 disaster mental health workers participated as a source of help and support: over 1.2 million New Yorkers received one or more face-to-face counseling sessions or public education services. Many users of these services indicated that their daily functioning had returned to pre-9/11 levels.

There is some evidence that alcohol consumption may have increased shortly after 9/11 (Vlahov et al., 2002, 2004a, 2004b). Others (for example, Knudsen et al., 2005), found no increase in alcohol consumption in their studies, however.

Studies have shown that merely watching the terrorist attacks repeatedly on television had a negative effect on psychological health. Gil-Rivas et al. (2004) found an increase in four acute stress symptoms in a sample of 142 adolescents indirectly exposed to the terrorist attacks through the media one year later. Parental support, positive affect and parenting self-efficacy (using data obtained separately from parents) were associated with greater positive affect in adolescents, suggesting that parents play an important role in adolescents' responses to traumatic national events. Others have also demonstrated that watching the attacks on the WTC on television had a negative effect on children and youth (Ahern et al., 2002; Gershoff and Aber, 2004; Gould et al., 2004). Watching people falling or jumping from the WTC towers had immediate short-term negative effects.

WITNESSING THE UNTHINKABLE

Shalev (2004) observed that some individuals, not under threat themselves but witnessing body parts or bodies burned beyond recognition, may experience more long-term distress than survivors of an attack. This notion might explain the reactions of those watching events on television or first responders such as police, firefighters and emergency service personnel.

LONG-TERM EFFECTS

It is likely that some survivors of 9/11 and other terrorist attacks, some first responders, and some soldiers fighting terrorism in Iraq, Afghanistan, Pakistan and other countries will show signs of PTSDs. The best research on US veterans of the Vietnam War (Dohrenwend et al., 2006) found that

19 percent had developed war-related PTSD during their lifetime and 9 percent were suffering from PTSD 11 to 12 years after the war. Hoge et al. (2006) reported that 19 percent of those serving in Iraq and 11 percent of those serving in Afghanistan reported symptoms of PTSD or other mental health problems. It has been estimated that 89 percent of those serving on these two fronts had been attacked or ambushed, and 95 percent indicated seeing dead bodies or human remains.

But there is some good news in several of the studies of survivors and witnesses of terrorist attacks. Most return to a life that was similar to their pre-event life. Bonanno (2004, 2005) has found that the vast majority of adults return to relatively stable patterns of healthy functioning along with a lasting capacity for positive emotions and growth. Interestingly, there were many ways to be resilient and some of these were not necessarily adaptive under normal circumstances. In conclusion, there is evidence of immediate negative effects on individuals affected directly or indirectly by a terrorist attack, but these seem to dissipate over time.

WHAT CAN BE DONE TO ASSIST INDIVIDUALS?

We can prepare employees of organizations for possible terrorist attacks by providing information resources and building evacuation routes. We can build better and safer buildings and retrofit older buildings to better withstand attacks. We can better prepare first responders with improved equipment, resources and counseling support. We can improve individual and organizational coping responses such as education, access to employee assistance programs (EAPs), communication systems in organizations and improved team functioning. And we can help individuals by offering education and training for improved coping and resilience. Finally, providing help and support, and encouraging employees to ask for help and support, would yield huge dividends to them.

EFFECTS OF TERRORISM ON ORGANIZATIONS

Terrorist attacks have been found to have some effects on organizations. These include psychological distress of employees and their families, disorganization of the workplace, unclear roles and responsibilities, more absenteeism and turnover, inability to concentrate on the job, and lower performance. These are in addition to the direct and indirect financial costs associated with death benefits, workplace injuries, counseling of staff, and the replacement of buildings and equipment (Inness and Barling, 2005).

FACILITATING ORGANIZATIONAL RECOVERY

The fire created by thousands of gallons of jet fuel pulled so much oxygen out of the air around the WTC that windows in nearby buildings blew out as the towers collapsed. The offices of Barron's Magazine, located on the 32nd floor of the World Financial Center shook and supplies and equipment flew out of the windows. Barron's employees left the building for their own safety. It took over a year to rebuild and refurnish their offices, but as employees hurried out of their building, they were already thinking about publishing the magazine on time. In fact a complete edition appeared only three days later. Not publishing it was never even considered.

Gittell et al. (2006), in a study of the airline industry following 9/11, found that some airlines rebounded faster and further than others. These differences were a function of several pre-9/11 factors: financial reserves, nature of their business model, and positive long-term employee relationships. Positive employee relationships supported the introduction of coping efforts by particular airlines (for example, Southwest). It is clear that some organizations rebounded quickly after 9/11 while others did not. Leadership quality, levels of organizational support and the health of the human system prior to the attack emerged as key elements here.

Sheffi (2005) considered the ways that organizations could recover from major disruptions, which included lowering their vulnerability and increasing their resilience. Organizations today are more vulnerable than ever. This results from the globalization of business, increasing dependence and interdependence with other organizations, longer supply lines, just-in-time (JIT) processes, increasing competition and higher levels of uncertainty across the board. Organizations need to focus on increasing both security (lowering the probability of an attack) and resilience (increasing resources to rebound quickly). Increasing security involves heightened defenses, including all employees in security initiatives, and increasing communication across various people, companies and agencies. Resilience involves creating redundancy in systems, developing back-up systems, allowing rapid movement of people, information and products from those parts of the organization that are still functioning, and developing a 'can-do' organizational culture that will get the job done under any circumstances. Argenti (2002) highlights the importance of putting the employee's interests above everything else in speeding up recovery.

Companies that rebounded quickly received tangible support from suppliers, competitors, regulators and customers following the attacks of 9/11.

Unfortunately, most companies are still conducting 'business as usual'. Since 9/11, few have thought about their vulnerabilities and changed in ways to make them better prepared, more secure and more resilient. Organizational preparation, according to Sheffi (2005) involves the following:

- building the business case for security;
- organizing for both preparation and resilience;
- prioritizing threats and vulnerabilities;
- reducing the probability of disruption;
- collaborating with other organizations, including government agencies, in sharing information;
- creating operational redundancies;
- strengthening relationships with suppliers and customers;
- increasing training so that employees can perform multiple jobs;
- developing and practicing emergency responses (for example, building evacuation routes); and
- developing a culture that emphasizes communication, teamwork, passion, performance, flexibility and getting the job done regardless of circumstances.

Most resilient organizations have cultures that have the following aspects (ibid.):

- an emphasis on results;
- teamwork and collaboration;
- informal networks and personal relationships;
- leadership at all levels;
- high levels of employee participation;
- employees that are engaged in their work;
- lots of communication; and
- an emphasis on innovation and flexibility.

Bachrach and Bamberger (2007), in a sample of NYC firefighters (n = 1600), examined both involvement in the terrorist incident and fire unit functions such as control and supervisor support climate on post-event negative emotional states. Data were collected almost 18 months after 9/11. They found that post-incident unit support and control climate had important protective functions for these first responders. Interestingly, the human system before the terrorist attacks contributed to the resilience of survivors of the attack, their organizations, and first responders.

WHY DON'T ORGANIZATIONS PREPARE?

Crisis management is 'an organization's pre-established activities and guidelines for preparing and responding to significant catastrophic events or incidents in a safe and effective manner' (Lockwood, 2005, p. 2). Lockwood contends that the human resources function should have a significant role in crisis management, given its responsibility for people, talent management, leadership development, safety and security, and communication planning. Lockwood reports that a recent Society for Human Resources Management survey found that as a result of 9/11, 56 percent of organizations created or revised their disaster preparedness plans, but 45 percent had not. Why don't organizations develop these plans? The Society offered the following reasons: organizations believe that it won't happen to them; other priorities are more important than crisis preparedness; they are unaware of risks to their organizations; and they ignore already available warning signs, and then rely on a weak and untested plan already available.

Mitroff and Alpsalan (2003) defined companies as crisis prepared ('proactive') or crisis prone ('reactive'). They examined several Fortune 500 companies and concluded that between 5 and 25 percent were crisis prepared. They found in their research that being prepared for a crisis had several benefits. Proactive companies experienced fewer crises, they stayed in business longer, they performed better financially, and they had better corporate reputations. Mitroff and Alpsalan identified three types of crisis: natural accidents (for example, fires, earthquakes), normal accidents (for example, *Exxon Valdez* oil spill, Three Mile Island nuclear plant), and intentional accidents (for example, deliberate evil actions such as terrorist bombings, kidnappings, and cyber attacks). Companies know how to deal with safety planning risk assessment and hazards – the first type of crisis; they can also prepare for the second type of crisis by obtaining information early from key actors; it is, however, harder for them to come to grips with the third type of crisis as it is hard to imagine just how evil some people can be.

Can organizations anticipate or predict crisis? Bazerman and Watkins (2004, p. 1) define a predictable surprise as 'an event or set of events that take an individual or group by surprise, despite awareness of all the information necessary to anticipate the events and their consequences'. They see 9/11 as reflecting an avoidable failure of leadership – it could have been anticipated or predicted – and offer suggestions on ways to improve recognition of impending crises and how to work toward preventing them. They first identify reasons why organizations do not see looming crises, covering some of the same ground as Lockwood (2005):

- leaders know that a problem or crisis exists that will not solve itself;
- the probability of a real disaster is increasing;
- solutions involve trade-offs between immediate costs and future benefits;
- the costs are certain, the benefits uncertain;
- the status quo seems more attractive than preparing for the uncertain future; and
- there is a vocal minority that benefits from the status quo.

As a solution they propose their RPM model involving three steps: enhance the Recognition of ongoing problems, Prioritizing problems so that one addresses and solves the most urgent, and Mobilizing support to head off the danger. Each step is then described in some detail with examples of how they would apply to actual events.

DEVELOPING ORGANIZATIONAL RESILIENCE BEFORE AND FOLLOWING A TERRORIST ATTACK

What do leaders and managers need to do to support recovery? Managers need to rebuild hope in people as employees lose a sense of hope in disasters such as 9/11. Employees turn to their leaders in times of crisis, and at such times everything gets magnified. How leaders behave is observed and remembered. Leaders need to instill trust by being trustworthy; this involves keeping promises, being open and honest, treating others with respect and dignity, caring for employees – the victims and survivors – and doing their share. Employees should be encouraged to seek help in times of need. People need help in times of trauma, yet most are reluctant to seek it. Leaders are particularly reluctant to seek help though they need it as much as anyone else, if not more, and employees look to their leaders for clues on how to respond. Organizational values are critical to healing; central here are valuing all employees as whole human beings, seeing the organization as a community, expressing a range of emotions during the tragedy (Mainiero and Gibson, 2003), and showing that the organization cares.

People are obviously confused following a terrorist attack. What is going on? How will we get through this? Did everyone survive? Managers can address these concerns by providing certainty about their job and their place in the organization, by maintaining personal connections to employees through being visible and accessible, by listening to staff, allowing for the expression of employee pain and suffering (ibid., 2003), focusing on the organization as a community, and indicating what the organization is doing to address problems and support high performance.

The routines of getting back to work as soon as possible are critical to recovery. Regular back-to-work routines create a sense of activity, purpose and effectiveness. Employees can get called to meetings to foster a sense of community. Regular communication channels can support help seeking and help giving. Formal and informal networks provide help, ideas, co-ordination, and information. Some employees behave courageously in times of crisis, thus fostering courage in others who are grappling with the same circumstances. Others feel inspired after seeing courageous acts by their leaders or fellow employees. Finally, organizations can provide employee assistance and counseling programs involving the family, an emergency response plan to speed up recovery, safety training, information on building evacuation routes, back-up communication and data systems and on-site street debriefing (Inness and Barling, 2005).

PREPARING FOR A TERRORIST BOMBING

Bombs are the most common method used by terrorists to create fear and inflict harm. About 70 percent of the terrorist attacks in the US between 1980 and 2001 were the result of bombings. There are some common-sense things to do in preparing for a terrorist bombing. These include: knowing your work, school or community disaster plan, identifying an alternative hospital within reasonable distance but not necessarily the closest to your work or school. If you think someone is going to set off a bomb, follow existing evacuation guidelines. If at work or school, leave the area immediately. If at home or in a public place, continue to follow existing evacuation guidelines that are in place. If at work or school, call 911 in the US, and the same if at home or in a public place. Then follow directions from people in authority such as the police, firefighters, emergency staff, first responders, the military or your company's supervisors.

We now have a greater understanding of building evacuation preparedness (Gershon et al., 2007). Evacuation leaders should be identified and trained for buildings 10 floors or higher, plans for assisting the disabled and disseminating information should be developed, and all employees should be knowledgeable about their building and aware of the exits.

MANAGING THE TERRORIST EVENT

Perliger et al. (2005), using analyses of suicide bombings by terrorists in Israel suggest a defensive method of dealing with such attacks in addition to the active defensive methods currently being advocated. How should authorities manage a terror event? Three factors appeared to be significant in a well-managed terrorist event: the clarity of the initial description of the event, internal coordination among units operating on the scene and external coordination among municipal agencies.

SECURITY IN THE GLOBAL BUSINESS ENVIRONMENT

Companies with a global workforce are paying greater attention to the risks faced by their employees, their liability, and ways of enhancing security practices. Some countries are unstable; others are in the process of becoming unstable. Companies need to know recent developments, where their employees are located, and to have an evacuation plan ready. Companies may also need to think about reducing international travel and overseas assignments. Problems such as social unrest, military conflicts, terrorist attacks and kidnappings exist. Companies continue to have employees in dangerous places such as the Middle East, South East Asia, and South and Central America.

There is some evidence that expatriates want more support and information from their head offices as they undertake assignments abroad. This should include security bulletins, contingency plans and emergency guidelines. Most expatriates indicated, however, that they would complete their current assignments and accept other expatriate assignments.

THE PRESENT VOLUME

This collection is about the impact of terrorism and terrorist attacks on employees of organizations, the leaders and management of these organizations and the organizations themselves. The events of 9/11 impacted on people, organizations and the economy as a whole. These effects continue today and concerns about future terrorist attacks will be with us for a long time. Some people and organizations showed resilience and returned to previous levels of functioning relatively quickly, others failed to do so. Reflecting the international dimensions of the topic, our authors come from five countries, four of which have suffered terrorist attacks in the past few years and all have had suspected or convicted terrorists within them.

Several books tackle terrorism at the political and cultural levels, its root causes and possible solutions, but relatively few focus on the organizational level. Terrorism is most likely to target organizations and the women and men who work in them to create maximum destruction and media attention. In addition, organizations, both in the private and public sectors, need to anticipate and prepare for potential attacks and respond if and when they are attacked. This collection focuses on the women and men who work in organizations, and those responsible for assisting them in times of emergency and terrorist crisis.

OVERVIEW OF THE CONTENTS

This volume is divided into three parts. Part I, containing three chapters, serves as an Introduction. In the first chapter, Ronald Burke sets the stage for those that follow. He begins by providing a brief overview of 9/11 on world events since these terrorist attacks, the effects of terrorist attacks and other disasters, both man-made and natural, on people in general, the impact of 9/11 on human resource management practices and organizations, individual coping and resilience, anticipating terrorist attacks and other disasters, improving responses to such attacks and disasters, and the implications of 9/11 and related terrorist events over the past seven years for organizations, their employees and their management. Terrorism is certain to be a worldwide threat and reality for years to come.

In Chapter 2, Jack Kondrasuk and Elizabeth Arwood offer a US analysis of terrorism. They first define terrorism, appreciating the existence of various definitions. They then provide a history of terrorism regarding the US: pre-9/11, 9/11, and post-9/11. Although the US had experience with terrorism previously, 9/11 dramatically changed the way the US now viewed it. The US is the number one terrorist target in the world. In response, the US has embarked on a war against terrorism. It has identified a small number of states it believes sponsor terrorism, terrorist organizations, individual terrorists and weapons used by terrorists. Kondrasuk and Arwood then review the effects of terrorism on US society in general and on the US government. They sketch out the future of terrorism in the US, concluding that it will last a long time. Changing the people likely to undertake acts of terror is seen as a promising avenue, but this is also a long-term strategy.

John Taylor (Chapter 3) writes about governments as actors in terrorism. He observes that governments are the targets of terrorism, the agents of terrorism and responsible for protecting their citizens from terrorism. Terrorism, at its root, results from anger, frustration, or resentments; government actions can cause terrorism, and harsh responses by governments to terrorist attacks may promote further violence. As a consequence; governments become the targets of terrorism. For terrorists, the government, business institutions, and innocent civilians blur into legitimate targets. Furthermore, governments create departments (for example, US

Department of Homeland Security) to both protect their citizens and fight against terrorism (the US war on terror). Taylor reviews the UK strategy which differs in significant ways from the US approach: the US sees the causes of terrorism as external; the UK sees the causes as internal. Defeating terrorism ultimately involves addressing its roots, and governments are the major actors in this effort.

Part II contains four chapters, and considers the impacts of terrorism. In Chapter 4, Kristen Walter, Brian Hall and Stevan Hobfoll use conservation of resources (COR) theory to understand the influence of mass casualty on individuals, groups and organizations. The key proposition in COR theory is that individuals and groups try to obtain, retain and protect the things they most value, termed 'resources'. Stress results then from the actual or threatened loss of resources or failure to gain resources following substantial investment. Resources, for them, are objective. When individuals or groups experience a traumatic event such as a terrorist attack and mass casualty, a loss of resources occurs. Managerial decision making is also affected by disasters such as terrorist attacks. These effects can be negative in many cases (for example, inaction, hostility to out-groups, risk aversions), influencing responses to these attacks. They then offer ways to best respond in these situations. These include promoting a sense of safety, calming, developing a sense of self- and collective-efficacy, connectedness and hope. Specific interventions are illustrated to provide concrete examples.

Ross Pastel (Chapter 5) focuses on the psycho-social impacts of biological warfare and bioterrorism following 9/11. Both use micro organisms or toxins to induce death and disease. These are used to attack a society by inducing fear and vulnerability among the population; the sending of anthrax through the US mail in October 2001, causing death, brought this to the forefront. The psycho-social impact of bioterrorism is a function of risk perception which is influenced by risk communication by leaders and information provided by the news media. Pastel reviews various bioterrorism agents associated with possible bioterrorist scenarios. Risk communication guidelines are reviewed and discussed. Bioterrorist attacks affect three groups: the community, healthcare workers, and patients and their families. Pastel then considers the effects of bioterrorism on each. Finally, he lays out some thoughts on how to best manage the effects of such attacks.

Arie Shirom and his colleagues (Chapter 6) summarize two studies investigating the effects of fear of terror on subjective and objective indicators of health. Both studies were carried out in Israel among adult samples. In the first study, objective exposure to terror did not predict self-reported health but fear of terror negatively predicted self-reported health for both females and males, The second study, using a subsample from the first study, considered low-grade inflammation as the objective health indicator. They found, in women only, a positive association between fear of terror and low-grade inflammation. Controlling several demographic and bio-medical variables, they conclude that fear of terrorism was prospectively associated with both subjective and objective indicators of health.

Hector González-Ordi and his colleagues (Chapter 7) examine the effects of the March 11, 2004 terrorist attacks at the Atocha train station in Madrid. This was the biggest terrorist attack in the history of Spain. Al Qaeda placed 10 bombs in four commuter trains during rush hour, killing 191 people and injuring 1800. Shortly after the attack, 2 million citizens of Madrid took part in a demonstration against terrorism. Surveys of Madrid residents showed acute stress symptoms following the attack, but most people were generally able to go about their daily lives shortly thereafter. The authors carried out three longitudinal studies of citizens in general, victims and their relatives, and emergency responders. Phone interviews were carried out with over 1500 people. About 10 percent of the general population had symptoms reflective of panic attack, women more than twice as likely to be affected as men. Interestingly, figures on symptom levels following the Madrid attack tended to be lower than those reported following the attacks on the WTC. Several variables (for example, sex, age, social support, stressors before the attack) predicted individual symptom levels following the attack in Madrid. Not surprisingly, victims and their families were more strongly affected than were the public at large, with first responders falling roughly in between these two groups. The authors conclude with an observation also made about the attacks in NYC following 9/11: although Madrid will never be the same, it emerged stronger following the attack.

Part III contains five chapters and focuses on what we know about building individual and organizational resilience. In Chapter 8, Michelle Slone and Anat Shoshani examine the interrelated function of the media and the viewing public in dealing with the psychological warfare of terrorism. The violence of terrorism is increasingly present in people's daily lives. Increased media exposure to terrorism increases anxiety, anger and fear. Negative stereotypes are also increased. Slone and Shoshani describe efforts to develop a preparatory intervention to help individuals deal with common responses to media coverage of terrorism involving high risk community sectors. Although evaluation data of the usefulness of such interventions is just emerging, their early results are promising. They conclude with their thoughts on regulating media exposure to terrorism. The public should be made aware of how media coverage evokes negative emotional reactions and helps to control their cognitions; the media needs to alert the viewing public to the potentially disturbing nature of their images. The power of the media in fact can be used to improve conflict resolution tactics among the viewers of terrorism.

Brian Flynn and Fred Lane (Chapter 9) integrate organizational and behavioral health concepts central to protecting, promoting, and sustaining the health of individuals and organizations in trying circumstances. Disasters such as terrorist attacks adversely affect both individuals and organizations. The authors first consider key factors necessary to sustain and develop organizations in normal as well as extraordinary circumstances. Leadership, particularly as it creates a culture, is central here. In addition, the overlapping roles and responsibilities of numerous stakeholders play a major part and are identified. Flynn and Lane then examine central concepts to achieve individual well-being. These include pre-event, during an event, and post-event responsibilities. They conclude with a review of concrete initiatives that unite organizations, leaders in organizations, and all employees around a shared value of promoting health and reducing negative consequences.

Parbudyal Singh (Chapter 10) tackles the challenges of maintaining work motivation during trying times. His review of many current theories of work motivation finds them only partially useful. Based on interviews with staff at a New York hospital after 9/11, along with archival data, press reports and recent literature, he distills important factors in motivating employees 'under attack'. This hospital lost some of its employees in the 9/11 attack as well as responding to the needs of others injured in the attack and their own workforce. Hospital employees continued their work during and after the attack, even working harder. Leadership, aided by the availability of a disaster plan, provided clear instructions and visible examples of commitment. Employees in the nursing profession generally are motivated to help in times of need. In addition, a sense of patriotism, coupled with a love for New York, supported high levels of work motivation. Singh suggests ways in which organizations can incorporate some of these messages in their own response to disasters.

Robyn Gershon and her colleagues (Chapter 11), using archival information and interviews with individuals who left the WTC after the terrorist attacks, distill lessons about employee responses to this workplace disaster and offer recommendations for workplace preparedness. Workplace readiness not only reduces injury and death but also preserves organizational assets and supports continuous operations. Information about individual responses to disaster is central to emergency preparedness. Individuals rarely panic during these emergencies and we have come to understand the conditions for panic. Emergency preparedness is the process of preparing for and recovering from disasters such as 9/11. The authors review key elements in emergency planning, including specific codes and standards. They conclude with findings from their WTC Evacuation Study. They collected data from over 1400 individuals, almost 10 percent of those who evacuated the two towers. These data identified individual, organizational and environmental factors that served as barriers to evacuation and progression. They conclude with specific recommendations for improving the evacuation of high-rise buildings. These are presented in the larger context of disaster mental health. Organizations have considerable leverage in addressing disaster mental health through providing social support and tangible services, employee assistance programs, and resilience training.

Finally, Jody Hoffer Gittell and her colleagues (Chapter 12) focus specifically on airline industry responses to 9/11. The airline industry faced devastating financial losses following these terrorist attacks. This chapter examines factors associated with the success of some airlines in rebounding quickly and the difficulties encountered by others. It considers employee relationships, financial reserves, and the underlying business model in 10 large airline companies. All three factors were important in maintaining productivity. Interestingly, employee relationships were central in the development and enactment of a viable business model. Some organizations do well in a crisis – particularly those having resilience. Contrasts of Southwest Airlines and US Airways are provided in some detail. A low-cost business model achieved through long-term positive employee relations, coupled with adequate financial reserves, allowed Southwest to return quickly to business as usual. The importance of positive employee relationships in this rebound is highlighted.

The world has become a less safe place over the past few years and threats of terrorism as well as actual terrorist attacks will be with us for some time. Organizations and their employees remain prime targets for such attacks. It is vital that organizations and their leadership be better prepared for and better able to recover should such attacks occur. We hope that this collection brings these issues to the forefront and strengthens the ability of organizations to anticipate, plan for, and respond to such disasters. We need resilient people and adaptive organizations.

NOTE

1. Preparation of this manuscript was supported in part by York University, Canada and Lancaster University, UK.

REFERENCES

Ahern, J., Galea, S., Resnick, H., Kilpatrick, D., Bucuvalas, M., Gold, J. and Vlahov, D. (2002), 'Television images and psychological symptoms after the September 11 terrorist attacks', *Psychiatry*, **65**, 289–300.

Argenti, P. (2002), 'Crisis communication lessons from 9/11', *Harvard Business Review*, December, 103–9.

Bachrach, S.B. and Bamberger, P.A. (2007), '9/11 and New York City firefighters' post hoc unit support and control climates: a context theory of the consequences of involvement in traumatic work-related events', *Academy of Management Journal*, **50**, 849–68.

Bazerman, M.H. and Watkins, M.D. (2004), *Predictable surprises: The Disasters You Should Have Seen Coming, and How to Prevent Them*, Boston, MA: Harvard Business School Press.

Berribi, C. and Klor, E. (2006), 'On terrorism and electoral outcomes: theory and evidence from the Israeli–Palestinian conflict', *Journal of Conflict Resolution*, **50**, 899–925.

Bonanno, G.A. (2004), 'Loss, trauma, and human resilience. Have we underestimated the human capacity to thrive after extremely aversive events?', *American Psychologist*, **59**, 20–28.

Bonanno, G.A. (2005), 'Adult resilience in the face of potential trauma', *Current Directions in Psychological Science*, **1**, 135–8.

Clauset, A., Young, M. and Gleditsch, K.S. (2007), 'On the frequency of severe terrorist events', *Journal of Conflict Resolution*, **51**, 58–87.

Dohrenwend, B.P., Turner, J.B., Turse, N.A., Adams, B.G., Koenen, K.C. and Marshall, R. (2006), 'The psychological risks of Vietnam for US Veterans: a revisit with new data and methods', *Science*, **313**, 979–82.

Felton, C.J. (2004), 'Lessons learned since September 11th 2001 concerning the mental health impact of terrorism, appropriate response strategies and future preparedness', *Psychiatry*, **67**, 147–52.

Galea, S., Vlahov, D.M., Resnick, H., Ahern, J., Susser, E., Gold, J. et.al. (2003), 'Trends of probable post-traumatic stress disorder in New York City after the September 11th terrorist attacks', *American Journal of Epidemiology*, **158**, 514–24.

Gershoff, E.T. and Aber, J.L. (2004), 'Assessing the impact of September 11th, 2001 on children, youth and parents: methodological challenges to research on terrorism and other normative events', *Applied Developmental Science*, **8**, 106–10.

Gershon, R.R.M., Qureshi, K.A., Rubin, M.S. and Ravais, V.H. (2007), 'Factors associated with high-rise evacuation: qualitative results from the World Trade Center Evacuation Study', *Prehospital and Disaster Medicine*, **22**, 165–73.

Gil-Rivas, V., Holman, E.A. and Silver, R.C. (2004), 'Adolescent vulnerability following the September 11th terrorist attacks: a study of parents and their children', *Applied Developmental Science*, **8**, 130–42.

Gittell, J.H., Cameron, K., Lim, S. and Rivas, V. (2006), 'Relationships, layoffs, and organizational resilience: airline industry responses to September 11', *Journal of Applied Behavioral Science*, **42**, 300–29.

Globe and Mail (2006), '50 changes 5 years after', Saturday, September 9, A5.

Gould, M.S., Munfakh, J.L.H., Kleinman, M., Lubell, K. and Prevenzano, D. (2004), 'Impact of the September 11th terrorist attacks on teenagers' mental health', *Applied Developmental Science*, **8**, 158–69.

Hoge, C., Auchterlonie, J. and Milliken, C. (2006), 'Mental health after deployment in Iraq and Afghanistan', *Journal of the American Medical Association*, **295**, 1023–36.

Inness, M. and Barling, J. (2005), 'Terrorism', in J. Barling, E.K. Kelloway and M.R. Frone (eds), *Handbook of Work Stress*, Thousand Oaks, CA: Sage, pp. 375–97.

Knudsen, H.K., Roman, P.M., Johnson, J.A. and Ducharme, L.J. (2005), 'A changed America? The effects of September 11th on depressive symptoms and alcohol consumption', *Journal of Health and Social Behavior*, **27**, 260–73.

Korten, D.C. (2006), *The Great Turning*, San Francisco, CA: Berrett-Koehler.

Lockwood, N.R. (2005), 'Crisis management in today's business environment: HR's strategic role', *Human Resources Magazine*, **50**, 1–10.

Mainiero, L.A. and Gibson, D.E. (2003), 'Managing employee trauma: dealing with the emotional fallout from 9-11', *Academy of Management Executive*, **17**, 130–43.

Mitroff, I. and Alpsalan, M. (2003), 'Preparing for evil', *Harvard Business Review*, April, 109–15.

Perliger, A., Pedahzur, A. and Zalmanovitch, Y. (2005), 'The defensive dimension of the battle against terrorism. An analysis of management of terror incidents in Jerusalem', *Journal of Contingencies and Crisis Management*, **13**, 79–91.

Shalev, A.Y. (2004), 'Further lessons from 9/11. Does stress equal trauma?', *Psychiatry*, **67**, 174–6.

Sheffi, Y. (2005), *The Resilient Enterprise*, Cambridge, MA: MIT Press.

Stein, B.D., Elliott, M.N., Jaycox, L.H., Collins, R.L., Berry, S.H., Klein, D.J. and Schuster, M.A. (2004), 'A national longitudinal study of the psychological consequences of the September 11, 2001 terrorist attacks: reactions, impairment, and help-seeking', *Psychiatry*, **67**, 105–16.

Vlahov, D., Galea, S., Ahern, J., Resnick, H., Boscarino, J.A., Gold, J., Bucuvalas, M. and Kilpatrick, D. (2004a), 'Consumption of cigarettes, alcohol, and marijuana among New York City residents six months after the September 11 terrorist attacks', *American Journal of Drug and Alcohol Abuse*, **30**, 385–407.

Vlahov, D., Galea, S., Ahern, J., Resnick, H. and Kilpatrick, D. (2004b), 'Sustained increased consumption of cigarettes, alcohol, and marijuana among Manhattan residents after September 11, 2001', *American Journal of Public Health*, **94**, 253–4.

Vlahov, D., Galea, S., Resnick, H., Ahern, J., Boscarino, J.A., Bucuvalas, M., Gold, J. and Kilpatrick, D. (2002), 'Increased use of cigarettes, alcohol, and marijuana among Manhattan, New York residents after the September 11th terrorist attacks', *American Journal of Epidemiology*, **155**, 988–96.

PART I

Introduction

1. International terrorism and threats to security: implications for organizations and management

Ronald J. Burke[1]

> Terrorism is the process of inducing fear in a civil population through violent acts that undercut trust and confidence, while creating a sense of personal vulnerability to random acts of evil.
>
> (Zimbardo, 2002, p. 16)

INTRODUCTION

The events of 9/11 will likely have effects that will last a long time. International terrorism had existed for years before that date and includes a previous attack on the World Trade Center, the bombing of the USS *Cole* in Yemen, attacks on the US military base in Saudi Arabia, and attacks on two US embassies in Africa. In addition, there have been successful terrorist attacks in Spain, the UK and Bali, among other countries. The destruction of the World Trade Center (WTC) in New York City (NYC), however, stands out as a critical event. It happened on US soil, was the largest terrorist attack, cost almost 3000 lives and destroyed a symbol of American capitalism and commerce.

This chapter includes the following content: a brief overview of the impact of 9/11 on world events since these terrorist attacks, the effects of terrorist attacks and other disasters on people, the impact of 9/11 on human resource management (HRM) practices and organizations, individual coping and resilience, organizational coping and resilience, anticipating disasters, improving responses to terrorist attacks and disasters, and implications of 9/11 and related events over the past six years for organizations, their employees and their management.

This chapter considers the effects of 9/11 on various parties: those directly or indirectly affected by 9/11 – individuals, families, organizations, and communities, those in first-responder roles, those responsible for helping the various parties get back on their feet – parents, supervisors and managers, HR professionals, counselors, and those responsible for preparing for the future – managers, risk assessors, and government policy makers.

What was uniquely important about 9/11? Several features stand out. There was global collaboration in the attack. The planning of the attack took place over several years. There was support from local people. The attacks emanated from 'friendly nations'. The attacks made use of advanced technology. The US was unprepared, though warning signs existed. The attacks were against the future (modernity), on a way of life. Finally, the attacks targeted a civilian population.

Since 9/11, there has been a continuation of terrorist attacks – not in the US but in the UK and elsewhere. There was a denial by the US of permission for Dubai to take over the ownership of seaports in the US. There has been increased border security resulting in greater costs and delays. We have seen heightened tensions between the Muslim and non-Muslim worlds through the publication of cartoons depicting the Prophet Muhammad in unflattering ways, raising the issue of Free Speech versus Blasphemy. Mobs in several countries attacked and destroyed the Danish and Norwegian embassies as a consequence. There have been continued violent insurgencies in Iraq and Afghanistan. There has been intensified conflict with Iran over its proposal to develop nuclear technologies, and increasing use of foreign aid as a bargaining chip in foreign policy. In late 2007, Iran captured 15 British sailors for purportedly entering Iranian waters with the intention of trying them on this charge. Escalating costs of the war on terror means taking resources away from pressing social issues in the US and elsewhere (for example, education, culture, the arts, poverty). Finally, the effects of international terrorism and 9/11 are compounded by other natural disasters that have hit the US and other parts of the world (tsunami, Katrina and other hurricanes, earthquakes and so on).

EMERGING ISSUES, CHALLENGES AND CONSEQUENCES

Tensions in the Middle East have driven oil prices up. The price of gasoline in North America rose dramatically in May 2004, hitting levels never before seen in the US. These increased costs have affected transportation costs and these latter increases are being passed on to consumers or borne by the providers of these services, cutting into their profits. The result is a jolt to economies in the developed and developing world. In June 2004, Alan Greenspan, US Federal Reserve Chairman, said that the risk of another large-scale terrorist strike in the US is the single biggest risk to the US economy.

Threats to security have increased levels of surveillance at airports and entry points to the US and many other countries. This has resulted in delays at airports and major border crossing points, slowing down the pace of business and tourist traffic flow. In addition, extra resources are being used in these efforts, increasing the costs of screening; these cost increases are again passed on to travelers and business or borne by various levels of government and eventually passed on to citizens through higher taxes or reductions of other services. Since 9/11, various countries have experienced anthrax-infected mail, letter bombs, bombs in packages and threats of bridge bombings and internet attacks.

Heightened fears have reduced the level of travel to and from various countries, resulting in a drop in air travel and tourism. Americans are traveling less, even to relatively safe destinations such as Canada. American attendance at Canadian stage theaters (Stratford Festival, Shaw Festival) dropped 20 percent in 2002. Terrorist attacks in Madrid, London, Bali, and the finding of explosives along parts of the French railway system have also reduced travel to these places (Ahmed, 2006).

The events of 9/11 have increased intergroup tensions within and across countries (Bar-Tal and Labin, 2001). Muslims and individuals of Arab descent have reported increased hostility from others in non-Muslim countries (Sheridan, 2006). There has been a heightened polarity between the Arab and the non-Arab world. The Muslim world reacted angrily to cartoons depicting the Prophet Muhammad in an unflattering way; to writers and filmmakers showing the Muslim world in a way they thought biased; and to the Pope's comments in September 2006 in which he cited an ancient quotation indicating that the Muslim religion was associated with aggression. In September 2006, a Berlin-based opera company canceled the showing of an opera that potentially depicted Muslims in a negative way, for fear of violence.

These factors operate to increase instability worldwide. Such instability increases the uncertainty of supply-chain management. The internationalization and globalization of business has resulted in greater interdependence of suppliers and consumers of these goods and services. International tensions and instability have required organizations to reconsider sources of their goods and services and to build in redundancies to guarantee supplies of these in turbulent times.

The need to proactively identify and capture terrorists and their allies has increased surveillance of citizens within North America and elsewhere by their own governments (Bennett and French, 2003). This has raised concerns about invasion of privacy, the arrest and detention of people on suspicion of supporting terrorism, and the weakening of due process. Flagrant examples of bending the rules supporting civil liberties to counter terrorism are becoming increasingly common.

The events of 9/11 have increased soul-searching by the major economic powers. Why does the Islamic world hate the United States, or the West more generally? What can be done about this? How stable are particular Islamic republics? This is also reflected in a loss of innocence among US citizens who for years had believed that they were respected throughout the rest of the world.

There is a new debate now in some countries about the role of violence and proactive attacks on potential or real terrorists and states thought to support or harbor terrorists. Some people believe that the use of violence to prevent terrorism is never justified; other people argue that there is a need for whatever means necessary, including violence, in order to reduce violence by terrorists in the long term.

This polarity is reflected in heightened beliefs by many in the Arab world of the decadence and greed of the Western world, and views by many in the West that the Arab world is corrupt, undemocratic and hypocritical – indicating that they support the West but also indirectly supporting hostility toward the West and terrorism.

The events of 9/11 took place over six years ago, and security has been improved in target countries. What is not clear is how much more prepared countries now are for such disasters. Canada created the Ministry of Public Safety and Emergency Preparedness in December 2003. Canadians should be worried: its Auditor-General issued a report in early 2004 indicating that Canada's anti-terrorism initiatives have been hindered by red tape, turf wars and bureaucratic incompetence. Information from surveys of first responders – those first on the scene in the event of a disaster – from 100 municipalities across Canada, indicated a sense that no one was in charge if a disaster struck, a condition similar to findings of the US 9/11 Task Force and analysis of the US government's response to Hurricane Katrina in September 2005.

The European Union (EU) made a decision in early 2004 to use the Canada–US border security methods following the bombings in Madrid. Police and intelligence services in Canada and the US have improved their information-sharing arrangements following 9/11.

Organizations are increasingly integrating responsibilities for materials sourcing, manufacturing, warehousing, transportation and customer service, termed 'supply chain management'. Border security issues complicate these linkages. Complex new border security arrangements are affecting supply chain management. Advance information must be provided to the US on all cargo entering or leaving their country by sea, air, truck or rail; similar measures have been implemented in Canada. Visitors to the US from particular countries must now be fingerprinted and photographed at airports and seaports. These measures add increased cost and complexity (more paperwork, driver overtime). Delays at the border can impact on just-in-time delivery schedules. Compounding these issues there was also disarray in the use of watch lists to identify terrorist suspects, mismanaged security alerts, delayed implementation of fingerprinting analysis technology, laxness in security screening at airports, and barriers to sharing intelligence between government agencies.

We need a strategic plan for counter-terrorism and security, better integration across government agencies, better analysis of terrorist threats, better use of technology to collect, analyze and share data, and the recruitment of talented men and women to work in intelligence and security functions.

There is the moral and ethical dilemma of whether Western-style democracies should resort to torture, assassination and the use of methods that violate their values and laws against enemies that do not uphold these values and laws in the war against terrorism. Ignatieff (2004) argues for a middle ground: societies may have to temporarily employ measures that run counter to their values and laws; these exceptions, however, must be justified and limited in their use. But can there ever be a 'just war'?

The war on terrorism is also having effects on target countries. Chief among these is the strengthening of executive power (the president, prime ministers) at the expense of the courts, legislatures and the media. The war on terrorism is fought in secret, making it difficult to observe and be monitored by these groups. Democracy itself may be at stake in some cases. Someone needs to guard the guards.

The terrorist attacks and the war on terror are having an influence on the internal politics of various countries. The bombing of the Madrid railway station affected the Spanish election in 2004. The war in Iraq influenced the 2006 House of Representatives election in the US. The war in Iraq led to reduced support for Prime Minister Tony Blair in the UK. Israeli elections have always been influenced by terrorist attacks.

Al Qaeda has begun to attack the economic infrastructure in Saudi Arabia. Their efforts have targeted the Westerners that Saudi Arabia depends on to keep its economy working and their oil flowing: 8.8 million foreigners live and work among the 17 million Saudis. On May 29, 2004, gunmen killed 22 people in the Saudi city of Khobar inside two oil industry office compounds.

There is an increasing consensus that the six-year war on terror has in fact made the world a far more dangerous place. Osama bin Laden's survival and the US-led invasion and occupation of Iraq have served as important recruiting factors for the Al Qaeda network. Following 9/11, Al Qaeda has claimed responsibility for attacks in Saudi Arabia, Indonesia, Turkey, the UK and Spain.

The threat of terrorism is obviously real. The state must ensure the safety of the general public. It is important that the state do so while protecting civil rights and liberties. Police and intelligence agencies must be accountable. Intelligence must not be obtained using torture. Targeted communities must be consulted; community outreach must be emphasized. Profiling must be based on information and specific behaviors of individuals.

My country (Canada) is thought by some to be doing a good job with immigration. Some Western European countries have an immigrant population that is failing to assimilate (culture clashes, surge of conservative Islam, citizens resent the welfare entitlements of newcomers). A backlash is building against these immigrants. The difficulty immigrants have in getting jobs is a major cause for concern. It is easier for immigrants to get jobs (poor ones in some cases) in North America than in some European countries. In addition, immigrants to North America tend to be more skilled and have other family members already there. Finally, North Americans, comprising immigrants historically, have little sense of ethnic or national identity.

According to the US State Department, the number of international terrorist attacks peaked at 666 in 1987 and fell to 348 in 2001 and 208 in 2003; the number of significant terrorist attacks in 2003 was 175, a 20-year high. Casualties from international terrorist attacks in 2001 were the highest – 3573 people killed compared with only 652 in 2003 (Gersham, 2004).

An examination of the anthrax attacks and other explosives sent through the post has identified procedures for effectively dealing with potentially dangerous materials, thoughts on how to view this new threat and plans for dealing with it (Babbs and O'Connor, 2003; Day, 2003).

In 2007, police in the UK issued a nationwide alert after linking seven letter bombs delivered over the previous three weeks (*National Post*, 2007). These attacks were not being treated as terrorism, but the motives were unclear.

International terrorism is on the increase in the Middle East and South Asia but has declined in most other parts of the world. Support for Al Qaeda and Osama bin Laden has dropped in the Western world. Media attention has made terrorism seem to be a larger concern than it should be. But the threat of terrorism is still real. Gathering and sharing information is critical to an effective anti-terrorism effort.

Reports from the 9/11 Commission (Kean and Hamilton, 2004) stressed the importance of understanding the organizational level – the way organizations operate, their structure and processes and their management practices and culture. The responses of employees are largely determined by their organization's policies, HRM practices and cultural norms. Blaming government or asking government alone to fix it is doomed to fall short.

Canadian banks came under pressure from the US to tighten regulations involving customers holding dual citizenship (immigrant clients), customers who are citizens of Canada or those on a US government blacklist. Some of these customers do not live in Canada while others do (*Globe and Mail*, 2007).

Can we ever go back to business as usual? Do we even want this? National priorities need to be reconsidered. Spending more money on defense and security means spending less money on health and education. The events of 9/11 have forced countries to change, eliminate denial, work together and build cooperation. Developed countries need to better understand the complaints, fears and challenges faced by developing countries. They need to capture the hearts and minds of their critics.

IMPACT ON PUBLIC ATTITUDES AND MORALE

Research studies of disasters in general have shown increases in psychological distress immediately following the disaster, with decreases over time. These disasters included the Oklahoma City bombing (Sprang, 1999, 2000, 2001; Benight et al., 2000), the nuclear accident at Three Mile Island (Collins et al., 1983), responses of Israeli children to scud missile attacks in the Persian Gulf War (Weisenberg et al., 1993), and responses of children to the Northridge California earthquake (Asarada et al., 1999).

Kisoc (2006) examined stress-related symptomatology of people in Turkey one year after two recent earthquakes (1999). Stress symptoms were common and severe among affected people. These were higher in women, those losing family members and other relatives in the earthquake (over 20,000 people died) and those having less education. Some stress symptoms declined 15–18 months after the earthquakes but were still evident.

Researchers have also devoted considerable effort to chronicling the impact of 9/11 on the general public. Fischer (2002) finds that the disaster research model of behavioral response to disaster fit events following 9/11. These include: fear (but not panic) of survivors, altruism from other citizens, shock and psychological dependence, live broadcast coverage by the media, local decision making, a reduction in crime and an overestimate of damage, injury and the death toll.

A survey of 407 psychologists reported that about half of a sample of their clients brought up the terrorist attacks during the selected counseling session (Daw, 2002a). More than 20 percent of clients mentioned specific trauma or anxiety reactions associated with the events of 9/11. In addition, focus groups conducted in October and November following 9/11 found that the attack caused fear, stress, anxiety and frustration and caused people to look inward and reassess their lives (Daw, 2002b).

Peterson and Seligman (2003) compared online e-survey responses to a questionnaire from individuals completing it before 9/11 and others completing it between September 12 and November 30 and found significant increases in six of 24 character strengths or virtues (love, gratitude, hope, kindness, spirituality and teamwork). These changes reflect what is now more important and suggest that other people rose in importance (see Wrzesniewski, 2002, for a description of how she changed after witnessing the WTC attack).

The American public did not respond with panic but instead with effective and adaptive action. Panic was prevented by providing timely and accurate information and instructions so that the public could make timely decisions. New heroes emerged, ordinary people, volunteers who made sacrifices for the well-being of others.

Six months after 9/11, Americans seemed to be getting back to 'normal', that is, the way things were on September 10. In the months following 9/11, the US was a wounded nation, having suffered the largest attack ever on its soil. In response, the US developed a common project and a higher cause. But the US is also a target. The citizens have generally rallied around President George W. Bush and he has performed in ways that have exceeded the modest expectations many people had for him.

But some things have changed. Americans have a higher opinion of how government can help than they had in previous decades. An increasing number of Americans have invested in family, neighbors and community. Many Americans are now more tolerant of bad news. But these increases may only be short-lived.

Liverant et al. (2004) studied anxiety responses and reactions to 9/11 among a sample of indirectly affected college students living in Boston. Anxiety was measured two and four months following 9/11, as were coping methods. About 20 percent of the first sample knew someone killed or injured in the 9/11 attacks. Furthermore,

- 44 percent saw the world as less safe after 9/11;
- 60 percent thought that another attack was likely;
- 83 percent thought that war was likely in the future;
- 68 percent reported being severely affected by 9/11;
- 67 percent felt severe anger about the attacks; and
- 19 percent changed their routine or travel plans as a result of 9/11.

Liverant et al. reported a significant decrease in anxiety over the two-month period. Those with a personal knowledge of a victim reported higher levels of anxiety. Some coping responses (avoidance coping – denial, mental disengagement, behavioral disengagement) were associated with higher anxiety levels.

Schuster et al. (2001) collected data within five days of 9/11 and reported that 90 percent of Americans indicated one or more distress symptoms to some degree and 44 percent indicated one or more stress symptoms to a considerable degree. A study conducted in the two weeks following 9/11 by Smith et al. (2001) reported that 50 percent of Americans reported feeling very nervous and tense following the tragedy, 46 percent reported feeling dazed and 50 percent reported having trouble sleeping. Galea et al. (2002) examined reactions to 9/11 among people in NYC directly and indirectly affected; 80 percent indicated that they had post-traumatic stress disorder (PTSD), and 10 percent depression five to eight weeks following 9/11. Silver et al. (2002) reported that 17 percent of the US population outside NYC indicated PTSD symptoms two months after 9/11, with 6 percent reporting these symptoms six months after 9/11. Taken together, these studies show that stress reactions to 9/11 were present in a substantial number of people indirectly exposed to the terrorist attack.

Olivas-Lujan et al. (2004) considered the impact of 9/11 on cultural values and levels of cosmopolitanism of US university students. Data were collected before 9/11 (April 11) and in December 2001 and March and April 2002. They report increases in authoritarianism and hierarchy/power distance and decreases in cosmopolitanism. In addition, students placed a lower value on variety, adventure and challenge and a higher value on security and stability in their ideal job following graduation.

Esses et al. (2002) found that attitudes toward immigration in the US and Canada became more negative following 9/11. Threats to a group are often associated with in-group favoritism and out-group bashing. This produced an increasingly shared social identity, higher authoritarianism and increased stereotyping of the out-groups.

Wrzesniewski (2002) reported an increase in people changing jobs or occupations following the WTC disaster in their search for greater meaning in their work. The disaster became a catalyst for an examination of what their work was contributing to the wider world.

Smith et al. (2001) report, based on a random telephone survey of 2126 US residents, that national pride, confidence in institutions, and faith in people and human nature had all increased – support for both the military and government was very high (confidence in). Citizen participation in civic (charitable actions) increased. Positive feelings seemed not to be affected (psychological well-being) by 9/11, but New Yorkers showed more negative emotion. Data were collected between September 13 and 27.

In 2002, Rasinski et al. (2002) re-interviewed the initial respondents. The second-wave data were collected 3–5 months following 9/11. Pride in the country continued to remain high; pride by New Yorkers was lower than the national average. Confidence in institutions dropped to levels approaching pre-9/11. Respondent anger declined, levels of positive emotion remained the same, while levels of negative emotion dropped. Psychological well-being improved.

An overview of the research on the impact of 9/11 on the attitudes and behaviors of the general public (see North and Pfefferbaum, 2002; Schlenger et al., 2002) suggests modest negative effects immediately following the terrorist attacks, with most people moving back to pre-9/11 levels within a few months (Knudsen et al., 2005). Individuals closer to NYC, Washington DC, and Pennsylvania indicated a larger impact than those living farther away. Individuals losing family, friends or property (or who knew someone who had) indicated a larger impact than those having no such loss. Women tended to have stronger emotional reactions to these terrorist attacks (anxiety, depression) than did men.

Impact of 9/11 on Children, Youth and Parents

The media, primarily the television networks, replayed footage of the airplanes flying into the WTC and the collapse of the towers in what seemed to be an almost endless stream of coverage. What effect does watching replays of the 9/11 attacks have on children, youth and parents?

Recent research has begun to address this question. A sample of these studies include: Ahern et al. (2002); Aber et al. (2004); Gil-Rivas et al. (2004); and Gould et al. (2004).

The findings support three conclusions. First, children throughout the US were affected to a large degree (for example, Schuster et al., 2001; Schlenger et al., 2002). Second, children and young people spending more hours watching coverage of the terrorist attacks were more negatively affected (Ahern et al., 2002). Third, these effects appeared to diminish over time (Gould et al., 2004). Parents, and their level of support and parental self-efficacy, played an important role in ameliorating adolescents' responses to 9/11 (Gil-Rivas et al., 2004).

Stein et al. (2004) report findings from a national longitudinal study of the psychological consequence of 9/11. They re-surveyed 355 people about two months after 9/11. Sixteen percent of adults had persistent distress reporting four or more substantial distress symptoms in both September and November. Adults with persistent distress also reported accomplishing less at work (65 percent), avoiding public gathering places (24 percent), consuming alcohol, medication or other drugs to sleep or feel better because of worries about terrorism (38 percent). Seventy-five percent talked with family and friends, but 43 percent indicated feeling unable to discuss these feelings because it made them feel uncomfortable. Few turned to counselors or general medical practitioners for information and help (11 percent).

Mainiero and Gibson (2003) report the results of a survey of employee reactions three months after the September 11th attacks. The findings show that while managers dealt with employees' physical safety concerns, they paid less attention to their emotional concerns. Three months after the event, common employee responses included fear, denial, and anger. Certain groups, women, employees with children, and those located closer to the crisis, were more likely to be emotionally affected. The authors suggest that managers should be trained to recognize and respond to employee trauma and offer suggestions on dealing with crisis events in ways that maintain employee morale.

Managers sometimes assume that events outside the organization will not affect employees within the organization. It is also likely that not all employees will be affected in the same way and to the same degree. Events outside the organization include: political instability around the world which has heightened following 9/11; negative views of the West, particularly the US and the UK; higher oil prices, increasing the costs of doing business; and new alliances against the West, the US in particular.

Mainiero and Gibson conducted an online internet survey of 5860 employees three months after 9/11. They asked whether employees were affected by 9/11, and if so, how? Were they more fearful; did they have difficulty concentrating; did they have less job satisfaction; did they limit their airline travel, delay a major purchase, or make changes to their values, goals and aspirations?

Women were more emotionally affected than men. The largest areas of change included more valued time spent with friends and family, less airline travel, changes to values, goals and aspirations and increased anxiety about personal safety.

What can managers do? Mainiero and Gibson suggest the following:

1. Demonstrate visible leadership, communicate, and be a role model.
2. Develop crisis contingency plans in advance of a disaster.
3. Clarify responsibilities for responding to the crisis.
4. Assess the overall emotional impact.
5. Create a positive climate for allowing employees to discuss their feelings.
6. Communicate to employees how business is affected by the crisis.
7. Address steps that the organization will take to improve security and manage employee risk.

8. Above all, be flexible.
9. Be prepared should another emergency occur.

Effects of the Palestinian Intifada

Since the beginning of the Al Aqsa Intifada in September 2000, Israel has had deaths – on a population size basis – equal to ten September 11ths. Zeidner (2006a) collected data from 770 Israeli adults (60 percent women) at the height of the Al Aqsa Intifada uprising (May–July 2002). Israeli women were more distressed by political violence than men, and appraised the crisis situation as more threatening and less manageable. Women also reported more somatic symptoms and more frequent PTSD symptoms than men did. Although both women and men used a mixture of coping strategies, women used both problem- and emotion-focused coping more than men did.

Hobfoll et al. (2006), in a study of 905 adult Jewish and Palestinian citizens of Israel, reported that exposure to terrorism was significantly related to greater loss *and* gain of psycho-social resources and to greater PTSD and depressive symptoms. Loss of psycho-social resources associated with terrorism was related to both PTSD and depression. The Israeli studies indicated that despite the distress and threats to safety, most Israelis adjusted with few mental health symptoms. But little is known about coping strategies likely to be effective in responding to the thrust of the actual terrorist attack.

Zeidner (2006b) reported that negative affectivity was strongly correlated with terror-related distress, appraisals, coping and mental health outcomes. Higher negative affectivity was correlated with seeing the threat as more stressful, more threatening and less controllable, with more emotion- and problem-focused coping and more reported symptoms.

Effects of the Madrid Terrorist Bombing

Miguel-Tobal et al. (2005) examined reactions to the March 11, 2004 terrorist attacks in Madrid, comparing districts in the general population (near the attack versus distant from the attack), victims and relatives and/or close friends, and emergency service personnel. They carried out three longitudinal studies at 5–12 weeks, 6–7 months and one year. Data were collected using interviews with 1589 people.

Women were more likely to indicate symptoms of panic attack than were men. About 11 percent of the general population reported such symptoms, 16.3 percent in the areas most directly affected by the attack. Women were more depressed than men; 8 percent of the general population were depressed, almost 10 percent in the areas most directly affected. Those directly involved in the attacks (victims, those with dead friends or relations, those with injured friends or relatives) scored higher on all mental health measures.

Miguel-Tobal et al. (2006) surveyed residents of Madrid via telephone interviews, 1–3 months after the terrorist train bombing to assess the prevalence of PTSD and major depression. They found that 2.3 percent reported PTSD symptoms and 8.0 percent reported symptoms consistent with major depression. The prevalence of PTSD was low, but depression prevalence was comparable to estimates following the 9/11 terrorist attack in NYC. They conclude that the magnitude of the attack may have affected the prevalence of PTSD in the two cities.

IMPACT ON HRM AND ORGANIZATIONS

Kondrasuk (2004) believes that 9/11 brought about changes in most areas of HRM; the 9/11 attacks had an immediate impact on people: shock, sorrow, outrage. However, with the passage of time, the initial effects declined, and life generally returned to pre-9/11 levels. Acts of discrimination and anger against Muslims and Arabs increased, people donated money to the victims of 9/11. Employees spent more time with their loved ones and at home. People drew closer together. Employers created crisis management teams and made disaster plans. The government tried to improve coordination among law enforcement and intelligence agencies. The economy suffered, particularly in the airline, hotel, and insurance industries. Border security was tightened, increasing both the cost and time required to move goods.

HRM effects included giving employees time off work if required, less business travel, the cancellation of meetings, and the collection of money for victims and their families. A few months after 9/11, stress symptoms appeared to increase. Employees tried to get a better balance between work and family (Leonard, 2002a, 2002b).

AON Consulting conducts research on employee attitudes in the US on a regular basis. Their surveys have two broad objectives: to examine the commitment of employees to their current employer and to identify what employers are doing to heighten commitment. AON Consulting undertook a planned survey in February 2001, but following the events of 9/11 decided to collect new data in October 2001 (AON Consulting, 2001). This period (2001) also coincided with a slowing American economy. They found that employee responses had generally not changed significantly. Some exceptions were observed, however. First, employees indicated a shift in work–life balance priorities, with more employees wanting to spend more time with their families. Second, employee commitment to their employers had increased. Third, employees had generally favorable views on how their employers had responded to the events of 9/11 as well as on the state of the economy. Fourth, employees expressed less confidence, however, about the future of their own employers.

Employees gave their employers their highest evaluation in terms of managing business issues, lower evaluations in providing for employee safety and effective leadership in helping their workforce deal with the attacks of 9/11 and their aftermath, and the lowest evaluations in helping employees cope with stress and anxiety arising from the terrorist attacks and future threats. Organizations directly affected by the attacks were likely to provide crisis counseling to their employees (Leonard, 2001a); employers less directly affected did less or nothing. Following 9/11, employees, had more positive perceptions of their organization's recognition of the importance of personal and family life.

Kondrasuk (2004) pays particular attention to the effects of 9/11 on HRM. Employers set up crisis management teams, and disaster plans were revisited or developed. Business travel decreased, and increased border security slowed the movement of goods. Bernasek (2002) assessed the extra financial burden of 9/11 at $151 billion a year.

HR managers were actively helping their employees. These efforts included allowing employees to delay or cancel traveling, or to take time off; providing financial aid to victims; canceling meetings; encouraging employees to help each other; and expecting tighter security.

Gorman and McKee (1990) describe the short- and long-term effects of an office building fire on First Interstate Bank. Short-term issues included pay, time keeping, expenses incurred as a result of the fire, parking, and transportation. Different employee groups were provided trauma counseling with a different emphasis (senior management versus first responders). They offer practical steps for preparing an organization for a crisis, which include ensuring that the organization has a business resumption/disaster recovery plan, a human system back-up, communication vehicles in place, Employee Assistance Programs and over-communicates.

Poe (2001) believes that the effects of 9/11 on the workplace may be long-lasting. She argues that although organizations helped their workforce immediately following the attack, they also need to attend to the long term. Leonard (2001a) noted a dramatic increase in the use of EAP services following the terrorist attacks. Leonard (2001b) also reported that organizations faced higher healthcare costs as a result of 9/11.

Leadership Responses to 9/11

Bligh et al. (2004) analyzed George W. Bush's speeches before and after 9/11. Following 9/11, these speeches referred more to the American people as a collective, included more patriotic and faith-based themes, were more aggressive and less ambivalent, but also tended to be slightly less optimistic. Bligh et al. undertook similar analyses of the media (newspapers, magazines, television news shows) and found generally similar results. After 9/11, President Bush achieved a balance between acknowledging the catastrophic events and painting a vision of a better future ahead.

Frohman (2006), and Frohman and Howard (2008), describe how an organization in Israel (headed by Frohman) continued working during the scud missile attacks of the first Gulf War. They offer three lessons learned: focus on long-term survival and long-term consequences, do the opposite of what people expect, and trust your instincts.

DIRECT AND INDIRECT EFFECTS ON ORGANIZATIONS

Impact on Property

The attacks on the WTC also destroyed or damaged cultural property such as buildings, artworks, historical and archeological artifacts, archives and libraries (Heritage Preservation, 2002). The WTC housed the records, archives and libraries of over 500 private and public sector organizations. Works of art worth an estimated value of $100 million were housed there (for example, Picasso, Hockney and Koenig). Damage to archival material also occurred at the Pentagon when American Airlines flight 75 crashed there.

Marketing

On a recent CNN television show, the correspondent interviewed a Pakistani small business man whose company produced a cola drink. Business was booming for him because Pakistanis were boycotting cola drinks made by US-based companies (Coke, Pepsi). Individuals in marketing, advertising and public relations are now concerned that their market research suggests that political developments, including opposition to the war in Iraq, are reducing the global appeal of US brands such as McDonald's and Microsoft. As a consequence, efforts are now being undertaken to reinvigorate the US brand by relaunching a campaign highlighting things made in the US.

Supply Chain Management

Sheffi (2001) notes two major impacts of international terrorism on supply chain management: operating under heightened security and dealing with the aftermath of terrorist attacks. These both imply less reliable delivery lead times and less certain demand forecasts. In addition, organizations may need to build in supply redundancies to take care of increasing uncertainty. Firms are vulnerable to attacks on their own facilities but also to attacks on their suppliers, customers, transportation providers and communication lines.

The EU has imposed new security regulations at its ports which took effect on July 1, 2004. These measures have increased costs and extended container loading times.

It has been estimated that the cost of border delays between the US and Canada is $8.3 billion per year. A simple piston used in a car engine travels back and forth across the Canada–US border four times before the completed vehicle gets shipped to retail market.

Insurgents in Iraq have frequently targeted and sabotaged vital oil pipelines. Delaying and interfering with Iraqi oil exports cripples the Iraqi economy and may, over the long term, increase oil prices worldwide. Infrastructure attacks (oil, power stations) and the assassination of employees in these sectors has a negative effect on Iraq's economic situation.

Economic Costs

The events of 9/11 have had significant short-term effects on some parts of the world. These include a loss of innocence, a heightened sense of vulnerability, the exploitation of freedom, grieving with the families of the deceased, compassion for those who suffered losses, confirming the centrality of the human condition, revitalizing bonds of social connection, recognizing the fragility of life and change events, and highlighting the need to work less, play more, connect more closely with family and friends, nourish one's roots, and seek meaning and spiritual values in one's daily transactions.

There are several ways in which terrorism can have an impact on a country's economy (Alavosius et al., 2002). Fielding (2004), using time-series data, examined the relationship of various dimensions of the Palestinian Intifada (for example, Israeli and Palestinian fatalities, number of GreenLine border closures) on 'capital flight', defined as the percentage of Israeli capital wealth held outside the country. The data indicated support for casual links going in both directions: more violence leads to more capital flight and more capital flight leads to more violence.

There have been other serious economic costs as well. The September 11 attacks on the US will cost property insurers over US$20 billion for claims on the WTC and the Pentagon. The insurance industry will likely pay out over US$40 billion when other costs of property damage and related charges such as business-interruption insurance liability insurance, workers' compensation, and life and health insurance losses are included (Leonard, 2001b).

INDIVIDUAL RESILIENCE

Who copes better with man-made or natural disasters? Friedberg et al. (2005) examined the relationship of forgiveness and ruminative tendencies and levels of trauma and stress experienced by NYC residents on the one-year anniversary of 9/11. Rumination predicted levels of trauma and perceived stress; lower levels of forgiveness predicted perceived stress but not trauma. Rumination mediated the relationship between forgiveness and perceived stress. Forgiveness served to buffer the effects of stress more in individuals low in rumination.

Zivotofsky and Koslowsky (2005) examined gender and coping strategies among individuals faced with a major perceived life-threatening stressor – the sniper(s) that terrorized the Washington, DC area in October 2002. Fourteen random shootings resulted in 10 deaths. Women's behavior was more affected than men's. Women reported greater use of both problem- and emotion-focused coping.

Assanangkornchai et al. (2004) found elevated levels of psychological distress in a sample of 590 Thai respondents following a severe flood in Hat Tai that killed at least 32 people, injuring about 1700 and causing severe property damage. Psychological distress was associated with respondents' perception of the severity of loss, ability to collect possessions, and indicating a 'negative' response to the flood.

Ben-Ezra et al. (2005) examined the immediate post-traumatic reactions of rescue personnel exposed to the Hilton hotel bombing in Sinai, Egypt on October 7, 2004. They found that rescue personnel having previous exposure to the same type of trauma had a lower level of PTSD symptoms than those who were exposed for the first time – an immunizing effect.

Bonanno et al. (2005) considered resilience among people in or near the WTC, considering specifically, trait self-enhancement, and the tendency toward overly positive or unrealistic self-servicing biases. Self-enhancement was associated with resilience right after 9/11, but to poorer social adjustment 18 months later.

ORGANIZATIONAL RESILIENCE

> When the worst happens, companies have to put their employee's interests above all else.
>
> (Argenti, 2002, p. 103)

Alpaslan and Mitroff (2004) examined the relationship between an organization's ethical orientation and crisis management initiatives before and after 9/11. Two ethical orientations were considered: a principles- or rule-based orientation and a consequences or cost–benefit approach. Organizations with the former valued a proactive approach to crises; organizations with the latter valued a reactive approach. Alpaslan and Mitroff found that the crisis management performance of proactive organizations was superior and they also experienced fewer crises than did organizations having a reactive approach. The authors propose that training managers consistent with a principles-based orientation will produce a more proactive and more effective approach to crisis management.

Buenza and Stark (2004) studied a Wall Street investment bank that lost its entire office and trading technology in 9/11. The trading staff survived but were forced to move operations to a temporary site in New Jersey. One year later the bank had returned to Wall Street, and restored its technology, without losing any of the traders that relocated to New Jersey. Traders had to deal with questions of uncertainty about the firm's future, the location of the firm should it survive, existential anxiety and their professional identities. This firm, prior to 9/11, had operated with a flat hierarchy. Groups of traders dealt with ambiguity, confusion and speed on a daily basis. The firm had HR policies that placed a high value on people. Immediately following the attack on the WTC, traders began calling each other to find out what was going on. The bank reopened immediately. The new offices were given the same labels as the old ones to restore meaning and order.

Buenza and Stark (2006) describe the actions of another investment bank adjacent to the WTC that moved to temporary quarters in New Jersey following the attack. Successful recovery following 9/11 involved a combination of 'planning and spontaneity, of redundancy and self-organization typical of firms with non-bureaucratic and non-hierarchical forms' (p. 1). Organizational responsiveness was less a function of contingency plans and hierarchical command structures than processes of self-organization and lateral coordination: easy access to managers, reduced emphasis on hierarchical status, emphasis on trust and sociability, physical proximity to colleagues and working collaboratively in networks. Six days after 9/11, this firm was trading again. No employee was hurt in the attacks. The human element was key in the recovery process – social ties and personal knowledge.

Dutton et al. (2002a, 2002b) describe the responses of Reuters America to the 9/11 attack. Reuters America had a data center in the WTC and four at the facilities in NYC as well as in Washington, DC and in the Sears Tower in Chicago. Employees were also at conferences and in meetings in NYC and elsewhere. Dutton et al. conducted interviews with 30 employees of Reuters America between October 19 and November 8 to chronicle the organization's responses. The following broad categories of action were observed:

1. *Turning chaos into order* A command center was established about an hour and a half after the attacks. The major priority was to find their people and continue to serve their clients. Those in the command center began to develop an approach for dealing with the situation. They found that efforts in 2000 to deal with the new millennium proved useful now.
   - Establish clear priorities People first, then customers, then the business.
2. *Searching for employees* They were able to locate all but six employees, those having died during the attack.
   - Caring for the families Managers contacted families and provided support and resources.
   - Mobilizing HR resources HR established a link with the command center, helped locate employees and families, and located counseling help. Within two days, Reuters had counselors on site and names of counselors were provided to employees as well.
   - Town hall meetings Three large town hall meetings were held on Friday September 21. Teleconferencing technology was used to allow employees across North America to take part.
3. *Technical recovery* Reuters made a commitment to getting their operations up and running within days. Customers were contacted to see whether they were alright and what Reuters could do to help. Customers were offered free access to previously paid-for services. An urgent need was to replace a data system lost in the WTC with a system located in another facility. Employees worked unusually long hours in difficult work settings to rebuild these systems. Individuals worked in client sites and provided services as fast as they could; cost was no object.

These responses by Reuters America reflected its culture; Reuters was committed to its people. The organization became flexible and creative in dealing with the crisis. Relationships between people became more open and trusting, both within Reuters and with clients. The importance of people was highlighted. The company was seen as more humane. The HR function gained in prominence. Reuters became more results oriented and less process oriented. Finally, Reuters developed procedures in the event that it faced another emergency.

Freeman et al. (2004) examined the losses and the recovery processes of Sandler, O'Neill & Partners, a financial services firm housed in the WTC but decimated in the 9/11 attacks. Sandler, O'Neill lost 39 percent of its workforce, two-thirds of its management committee and its entire office. Yet one year later, the firm had recovered and was performing at an even higher level. Freeman et al. identify Sandler, O'Neill's use of moral purpose as the key element in its recovery. This sense of moral purpose unleashed motivation, made use of outside help and tapped high levels of physical and psychological resources. Leadership played a key role in handling anxiety and grief and the culture encouraged and supported team work and self-management (Barbash, 2003). Freeman et al. identified nine other organizations that suffered large numbers of deaths, and while none has become as successful post-9/11 as Sandler, O'Neill, all have recovered or are recovering. Situational factors also played a role in Sandler, O'Neill's recovery. This firm had not suffered any previous attacks or crises. The attack was over quickly. The attackers were not known to the firm. The environment supported Sandler, O'Neill; they were attacked because they were a symbol of success. Finally, the firm was healthy prior to 9/11. The sense of moral purpose (survival, at war, don't give in to terrorists, take care of the families) provided the motivation, but others outside the organization were sympathetic to Sandler, O'Neill (gave them better deals) and the employee deaths created opportunities for new leadership and change. Resources also played a role in how Sandler, O'Neill responded. These included cognitive abilities (knowledge and experience), financial resources, an organic structure, employee attitudes (excellence, responsibility, helping others) and effective management and containment of anxiety and grief. But a sense of purpose was the key.

Gittell et al. (2006), using data from the US airline industry, found that the maintenance of positive relationships with employees and financial reserves contributed to organizational resilience following 9/11. Financial reserves allowed some airlines to retain staff (positive employee relationships) instead of dismissing them, and that contributed to airlines' recovery following the attacks. The data were collected from nine airlines, using publicly available sources. Financial resources included low levels of debt and high levels of cash on hand. Layoffs were measured using airline announcements. Performance recovery was measured by comparing stock prices at four different times following 9/11 with stock prices on September 10.

Kelly and Stark (2002) interviewed executives and employees in financial services firms with offices in the WTC and adjacent buildings that began trading less than a week after the attack. They found that strong personal ties among employees, lateral self-organization in the form of teams and non-hierarchical relations were critical to the recovery process.

Although working under stress, fear and grief, the achievements of women and men in the days and months following 9/11 have been impressive. The consensus from those in the trenches was that 'people' made the difference. Interestingly, the use of technology helped with communication. In addition, factors that prepared the firms for the uncertainty and chaos that followed 9/11 were those factors that made these firms competitive in the marketplace. The events of 9/11 are requiring organizations to rethink their resources, locations, strategy and management structures.

Kendra and Wachtendorf (2003a, 2003b) studied the reconstitution of the Emergency Operations Center (EOC) following its destruction in the WTC attack, to better understand resilience. Although the physical EOC facility was destroyed in the collapse of the WTC, the organization that had been created to manage crises in NYC continued. This organization called on the resources of NYC and surrounding communities, states and the federal government. Human relationships that existed before 9/11, organizational response patterns and assigned roles contributed to EOC resilience following the attack. The research team began fieldwork within two days after the attack, and continued for two months.

Resilience can be defined as 'an ability to sustain a shock without completely deteriorating; that is most conceptions of resilience include some idea of adapting to or "bouncing back" from a disruption' (Kendra and Wachtendorf, 2003b, p. 41). Resilience is still best thought of as an art as well as the application of scientific knowledge and techniques (Sutcliffe and Vogus, 2003).

Organizational features associated with resilience include redundancy, the capacity for resourcefulness, effective communication and the capacity for self-organization in the face of unanticipated and huge demands. Resilience embodies attitudes about useful actions; it is about coping skills, creativity, improvisation, established roles in a social system, wisdom and straight talk, and rapidity, among others.

Argenti (2002), based on interviews with managers of their experiences and responses to the events of 9/11, distilled five lessons for any company facing a crisis that impacts on its employees' confidence and motivation. Some involved preparation, having plans and mechanisms for emergency responses; others involved responding to the unexpected:

1. Get on the scene. Managers must be visible, need to indicate that dealing with the anxiety and distress of staff is their top priority, need to lead, and provide a sense of calm, reason and humanity.
2. Use communication channels carefully and creatively. See the mass media as allies.
3. Stay focused on the business. Focus on the work, capitalize on employees' needs to help operate with both head and heart.
4. Have a plan in place. Identify contingency work sites, set up communication channels.
5. Improvise, but from a strong foundation the firm's guiding values and principles were the best basis for guiding employee actions. Imbued with these, managers and employees will do the right thing.

Sheffi (2001, 2005, 2006) believes that organizational resilience is a supply chain issue. He and his colleagues studied how companies rebound – or fail to rebound – after a low probability/high impact event such as a terrorist attack, examining dozens of organizations. Their major conclusion was that an organization's ability to return to business depends more on decisions made before the event rather than during and after. This involves organizational culture, communication and flexibility. Companies need to integrate corporate risk management into their business strategies. Parts and factories should be interchangeable (identical).

Resilient companies communicate obsessively. Decision-making power resides at low organizational levels. People have a passion for their work. They know the mission of their organization and buy into it. Resilient organizations are flexible and agile. Companies in fast-changing industries are more likely to be resilient.

Buenza and Stark (2003) conclude that successful recovery in the financial services companies they studied was a combination of planning and spontaneity, of redundancy and self-organization, typical of firms with non-bureaucratic and non-hierarchical forms. This hinged on a socio-technical network of relations, connections, policies and time-sensitive data that could not be seen from the outside, distributed intelligence being a key part in all of this.

Barbash (2003) looked back at the 9/11 survivors of Cantor Fitzgerald, and concluded that working to rebuild the firm has been the key to healing.

Evacuating Buildings

Gershon (2006) studied individual organizational and structural factors on evacuation preparedness, and the evacuation process and decision making for survivors of the WTC attack. In the absence of leadership or direction, many people delayed their decision to evacuate, or actually froze and did not respond to messages. People with disabilities were dependent on others for help, sometimes slowing movement on the staircases. Evacuees were unfamiliar with the building, and unaware of the fire exits. Once off their own floor, they became spatially disoriented. Once outside the building, evacuees were unsure of how to vacate the immediate area. People were relieved when they met members of the uniformed services (fire, police) in the stairs and lobby and followed their direction. Natural leaders emerged and helped people evacuate.

Gershon recommends the following:

1. Evacuation leaders should be identified and trained for all buildings 10 floors or higher.
2. Plans for the evacuation of the disabled or infirm should be made, identifying assistants, elevator use and so on.
3. All employees should be fully oriented to the building, including all exits.
4. Pre-planning is needed for mass evacuation from urban areas.

ANTICIPATING AND PLANNING FOR POSSIBLE TERRORIST ATTACKS

Watkins and Bazerman (2003) suggest that there are signs, if anybody had noticed, that would have predicted 9/11. These included Islamic militants willing to become martyrs for their cause; increasing hostility to the US; the 1993 car bombing of the WTC; the failed hijacking of an Air France plane in 1995 planned to hit the Eiffel Tower; a failed 1995 Islamic plot to hijack US commercial planes over the Pacific Ocean; plans to crash a light plane filled with explosives on to the CIA's headquarters; and several US government reports indicating lack of airport security. Yet despite these signals, no precautionary measures were undertaken. Watkins and Bazerman attribute this to lapses in recognition of the potential disaster, low priority given to airline security, and the resulting failures by the airlines to mobilize resources to increase airline security. However, some surprises really are unpredictable.

Watkins and Bazerman identify three kinds of barriers to identifying predictable surprises: psychological, organizational and political. They propose a three-stage process for improving an organization's ability to predict surprises: recognition, prioritization and mobilization:

1. Recognition – is the organization scanning the environment for emerging threats? Leaders need to pay attention.

2. Prioritization – did the organization give priority, based on cost–benefit analyses, to threats having the highest costs? Serious threats have to be identified.
3. Mobilization – did the organization take precautionary measures equal to the risks involved? Leaders need to respond effectively and appropriately.

Psychological barriers involve cognitive biases (for example, giving more weight to evidence supporting our position, see things as better than they are, try to maintain the status quo), which make it harder to anticipate surprises. *Organizational barriers* involve silos and fragmented information that make it difficult to get a complete understanding. *Political barriers* are political vulnerabilities (for example, imbalances of power) within an organization.

Watkins and Bazerman offer the following practical advice (pp. 79–80) to improve the prediction of 'predictable surprises':

1. Ask yourself and colleagues, 'What predictable surprises are currently brewing in our organization?'.
2. Use scenario planning and risk assessment tools. Include outsiders in these exercises.
3. Create cross-functional teams.
4. Use action learning groups of high potential managers to collect information on and analyze business challenges.
5. Develop mechanisms and rewards to support information sharing and common objectives.
6. Develop informal and formal networks to influence potential decisions.

IS PLANNING AND PREPARATION REALLY POSSIBLE?

McConnell and Drennan (2006) think that planning and preparing for crises may in fact be 'mission impossible'. Crises and disasters are low probability events that place huge demands on resources and have to compete with other priorities. It is difficult to order possible threats. Planning for crisis requires integration and synergy across various networks that are usually fragmented. Finally, planning and preparation requires extensive and costly training and exercises that typically only produce a symbolic or surface appearance of readiness.

Boin and Smith (2006) emphasize a new relationship of the private sector which operates critical infrastructures and public management. Thus, private and public sectors must together deal with the prevention of attacks, effective communication of information across organizational boundaries and occurrence of crisis decision making – all in all a complex task.

IMPROVING RESPONSES TO DISASTERS

There appears to be common shortcomings in the nature of the responses when disasters strike. Gheytanaki et al. (2007) identified a dozen major failures that have occurred in major disasters including Hurricane Katrina:

- lack of efficient communication;
- poor coordination plans;
- ambiguous authority relationships: who is in charge;
- who should be in charge;
- counter-terrorism versus all-hazards response;
- ambiguous training standards and lack of preparation;
- limited learning in 'lessons learned';
- performance assessment not integrated into the process;
- the role of socio-economic and racial factors in responses;
- rumor and chaos;
- personal and community differences in levels of preparedness; and
- different views on disaster mental health and the role of mental health professionals.

Most writing to date has focused on the final stage of a disaster – dealing with trauma – rather than on prevention. There is material on a variety of sources that address failures of communication, coordination, planning, and command. We know something about the ways individuals behave in disasters. Basic research has been carried out for some time on learning, training and performance assessment. Progress is also being made in the training of disaster mental health professionals. Applying those resources to prevention – as well as disaster response – is long overdue.

PREPARING FOR THE FUTURE: OUR CHILDREN'S CHILDREN'S WAR

Canton (2006) lists terrorism as one factor that will shape the world over the next 20 years. He believes that terrorism is increasing and that decreasing the gap between the developed nations (the haves) and the developing nations (the have-nots) is crucial. He concludes that if this gap widens, terrorism will increase; poverty and terrorism go together:

- Crime, drug and terrorist networks exploit impoverished war zones in Colombia, Somalia, Chechnya and Bosnia.
- Terrorists control trade in many conflict-ridden zones in Africa, Asia and the Middle East.
- Al Qaeda recruits among the poor in nations such as Malaysia, Iraq, Iran, Sierra Leone, Indonesia, Afghanistan and Pakistan.
- Al Qaeda also recruits from the West's Muslim community in the US and Europe, as evidenced by recent bombings, arrests, and failed plots.

The events of 9/11 have dramatically changed America's (and many parts of the world's) vision of the future by raising fear and uncertainty to new levels. Thus, global security and terrorism became important priorities following 9/11. The forces of globalization have had a significant influence on the world today, in some cases magnifying the problems. These forces include linking the economies of various nations into an interconnected network, increased culture clash between nations, and has influenced terrorism, potentially increasing or decreasing terrorism levels.

Canton believes that the Third World War has already started, the first shot being 9/11. A new set of risks and threats are now upon us. Unfortunately most of the developed world is still unprepared for terrorism. It is hoped that the chapters that follow will naturally identify the most pressing needs here and point to necessary individual, organizational and national initiatives to address them (Felton, 2004; Baer et al., 2005).

NOTE

1. Preparation of this manuscript was supported in part by York University, Canada. Louise Coutu prepared the manuscript.

REFERENCES

Aber, J.L., Gershoff, E.T., Ware, A. and Kotter, J.A. (2004), 'Estimating the effects of September 11th and other forms of violence on the mental health and social development of New York City's youth: a matter of context', *Applied Developmental Science*, **8**, 111–29.

Ahern, J., Galea, S., Resnick, H., Kilpatrick, D., Bucuvalas, M., Gold, J. and Vlakov, D. (2002), 'Television images and psychological symptoms after the September 11 terrorist attacks', *Psychiatry*, **65**, 289–300.

Ahmed, M. (2006), *The London Bombings: An Independent Inquiry*, London: Gerard Duckwater.

Alavosius, M.P., Braksick, L.W., Daniels, A.C., Harshbarger, D., Houmanfar, R. and Zielstra, J. (2002), 'The impact of terrorism on the US economy and business', *Journal of Organizational Behavior Management*, **22**, 3–26.

Alpaslan, M.C. and Mitroff, I.T. (2004), 'Bounded morality: the relationship between ethical orientation and crisis management before and after 9/11', in M.A. Rahim, K.D. Mackenzie and R.T. Golembiewski (eds), *Current Topics in Management*, Stanford, CT: JAI Press, pp. 137–55.

AON Consulting (2001), *United States back@work*, Chicago, IL: AON Consulting.

Argenti, P. (2002), 'Crisis communication: lessons from 9/11', *Harvard Business Review*, December, 103–9.

Asarada, J., Glynn, S., Pynoos, R.S., Nahum, J., Guthrie, D., Cantwell, D.P. and Franklin, S. (1999), 'When the earth stops shaking: earthquake sequelae among children diagnosed for pre-earthquake psychopathology', *Journal of the American Academy of Child and Adolescent Psychiatry*, **38**, 1016–23.

Assanangkornchai, S., Tangboonngam, S. and Edwards, J.G. (2004), 'The flooding of Hat Tai: predictors of adverse emotional responses to a natural disaster', *Stress and Health*, **20**, 81–9.

Babbs, C. and O'Connor, B. (2003), 'Dealing with the threat of an attack through the post using biological agents: the UK experience', *Journal of Contingencies and Crisis Management*, **11**, 118–23.

Baer, M., Heron, K., Morton, O. and Ratliff, E. (2005), *Safe: The Race to Protect Ourselves in a Newly Dangerous World*, New York: HarperCollins.

Barbash, T. (2003), *On Top of the World: Cantor Fitzgerald, Howard Lutnick, and 9/11: A Story of Loss and Renewal*, New York: HarperCollins.

Bar-Tal, D. and Labin, D. (2001), 'The effect of a major attack on stereotyping: terrorist attacks in Israel and Israeli adolescents' perceptions of Palestinians, Jordanians and Arabs', *European Journal of Social Psychology*, **31**, 265–80.

Ben-Ezra, M., Essar, N. and Saar, R. (2005), 'Post-traumatic reactions among rescue personnel 96 hours after the Hilton Hotel bombing in Sinai: the effect of previous exposure', *Stress and Health*, **21**, 269–72.

Benight, C.C., Freyaldenhoven, R.W., Hughes, J., Ruiz, J.M. and Zoschke, T.A. (2000), 'Coping, self-efficacy, and psychological distress following the Oklahoma City bombing', *Journal of Applied Social Psychology*, **30**, 1331–44.

Bennett, C. and French, M. (2003), 'The state of privacy in the Canadian State: fallout from 9/11', *Journal of Contingencies and Crisis Management*, **11**, 2–11.

Bernasek, A. (2002), 'The friction economy', *Fortune*, **145**, 104–12.

Bligh, M.C., Kohles, J.C. and Meindl, J.R. (2004), 'Charting the language of leadership: a methodological investigation of President Bush and the crisis of 9/11', *Journal of Applied Psychology*, **89**, 562–74.

Boin, A. and Smith, D. (2006), 'Terrorism and critical infrastructures: implications for public–private crisis management', *Public Money and Management*, No. 1, 295–304.

Bonanno, G.A., Rennicke, G. and Dekel, S. (2005), 'Self-enhancement among high-exposure survivors of the September 11th terrorist attack: resilience or social maladjustment?', *Journal of Personality and Social Psychology*, **88**, 984–98.

Buenza, D. and Stark, D. (2003), 'The organization of responsiveness: innovation and recovery in the trading rooms of Lower Manhattan', *Socio-Economic Review*, **1**, 135–64.

Buenza, D. and Stark, D. (2004), 'Survival and sensemaking: organizational resilience in Wall Street Trading Room after 9/11', paper presented at the Annual Meeting of the Academy of Management, New Orleans, August.

Buenza, D. and Stark, D. (2006), *Trading Sites – Destroyed, Revealed, Restored*, New York: Social Science Research Council.

Canton, J. (2006), *The Extreme Future: The Top Trends that Will Reshape the World for the Next 5, 10 and 20 Years*, New York: Dutton.

Collins, D.L., Baum, A. and Singer, J.E. (1983), 'Coping with chronic stress at Three Mile Island: psychological and biochemical evidence', *Health Psychology*, **2**, 149–66.

Daw, J. (2002a), 'Practice net takes a snapshot of psychology practice after Sept. 11', *Monitor on Psychology*, **33**, 14.

Daw, J. (2002b), '"Window to self-discovery" opened after Sept. 11', *Monitor on Psychology*, **33**, 15.

Day, T.G. (2003), 'The Autumn 2001 anthrax attack on the United States Postal Service: the consequences and response', *Journal of Contingencies and Crisis Management*, **11**, 110–17.

Dutton, J.E., Quinn, R. and Pasick, R. (2002a), *The Heart of Reuters – Part A*, Ann Arbor, MI: Center for Positive Organizational Scholarship, Ross School of Business, University of Michigan.

Dutton, J.E., Quinn, R. and Pasick, R. (2002b), *The Heart of Reuters – Part B*, Ann Arbor, MI: Center for Positive Organizational Scholarship, Ross School of Business, University of Michigan.

Esses, V.M., Dovidio, J.F. and Hodson, G. (2002), 'Public attitudes toward immigration in the United States and Canada in response to the September 11, 2001 attack on America', *Analysis of Social Issues and Public Policy*, **2**, 69–85.

Felton, C.J. (2004), 'Lessons learned since September 11th 2001 concerning the mental health impact of terrorism, appropriate response strategies and future preparedness', *Psychiatry*, **67**, 147–52.

Fielding, D. (2004), 'How does violence conflict affect investment location decisions? Evidence from Israel during the Intifada', *Journal of Peace Research*, **41**, 465–84.

Fischer, H.W. (2002), 'Terrorism and 11 September 2001: does the "behavioral response to disaster" model fit?', *Disaster Prevention and Management*, **11**, 123–7.

Freeman, S.F., Hirschhorn, L. and Matty, M. (2004), 'Organizational resilience and moral purpose: Sandler O'Neill & Partners, L.P. in the aftermath of September 11, 2001', paper presented at the Annual Meeting of the Academy of Management, New Orleans, August.

Friedberg, J.P., Adonis, M.N., Von Bergen, H.A. and Suchday, S. (2005), 'September 11th related stress and trauma in New Yorkers', *Stress and Health*, **21**, 53–60.

Frohman, D. (2006), 'Leadership under fire', *Harvard Business Review*, December, 124–31.

Frohman, D. and Howard, R. (2008), *Leadership the Hard Way*, San Francisco, CA: Jossey-Bass.

Galea, S., Ahern, J., Resnick, H., Kilpatrick, D., Bucuvalas, M. and Gold, J. et al. (2002), 'Psychological sequelae of the September 11th terrorist attacks in New York City', *New England Journal of Medicine*, **346**, 982–7.
Gersham, J. (2004), 'A secure America in a secure world. Task Force on Terrorism', September 9, Foreign Policy in Focus, www.fpif.org, p. 8.
Gershon, R. (2006), 'World Trade Center Evacuation Study', paper presented at the Work, Stress and Health 2006 conference sponsored by the American Psychological Association, Miami, FL, March.
Gheytanaki, A., Joseph, L., Gierlach, E., Kimpara, S., Housley, J., Franco, Z.E. and Bentler, L.E. (2007), 'The dirty dozen: twelve failures of the Hurricane Katrina response and how psychology can help', *American Psychologist*, **62**, 118–30.
Gil-Rivas, V., Holman, E.A. and Silver, R.C. (2004), 'Adolescent vulnerability following the September 11th terrorist attacks: a study of parents and their children', *Applied Developmental Science*, **8**, 130–42.
Gittell, J.H., Cameron, K., Lim, S. and Rivas, L. (2006), 'Relationships, layoffs and organizational resilience', *Journal of Applied Behavioral Science*, **42**, 300–329.
Globe and Mail (2007), 'Bank rules tightened after 9/11', Toronto, January 18, A9.
Gorman, L. and McKee, K.D. (1990), 'Disaster and its aftermath', *HR Magazine*, **35**, 54–8.
Gould, M.S., Munfakh, J.L.H., Kleinman, M., Lubell, K. and Provenzano, D. (2004), 'Impact of the September 11th terrorist attacks on teenagers' mental health', *Applied Developmental Science*, **8**, 158–69.
Heritage Preservation (2002), *Cataclysm and Challenge: Impact of September 11, 2001 on Our Nation's Cultural Heritage*, New York: Heritage Preservation.
Hobfoll, S.E., Canetti-Nisim, D. and Johnson, R.J. (2006), 'Exposure to terrorism, stress-related mental health symptoms, and defensive coping among Jews and Arabs in Israel', *Journal of Consulting and Clinical Psychology*, **74**, 207–18.
Ignatieff, M. (2004), *The Lesser Evil*, Princeton, NJ: Princeton University Press.
Kean, T.H. and Hamilton, L.H. (2004), *The 9/11 Report: The National Commission on Terrorist Attack from the United States*, New York: St. Martin's Press.
Kelly, J. and Stark, D. (2002), 'Crisis, recovery, innovation: learning from 9/11', *Environment and Planning A*, **34**, 1523–33.
Kendra, J. and Wachtendorf, T. (2003a), 'Creativity in emergency response to the World Trade Center disaster', in J.L. Monday (ed.), *Beyond September 11: An Account of Post Disaster Research*, Special Publication 39, Boulder, CO: Institute of Behavioral Science, National Hazards Research and Applications Information Center, University of Colorado, pp. 37–53.
Kendra, J.M. and Wachtendorf, T. (2003b), 'Elements of resilence after the World Trade Center disaster: reconstituting New York City's Emergency Operations Center', *Disasters*, **27**, 37–53.
Kisoc, I. (2006), 'Stress symptoms of survivors of the Marmora Region (Turkey) earthquakes: a follow-up study', *International Journal of Stress Management*, **13**, 118–26.
Knudsen, H.K., Roman, P.M., Johnson, J.A. and Ducharme, L.J. (2005), 'A changed America? The effects of September 11th on depressive symptoms and alcohol consumption', *Journal of Health and Social Behavior*, **46**, 260–73.
Kondrasuk, J.N. (2004), 'The effects of 9/11 and terrorism on human resource management: recovery, reconsideration and renewal', *Employee Responsibilities and Rights Journal*, **16**, 25–37.

Leonard, B. (2001a), 'EAPs pushed to the limit following the terrorist attacks', *HR Magazine*, **46**, 25.

Leonard, B. (2001b), 'Employers' healthcare costs may rise even more because of the attacks', *HR Magazine*, **46**, 28.

Leonard, B. (2002a), 'After September 11, a changed workforce', *HR News*, **21**, 1–7.

Leonard, B. (2002b), 'Anniversary of a tragedy', *HR Magazine*, **47**, 44–6.

Liverant, G.I., Hoffman, S.G. and Litz, B.T. (2004), 'Coping and anxiety in college students after the September 11th terrorist attacks', *Anxiety, Stress and Coping*, **17**, 127–40.

Mainiero, L.A. and Gibson, D.E. (2003), 'Managing employee trauma: dealing with the emotional fallout from 9-11', *Academy of Management Executive*, **17**, 130–43.

McConnell, A. and Drennan, L. (2006), 'Mission impossible? Planning and preparing for crisis', *Journal of Contingencies and Crisis Management*, **14**, 59–70.

Miguel-Tobal, J.J., Cano-Vindel, A., Gonzales-Ordi, H., Iruarrizaga, I., Rubenstein, S., Viahov, D. and Galea, S. (2006), 'PTSD and depression after the Madrid March 11, train bombings', *Journal of Traumatic Stress*, **19**, 69–80.

Miguel-Tobal, J.J., Vindel, A.C., Iruarrizaga, I., Ordi, H.G. and Galea, S. (2005), 'Psychopathological repercussions of the March 11 terrorist attacks in Madrid', *Psychology in Spain*, **9**, 75–80.

National Post (2007), 'Letter bombs have UK on high alert', Toronto, February 8, A14.

North, C.S. and Pfefferbaum, B. (2002), 'Research on the mental health effects of terrorism', *Journal of the American Medical Association*, **288**, 633–6.

Olivas-Lujan, M.R., Harzing, A.W. and McCoy, S. (2004), 'September 11, 2001: two quasi-experiments on the influence of threats on cultural values and cosmopolitanism', *International Journal of Cross-Cultural Management*, **4**, 211–28.

Peterson, J. and Seligman, M.E.P. (2003), 'Character strengths before and after September 11', *Psychological Science*, **14**, 381–4.

Poe, A.C. (2001), 'Long after the initial shock of the September terrorist attacks has worn off, employees may need your continuing support. Aftershocks of the 11th', *HR Magazine*, **46**, 46–51.

Rasinski, K.A., Berktold, J., Smith, T.W.T. and Albertson, B.L. (2002), *America Recovers: A Follow-up to a National Study of Public Response to the September 11th Terrorist Attacks*, Chicago, IL: National Opinion Research Center.

Schlenger, W.F., Caddell, J., Ebert, L., Jordan, B., Rourke, K.M., Wilson, D. et al. (2002), 'Psychological reactions to terrorist attacks: findings from the National Study of America's reactions to September 11', *Journal of the American Medical Association*, **288**, 581–8.

Schuster, M.A., Stein, B.D., Jaycox, L.H., Collins, R.L., Marshall, G.N., Elliot, M.N. et al. (2001), 'A national survey of stress reactions after the September 11, 2001, terrorist attacks', *New England Journal of Medicine*, **345**, 1507–12.

Sheffi, Y. (2001), 'Supply chain management under the threat of international terrorism', *International Journal of Logistics Management*, **12**, 1–11.

Sheffi, Y. (2005), *The Resilient Enterprise: Overcoming Vulnerability for Competitive Advantage*, Cambridge, MA: MIT Press.

Sheffi, Y. (2006), 'Yossi Sheffi: the thought leader interview', *Strategy and Business*, Spring, 93–101.

Sheridan, L.P. (2006), 'Islamaphobia pre- and post-September 11th, 2001', *Journal of Interpersonal Violence*, **21**, 317–36.

Silver, R.C., Homan, E.A., McIntosh, D.N., Poulin, M. and Gil-Rivas, V. (2002), 'Nationwide longitudinal study of psychological responses to September 11th', *Journal of the American Medical Association*, **288**, 1235–44.
Smith, T.W., Rasinski, K.A. and Tace, M.T. (2001), 'America rebounds: a national study of public response to the September 11th terrorist attacks', October 25; National Opinion Research Center, University of Chicago, http://www.norc.uchicago.edu/projects/reaction/index.asp, September 12, 2004.
Sprang, G. (1999), 'Post-disaster stress following the Oklahoma City Bombing: an examination of three community groups', *Journal of Interpersonal Violence*, **14**, 169–83.
Sprang, G. (2000), 'Coping strategies and traumatic stress symptomatology following the Oklahoma City bombing', *Social Work and Social Sciences Review*, **8**, 207–18.
Sprang, G. (2001), 'Vicarious stress: patterns of disturbance and use of mental health services by those indirectly affected by the Oklahoma City bombing', *Psychological Reports*, **89**, 331–8.
Stein, B.D., Elliott, M.N., Jaycox, L.H., Collins, R.L., Berry, S.H., Klein, D.J. and Schiater, M.A. (2004), 'A national longitudinal study of the psychological consequences of the September 11, 2001 terrorist attacks: reactions, impairment and help-seeking', *Psychiatry*, **67**, 105–17.
Sutcliffe, K.M. and Vogus, T. (2003), 'Organizing for resilience', in K.S. Cameron, J.E. Dutton and R.E. Quinn (eds), *Positive Organizational Scholarship: Foundations of a New Discipline*, San Francisco, CA: Berrett-Koehler, pp. 94–110.
Watkins, M.D. and Bazerman, M.H. (2003), 'Predictable surprises: the disasters you should have seen coming', *Harvard Business Review*, March, 72–80.
Weisenberg, M., Schwarzwald, J., Waysman, M., Solomon, Z. and Klingman, A. (1993), 'Coping of school-age children in the sealed room during scud missile bombardment and postwar stress reactions', *Journal of Consulting and Clinical Psychology*, **61**, 462–7.
Wrzesniewski, A. (2002), 'It's not just a job, shifting meanings of work in the wake of 9/11', *Journal of Management Inquiry*, **11**, 36–58.
Zeidner, M. (2006a), 'Individual differences in psychological reactions to terror attack', *Personality and Individual Differences*, **40**, 771–81.
Zeidner, M. (2006b), 'Gender group differences in coping with chronic terror', *Sex Roles*, **54**, 297–310.
Zimbardo, P. (2002), 'President's column', *Monitor on Psychology*, **33**, 16.
Zivotofsky, A.Z. and Koslowsky, M. (2005), 'Gender differences in coping with the external major stress of the Washington, DC sniper', *Stress and Health*, **21**, 27–31.

2. A US analysis of terrorism

Jack N. Kondrasuk and Elizabeth Arwood[1]

INTRODUCTION

The morning of 11 September 2001 seemed like a typical morning to most citizens of the United States (US). Businesses and people were going along at their usual, hectic pace. Then at 8:46 a.m. Eastern Standard Time a plane crashed into the North Tower of the World Trade Center in New York City. About 15 minutes later another hijacked commercial airliner crashed into the South Tower. News media across the US began reporting these horrific and unexplainable events. People were transfixed by the fire and smoke bursting out of the Twin Towers as viewed on their TV screens. The US populace was accustomed to seeing violence on TV programs and knew it was not real. But it was real. We saw the desperation and horror of people trying to escape the infernos by jumping out of windows of the Twin Towers, of people running for their lives through the billowing smoke and flying debris as the South Tower completely collapsed and disintegrated at 10:05 a.m. Resembling a mirror image, the North Tower collapsed at 10:28 a.m. The firefighters and police officers attempted to evacuate the employees of the buildings. Many of these public servants, as well as bystanders watching in disbelief, were engulfed in the infernos and buried alive. At 9:45 a.m. the headquarters of the US military, the Pentagon, was directly hit with a third commercial airliner also loaded with fuel and used as a flying bomb. A fourth hijacked commercial airliner crashed into a field in Pennsylvania at 10:10 a.m. as the passengers fought the hijackers and caused the abortion of the plan to crash it as a bomb into an important government headquarters target in Washington, DC. The nation was paralyzed in shock as about 3000 relatives, friends, and fellow citizens were killed on that morning. That was what brought America together and into the 'war on terrorism'.

Terrorism is a widespread problem throughout the world (Kondrasuk et al., 2005). It occurred in 54 countries in the world from May 2003 to May 2004 and on all but one continent (Antarctica) (Terrorism Research Center, 2007). There were 20 other countries that had religious militant groups that could easily have caused terrorist events (Deming, 2004). Additional countries, such as South Africa, were not listed in any of the aforementioned groupings, but they are also involved in the war on terrorism by arresting suspected terrorists within their borders (Kondrasuk et al., 2005). While terrorism is a global issue, this chapter will focus on terrorism from the viewpoint of the US. The people of the United States of America knew of terrorism before '9/11' (as that fateful day was termed). Before 9/11, the majority of the US population held more of an intellectual and remote view of terrorism. The events of 9/11 drastically changed the way Americans viewed terrorists and terrorist attacks. As a reaction to 9/11, the US pulled together and focused on the self-proclaimed 'war on terrorism'. The US saw the resultant invasions of Afghanistan, and later Iraq, as necessary to avenge the 9/11 attacks and to protect the homeland. The continued occupation of Iraq and subsequent US casualties have eroded US support for the country's involvement even though US protection agencies have thwarted potential terrorist attacks. The definition of the 'US view of terrorism' varies depending on the time period – whether before 9/11, at 9/11 and immediately after, or post-9/11 through to the present time. Over this time the mindset of the American people has continued to evolve regarding 'terrorism'.

PURPOSE OF CHAPTER

This chapter will seek to explain the US view of terrorism and its evolution. It will analyze how the US tends to view the definition of terrorism, the history of terrorism, the present situation, and the future of terrorism for the US. The US view can be compared with and contrasted to views of terrorism from various parts of the world as expounded in other chapters in this book, but it is probably essential to consider the US view, as the US has been considered the number one terrorist target in the world (National Commission on Terrorism, 2003) and the premier economic and military power (Wikipedia, 2004). (Note: this chapter is a general media and research-based view and not an official government pronouncement.)

DEFINITION OF TERRORISM

'Terrorism' is difficult to define. A 'terrorist' is variously seen as a murderer, an insurgent, a guerilla, a militant, a revolutionary, a rebel, a warrior, a soldier, or a freedom fighter. In fact, it is common in the US media to refer to US opponents in Iraq as 'insurgents' if their group is entirely from Iraq but 'terrorists' if their members come from outside Iraq (Ifel, 2004). The US Code for defining an act of terrorism is

> an activity that – (A) involves a violent act or an act dangerous to human life that is a violation of the criminal laws of the United States or any State, or that would be a criminal violation if committed within the jurisdiction of the United States or of any State; and (B) appears to be intended (i) to intimidate or coerce a civilian population; (ii) to influence the policy of a government by intimidation or coercion; or (iii) to affect the conduct of a government by assassination or kidnapping. (Chomsky, 2006)

However, the US State Department concludes that there is no one universally agreed-upon definition for terrorism. The State Department itself, for explanatory and communication purposes, defines 'terrorism' as 'premeditated, politically motivated violence perpetuated against non-combatant targets by sub-national groups or clandestine agents, usually intended to influence an audience' (Kellerhals, 2008). A dictionary definition of terrorism is 'violence or the threat of violence especially bombing, kidnapping, and assassination carried out for political purposes' (Microsoft, 2007). For the purposes of this chapter, terrorism will be defined as: (i) a non-military group of people united in a common political, religious, or ideological cause, (ii) who operate in a clandestine way, (iii) without publicly known headquarters, (iv) committing or threatening to commit acts of significant violence, (v) against those who oppose their views, and (vi) are mainly civilians. Terrorists (vii) do not abide by civil laws or formal rules of conduct such as the Geneva Convention, (viii) try to produce fear in others, and (ix) seek to influence others to take a particular political course of action. Very importantly, (x) they are based outside the US and could be referred to as 'Foreign Terrorist Organizations' (FTOs). This definition excludes the Ku Klux Klan, white supremacy groups, and workplace violence in the US (Kondrasuk, 2004).

HISTORY OF TERRORISM REGARDING THE US

Pre-9/11

Although there may have been terrorist-type activities conducted during the Revolutionary and Civil Wars in the US, the perpetrators were usually from an army fighting the US or groups within the US rather from non-military groups based outside. The first modern recorded terrorist activity in the US was a bomb planted in New York airport luggage in 1975. As a precursor to the later 9/11 attack, al Qaeda operatives conducted a truck bombing attack on New York's World Trade Center in 1993 where six people were killed and thousands injured. Probably due to the relatively small damage to the structure, it made the news but not the major headlines in the US. The Oklahoma City Federal Building bombing in 1995 did make the headlines. Initially thought to be conducted by foreign terrorists, later investigations showed a US extremist group to be the culprits (Terrorism Research Center, 2007).

Various terrorist events have occurred at US interests outside of US soil. There were two attacks on US embassies in 1983. In 1985 one American man was killed when the *Achille Lauro*, an Italian cruise ship, was hijacked by the Palestine Liberation Front. The 1988 bombing of a US airline Pan Am flight over Scotland resulted in the deaths of 270 passengers, including 189 Americans. Two men accused of being Libyan intelligence agents were eventually charged with planting the bomb (Kushner, 2003). The Japanese Red Army (a terrorist group) car bombed the United Service Organizations Club in Naples, Italy in 1988 (ibid.). The Khobar Towers complex near a US military complex in Saudi Arabia was bombed in 1996. The US embassies in Nairobi, Kenya and Dar es Salaam, Tanzania were both bombed in 1998 by al Qaeda terrorists (Terrorism Research Center, 2004). Al Qaeda also attacked a US Navy destroyer, the USS *Cole*, in the Persian Gulf in 2000 (Infoplease Staff, 2007). Then came the events of 11 September 2001 and everything changed for the US.

9/11

The September 11th attacks penetrated into the daily lives of nearly every US citizen. The initial media headlines stated that over 10,000 people were killed that morning; the final count was about 3000 dead (September 11 News, 2004). Ironically, it was on 9/11 that the majority (66 percent) of the population had the greatest confidence that the government could protect them from future terrorist attacks (Polling Report, 2007, ABC News/ Washington Post Poll, July 11–15, 2002). In addition to the astounding number of deaths caused by the attacks of 9/11, the events also had a significant economic impact. Thirty percent of the office space in Lower Manhattan, the third largest business district in the US, was damaged or destroyed and the fledgling recession at the time was spurred to greater growth (Wikipedia.com).

Public opinion polls have shown the significance of the September 11th attacks. Surveys showed that respondents were more likely to view the 9/11 attacks as more serious (35 percent) than Pearl Harbor (14 percent). In fact, 46 percent of people surveyed judged the 9/11 attacks as the most significant event of their lives (Polling Report, 2007, Pew Research Center for the People & the Press survey, August 9–13, 2006). Almost all (98 percent) of the respondents in another poll could recall where they were when they first heard about the 9/11 attacks. In August 2006, nearly five years after the attacks, 85 percent of the respondents still thought about 9/11 at least a few times a month. Over 90 percent believed that 9/11 had changed life in the US, but fewer (66 percent) believed it changed their own lives some or a great deal (Polling Report, 2007, Time/Discovery Channel Poll, August 22–24, 2006). Between 53 and 68 percent of the respondents believed that the 9/11 attacks were motivated in a major way by US ties to Israel in the Palestinian situation, resentment of US military and economic power, and/or opposition to the US military presence in the Persian Gulf region (Polling Report, 2007, Newsweek Poll, September 20–21, 2001).

The terrorist attacks on 9/11 made many changes in US citizens' daily lives. The population felt very vulnerable, which brought diverse groups together in a way not seen before. Symbols of American nationalism were evident in forms such as US flags proudly flown on buildings, to vehicle bumper stickers with slogans such as 'These Colors Don't Run' (Neier, 2004). Because of strong anger against the suspected perpetrators of the 9/11 attacks, there were two exceptions to the commingling of the US population. Those of Arab extraction and Muslim faith faced increased discrimination in the form of burning mosques, refusing employment, or even death threats. The government stepped in to threaten stern prosecution of any discrimination against Arabs or Muslims. Ironically, at the same time, some people in the US became more inquisitive about Islam and its teachings, and many citizens purposely sought closer ties to Arabs and Muslims to counteract the hate shown by others.

In addition to the change in public sentiment, the 9/11 attacks produced notable effects on American organizations and their employees. People avoided travel by public transportation, especially by airplane. The US travel industry, the insurance industry, the economy, and the stock market were damaged by the attacks (Kondrasuk, 2005). For instance, since September 2001, the airline industry has lost $40 million, 155,000 employees, and has taken 816 planes out of service (Hutson, 2006). The attacks brought the New York Stock Exchange to an abrupt halt and exacerbated a fledgling recession. Employees, wanting to spend less time at work and more time with their families, still became more loyal to their employers (Burke, 2005). The workforce witnessed little change in employee job satisfaction. Despite the preservation of job satisfaction, immediately following September 11, many Americans suffered symptoms of stress and other deleterious health effects. Many people had increased anxiety and avoided tall buildings and crowds. A few days after September 11, a survey showed that 90 percent experienced one or more symptoms of stress (Schuster, 2001). Two months later another poll was taken, showing that 16 percent of adults surveyed had persistent substantial symptoms of stress, and of those adults, 65 percent accomplished less at work. To this day, many

Americans continue to experience terrorism-related stress and disruption to their daily lives (Stein, 2004). Furthermore, 9/11 made the US want revenge!

Post-9/11

After seeking information on the perpetrators of 9/11 and options for dealing with them, President George W. Bush declared 'war on terrorism'. (Incidentally, the term has changed from the government's terming it the 'war on terrorism' to the 'war on terror' to the 'global war on terror' as abbreviated by GWOT.) This was the precursor to attacking the Taliban in Afghanistan in 2001 and seeking al Qaeda and its leader, Osama bin Laden. Polls of a cross-section of US citizens tended to show general support for the foray into Afghanistan. From October 2001 to April 2002, a relatively steady 87–91 percent of the US respondents approved of the US military incursion into Afghanistan (Polling Report, 2007, CBS News Poll, April 1–2, 2002). The US was able to quickly overthrow the Taliban government, a supporter of al Qaeda, in Afghanistan. Although al Qaeda bases and training camps were destroyed, the top leadership, including Osama bin Laden, evaded death or capture. Most US citizens (70 percent) felt that the war in Afghanistan was successful for the US (Polling Report, 2007, ABC News Poll, October 2–6, 2002).

Later, President Bush would order an attack on Iraq, allegedly because it had weapons of mass destruction (WMD) and to continue the war on terrorism. In 2002–03 the US sought United Nations support for invading Iraq. Gaining some, but not full UN support, the US, with strong support from its close ally the UK and other coalition force members, attacked Iraq in March 2003. This preemptive strike on Iraq was not strongly backed by the US population as 45 percent thought it was a good thing but 43 percent thought it was a bad thing (Polling Report, 2007, Knight Ridder poll, January 3–6, 2003). After the two-week initial sweep and take-over of Iraq by US and 40 other countries (the 'coalition forces'), guerrilla warfare ignited against the occupying coalition forces. This warfare was initially termed an 'insurgency' and conducted primarily by the 20 percent Sunni Muslim minority (former Iraqi President Saddam Hussein's political party when he was President). There is evidence that Saddam Hussein preplanned the insurgency before the war even started (Pound, 2004). The insurgency was assisted by foreign fighters (mainly Muslims from neighboring Arab countries, termed 'terrorists' by the US media) (Jack, 2004; Stojanovic, 2004; Vick, 2004). Iraq's Shiite Muslim (60 percent) majority was not initially seen as being involved in the physical insurgency – only seeking political power (Stojanovic, 2004; Vick, 2004). Some view the insurgency and terrorism in Iraq as part of an Islamic Jihad to make Iraq an Islamic state and training ground for like-minded fighters (Ware, 2004). In 2004, Abu Musab al-Zarqawi, a Jordanian citizen, declared his allegiance to al Qaeda and changed the name of his group to al Qaeda in Iraq. Subsequently, Osama bin Laden confirmed al-Zarqawi as his leader of al Qaeda in Iraq (CNN, 2004). The initial insurgency then evolved into a civil war between Shiite and Sunni Muslim factions in Iraq. This was further complicated by Muslim fighters infiltrating into Iraq to join in the fighting and by neighboring countries (such as Iran and Syria) supplying and/or allowing military goods to be smuggled into Iraq for military use.

The deaths and injuries of military personnel in Iraq and the involuntary extensions of soldiers' tours of duty have adversely affected enlistments in the US National Guard and Reserve and soured the US population to US involvement in Iraq. Surveys showed that the occupation of Iraq by US and coalition forces has affected the perceived threat of terrorist attacks in the US and the public view of the war on terrorism. Slightly over 40 percent of those polled think that the actions have made the US safer, while more than 50 percent think that it has increased US risk (Polling Report, 2007, Time Poll, November 29–December 1, 2005). In 2002, 60 percent of the US population believed that terrorists could always find some way to attack the US, while 88 percent believed that al Qaeda already had terrorists in the US that could launch another major attack (Polling Report, 2007, CNN/USA Today/Gallup Poll, September 2–4, 2002).

PRESENT US TERRORIST SITUATION

The official view of the US is that 'America is at war with a transnational terrorist movement fueled by a radical ideology of hatred, oppression, and murder' (White House, 2006). It states that the basic strategy to win the war on terror requires the US to:

1. develop more democracies in the world – especially in troubled, volatile areas;
2. prevent attacks by terrorists;
3. keep WMD from use by terrorists; and
4. keep rogue states from supporting or harboring terrorists.

Measured by surveys of the population, there has been a decrease in people's beliefs that there will be a future terrorist attack in the US. This information counters the belief that the US and its allies are having less success in the war against terrorism (Polling Report, 2007, CNN Research

Corporation Polls, January 19–21, 2007). People (41–54 percent) in the US generally believe that neither the US and its allies nor the terrorists are winning the war on terror (ibid.). However, if forced to choose, more believe that the US is winning (46 percent) than losing (33 percent) (Polling Report, 2007, Pew Research Center for the People & the Press survey, November 1–4, 2006). The reader can go to online sources for up-to-the-minute information about terrorism from a US perspective such as government websites (for example, www.state.gov), media (http://abcnews.go.com/US/Investigation/story?id=90203&page=1), US surveys (http://www.pollingreport.com/terror.htm), and private organizations specializing in terrorism information (for example, www.terrorism.com).

Terrorists and Goals

The US government officially lists countries in the world that are believed to sponsor terrorism. Although Afghanistan was removed from this list in 2002, Iraq in 2004, and Libya in 2006, the following remain: North Korea, Iran, Sudan, Cuba, and Syria. Various sanctions and other pressures are applied to these countries to stop their support of terrorism (State Department, 2007b). Iran was viewed by the US State Department as 'the world's pre-eminent state sponsor of terrorism' (Jack, 2004, p. 36). Supposedly, in 2003 Iran placed a $500 bounty on each US soldier killed and $1000 for each tank destroyed in Iraq, and set up a special group to assassinate US Coalition Provisional Authority administrator, Paul Bremer. Iran provided training opportunities, money, supplies, and entry to Iraq for terrorist groups such as Ansar al-Islam. Iran supposedly provided training and funds to anti-US Shiite cleric Moqtada al-Sadr and his Mahdi Army militia, Hezbollah, Ansar al-Islam, and al Qaeda (Jack, 2004).

The US State Department has developed a list of 42 presumed dangerous terrorist organizations that are based outside the US. These FTOs remain on the list until removal by an Act of Congress, a court order, a successful petition by the FTO in question (after being on the list for at least two years), or removal by the Secretary of State because of changed conditions. The State Department looks at the number of actual terrorist attacks committed by the designated group and the potential and intent for future terrorist attacks (State Department, 2007a). The main terrorist organization threatening the US at the present time is officially seen as al Qaeda – the one behind the 9/11 attacks (State Department, 2007c). It is seen by the US as a network rather than a single group; in Arabic, its name means 'the base'. It was considered stronger in 2004 than it was in 2001 even though the US has tried to destroy its leadership and training facilities in

Afghanistan. Osama bin Laden continues to lead al Qaeda even though he has a $25 million bounty on him and is constantly pursued. He is presumed to be living in the rugged mountains between Pakistan and Afghanistan. The State Department expects the US to be confronting al Qaeda for decades into the future (State Department, 2007c).

Who are the individual terrorists? The official US view states that terrorists believe they have no stake in the existing order, feel alienated, believe their grievances can be blamed on others, succumb to false and misleading information, thrive on conspiracy theories, and believe their ideology justifies killing others to achieve it. They are likely to join others of like mind in terrorist cells which may or may not belong to some of the previously mentioned terrorist organizations (White House, 2006). The government and the US populace also believe that there is a religious aspect that defines terrorists although there are two different views on this. On one side is the view that there is a global Islamic Jihad to install Islamic states in the world. This view says that the insurrections and terrorism around the world are the result of a groundswell of desire by Muslims as the world's fastest growing religion. In Kakutani's (2004) review of the book *Why the West is Losing the War on Terror*, the book's author believes the West looks at al Qaeda as a terrorist group, but it is really a group leading a worldwide Islamic insurgency. The other viewpoint says that the insurrections and terrorism are the result of a few Muslim extremists and are contrary to the view of the majority of the world's 1.3 billion Muslims and 7 million Muslims in the US (*Why Islam?*, 2005). Proponents of this view contend that the Qur'an forbids Muslims to kill non-combatants, forbids forcing others to adhere to Islam, and that Jihad mainly deals with the inner personal struggle to submit to God (ibid., 2004).

How do Americans view Muslims? The population believes that terrorist groups like al Qaeda have the ultimate goal of destroying the US (48 percent), taking over the Middle East for themselves (26 percent), or both (15 percent) (Polling Report, 2007, CBS News/New York Times Poll, September 15–19). Forty-two percent of Americans believe that Muslims around the world are sympathetic to terrorists, while 22 percent believe that Muslims are sympathetic towards the US regarding the war against terrorism (Polling Report, 2007, Fox News/Opinion Dynamics Poll, February 28–March 1, 2006). While 82 percent of the poll respondents stated that Muslim extremists would attack the US in the near future (Polling Report, 2007, Newsweek Poll, August 2–4, 2005), most believed that it is a small group of Muslim extremists – not Muslims in general – that are the terrorists (Polling Report, 2007, Pew Research Center for the People & the Press survey, July 7–10, 2005/Polling Report, 2007, Newsweek Poll, January 31–February 1, 2002).

Again, it was a situation of the respondents viewing Muslims and Arab Americans more positively than they thought others in the US viewed them. Most (62 percent) of the respondents believed that Arab Americans were *not* more sympathetic to terrorists than other US citizens (Polling Report, 2007, CBS News/New York Times Poll, September 20–23, 2001). However, a couple of weeks after 9/11, 90 percent of the respondents thought that it was very or somewhat likely that Arab Americans, Muslims and immigrants from the Middle East would be singled out unfairly by people in the US; 55 percent believed that Arabs themselves, around the world, had a negative view of the US (Polling Report, 2007, ABC News Poll, October 8–9, 2001); about 65 percent believed that the US war on terror would evolve into a world war with the US/allies on one side and the Arabs/Muslims on the other (ibid.); but 88 percent believed that it was important that the US get the support of as many Arab and Islamic countries as possible (Polling Report, 2007, Harris Poll, September 19–24, 2001).

Weapons Used By Terrorists

The main types of weapons employed by terrorists now are explosives (for example, suicide bombers, explosives in cars and trucks rammed into targets, and remote control detonation devices) and guns (for example, snipers, automatic weapons). Chemical (for example, sarin), biological (for example, anthrax), informational (computer hacking), and radiological (for example, dirty bombs) weapons are considered definite threats but so far have not been used extensively, and nuclear weapons (for example, atomic or hydrogen bombs) are very expensive and difficult to make. All weapons depend on the element of uncertainty leading to fear and in effect are concomitant with psychological weapons. More details on terrorist weapons can be found elsewhere (Kondrasuk, 2005).

Terrorist Attacks on US Soil

Between 9/11 and the present time, the US has encountered some actual and numerous attempted attacks by terrorists. There have been letter and package bombs sent to influential leaders – especially to congressional leaders. Anthrax-tinged letters caused some casualties. The perpetrators of the letter bombs and the anthrax infections were never found. Terrorists were caught on the West Coast before they could employ chemical attacks or 9/11-style airplane bombs on commercial towers in Los Angeles. There were warnings that West Coast bridges would be blown up, but none were. An al Qaeda operative was caught trying to detonate a bomb in his shoe while on a flight over the Atlantic Ocean. In the summer of 2004, an al Qaeda operative was accused of trying to blow up apartment buildings and setting off a dirty bomb (Hanley, 2004). During the fall of 2004, Osama bin Laden's videotape threatened 'to inflict catastrophic effects' on the US before the 2004 presidential election (Johnston and Stout, 2004). Following these events, the main topic in those presidential debates focused on terrorism (Terrorism Research Center, 2007).

Terrorist Attacks Overseas

US embassies overseas have been threatened and some actually attacked. In 2004, the US closed its consular offices in Bombay, India and warned US citizens against a terrorist attack (Reuters, 2004). There was an attack on the US consulate in Saudi Arabia in 2004 (CNN, 2004). Also in 2004, a bin Laden videotape asked followers to stop the flow of oil to the West (ibid.). Others who have helped the US are more likely to become terrorist targets themselves. For instance, the UK, Spain, Italy, the Philippines, and Poland have had their citizens in Iraq taken hostage by insurgents and terrorists for the stated purpose of coercing those countries to withdraw their troops from Iraq. The UK, a staunch ally of the US, suffered a significant terrorist attack of its subway and bus system in what the British refer to as '7/7' as it occurred on 7 July 2005. Many people were injured and 56 were killed (Wikipedia, 2007). It has also affected relations with Mexico, with increased security at the countries' border to stop undocumented people – including possible terrorists – from entering the US.

Effects of Terrorist Attacks

On society in general

Americans have tended to believe that terrorism is a threat to the country but less so to them personally. Over time, the majority of the US population have not felt that terrorist attacks will occur in their communities. In fact, from 2002 to 2005, 82–90 percent of the respondents believed that it was not too likely or not likely at all that a terrorist attack would occur in their community (Polling Report, 2007, CNN/USA Today/Gallup Poll. January 7–9, 2005) even though 72–90 percent of respondents in another national survey believed that terrorist attacks will occur somewhere else in the US during a similar time period (Polling Report, 2007, Time/CNN Poll, September 3–4, 2003). In 2003, 34 percent of the US population thought that terrorism was a threat to their lives, but few (24 percent) did anything about it and most (90 percent) went on with their lives as usual according to polls (Polling Report, 2007, Fox News/Opinion Dynamics Poll,

February 11–12, 2003). However, the views of safety vary politically. The percentages of those feeling much or somewhat safer after 9/11 is 75 percent all, 90 percent Republicans, and 69 percent Democrats (Polling Report, 2007, University of Pennsylvania National Annenberg Election Survey, March 1–15, 2004).

Terrorism has significantly affected US politics. The public's view of terrorism and homeland security was a significant issue in the 2004 national election. Terrorism became the main and most important focus of the 2004 presidential election campaign (Nagourney, 2005). According to a CNN poll, 49 percent of voters believed that terrorism was an extremely important campaign issue (Polling Report, 2007, CNN Poll, September 22–24, 2006). Since 2004, the US view of how the war on terrorism is being handled has changed dramatically as measured by public polls. In early 2004, public opinion regarding President Bush's handling of terrorism and homeland security remained strong, with a public approval of 65 percent (Polling Report, 2007, Newsweek Poll, January 18, 2007). According to the same poll, only 41 percent now approve of the way Bush is handling terrorism and homeland security. This change in public sentiment since 2004 may be the result of new intelligence information that Iraq may not have been as large a threat as previously thought, especially with regard to WMD, and the increasing number of fatalities of US military personnel.

On the government

Terrorism has drastically affected the US government and resulted in many changes in how the government functions. Officially, the government has listed priority actions in the GWOT as: (i) preventing terrorist attacks (by destroying terrorist leaders, followers, weapons, funding, communications, travel, and defending possible targets); (ii) denying WMD to terrorists and their supporters; and (iii) denying terrorists the control, support, or sanctuary of rogue states such as Iran and Syria (White House, 2006). The government has developed the US Counterterrorism Team to integrate all the government anti-terrorist operations. It consists of the White House, the Department of State, the Department of Defense, the Department of the Treasury, the Department of Justice, the Department of Homeland Security, the Central Intelligence Agency (CIA), the Office of the Director for National Intelligence, the Agency for International Development, and the National Counterterrorism Center (State Department, 2007b). The Counterterrorism Office of the State Department set policies for dealing with terrorists, such as make no concessions and set no deals, bring terrorists to justice, focus on and change countries sponsoring terrorism, and collaborate with and support allies on counterterrorism (State Department,

2007d). The Secretary of State provides Congress with an annual 'Pattern of Global Terrorism' report (State Department, 2007d).

Many resources are spent on anti-terrorism efforts. The intelligence budget (other than tactical military intelligence) was increased from $15.5 billion in 1998 to $44.4 billion in 2005 (Kaplan and Whitelaw, 2006). However, Bush's budget proposal released on 5 February 2007 revealed a scheme to reduce the budget for Homeland Security by 8 percent to $34.6 billion (Rosenberg, 2007). As an unclassified example of where such money goes, the 2003 Homeland Security budget was: $14 billion for border and port security, $5 billion for aviation security, $4 billion for immigration enforcement, $3 billion for biodefense, and $1 billion for infrastructure protection (Magnusson et al., 2004). The Department of Homeland Security has conducted Top Officials National Exercise Plan (TOPOFF) exercises for private businesses and government organizations to practice responding to terrorist attacks in the US (Phillips, 2006). The first TOPOFF exercises in 2000 simulated a radiological bomb detonation in Washington, a mustard gas attack in Portsmouth, NH and a pneumonic plague outbreak in Denver. These exercises were conducted again in 2003, 2005 and 2006 (ibid.).

The government has immersed itself fully into the transportation infrastructure. The government has taken control of airport security with the advent of federal-level employees – the Transportation Security Administration (TSA). Laws and airline changes added incognito armed sky marshals on some flights, allowed pilots to carry guns, provided TV systems in cockpits to survey the entire cabin, installed bullet-proof doors to cockpits, and set up no-fly zones over sensitive areas (Access Control & Security Systems, 2002). Although less has been done than at airports, the government has also increased restrictions on ports and shipping. Analysts believe that nuclear weapons could fairly easily be smuggled into the US through one of its 361 commercial shipping ports. In fact, bowing to pressures from Congress and the public, the Bush administration backed down from allowing a United Arab Emirate company to control shipping at a US port. A survey at the time showed the public outcry over having outside influences, especially Arab, control any US shipping; poll respondents (70 percent) stated that the UAR company should definitely not manage any US ports (Polling Report, 2007, ABC News/Washington Post Poll, March 2–5, 2006). In fact, 42 percent in one poll said that it was never OK for a company outside the US to manage a US port, even if the company was in a country that was an ally (Polling Report, 2007, CBS News Poll, March 9–12, 2006). The Customs & Border Protection division of Homeland Security told Congress that it recorded 318,000 alarms from port sensors from 2004 to 2007, yet none of those alarms constituted a serious threat.

Newer devices to distinguish harmless radiation from significant threats are now becoming available and are being installed (Pressman, 2007). In contrast, little has been done for other transportation areas such as railroads or buses.

Some counter-terrorism activities are not open to public perusal and cannot be described in this chapter. However, hints of these activities sometimes leak out. For instance, the CIA has apparently used private jets in 'extraordinary rendition' operations to remove suspected terrorists from a given country to another country for in-depth questioning and possible torture to extract information about their terrorist activities (Rogoway, 2004).

Various laws have been passed since 9/11 in the GWOT. Probably the most prominent was the 2002 USA Patriot Act (Uniting and Strengthening America by Providing Appropriate Tools Required to Intercept and Obstruct Terrorism). It was designed to legally increase surveillance, interrogation, and detention of possible terrorist suspects. The Patriot Act was revised (with anti-money-laundering and greater national and international data-gathering provisions) and renewed in 2006. Other anti-terrorist laws were passed subsequent to 9/11. They included the New Customs Trade Partnership Against Terrorism Act (Customs TPAT) which enhanced cargo security throughout the entire import–export process; the Container Security Initiative of 2002 which deployed American inspectors to dozens of foreign ports where cargo is screened before entering the US; the 2002 Aviation and Security Transportation Act, which authorized the national government to control all airport security and create the TSA; and the Intelligence Reform and Terrorism Prevention Act of 2004 (IRTPA) which revised the procedures for designating FTOs (State Department, 2007a). Other anti-terrorism laws have been passed more recently. The Safe Port Act (2006) requires improved technology for inspecting all cargo containers entering the 22 largest US ports and requires the Department of Homeland Security to provide plans to ensure that shipping continues even in the event of a terrorist attack. The Department of Homeland Security Appropriation Act of 2006 provides $33.8 billion to help secure the homeland. It requires nuclear detection equipment at US ports, raises security standards at chemical plants, safeguards American cities against WMD, and strengthens the capabilities of the Federal Emergency Management Agency. The Military Commission Act of 2006 allows the CIA to continue its present procedures for questioning key terrorist leaders and operatives without civil rights abuse lawsuits filed by suspected terrorists. The Secure Fence Act of 2006 authorizes more checkpoints, lighting, fencing, and technological monitoring along the US–Mexican border to help prevent people from entering the US illegally (*Jurist*, 2007).

The public, Congress, and the Executive branch in the US government are now grappling with whether it is better to fight terrorism overseas or become more isolationist in policy and stay home to protect US borders. In a 2006 survey, 47 percent of the respondents said that terrorism should be confronted overseas versus 45 percent who said to stay home and focus here. However, these numbers diverge on political lines as 69 percent of the Republicans versus 29 percent of the Democrats surveyed took the position that the US should confront terrorists overseas (Polling Report, 2007, CBS News/New York Times Poll, September 15–19, 2006). In the meantime, the Executive branch continues to work on the current challenges in Iraq. One of the main issues is bringing political stability to the region by promoting democracy. It is still being debated within the government and the public as to how long the US will deploy troops in Iraq.

On businesses and other non-government organizations: impacts and actions

An official of the Department of Homeland Security has stated that private corporations have an advantage over the government in preventing terrorism because companies can make quicker decisions and thus act much faster (Phillips, 2006). The obvious effects of the threat of terrorism on business are increased resources spent on security. For example, airports have had to make tremendous changes, at significant cost, to serve their customers. Six months after September 11, 156 terminal and concourse evacuations occurred in US airports, which caused 2395 flights to be delayed or cancelled (Rupp et al., 2003). Previously exempt businesses such as sports stadiums must spend more on increased security in the form of security guards and surveillance cameras because they are now viewed as soft targets (high potential casualties or economic loss with easy terrorist access) (Appelbaum, 2005). The financing of terrorism through coupon fraud, credit card scams, money laundering, and donations to Islamic charities is an additional concern for business organizations and for the strength of the economy as a whole. Also, increased restrictions have made it more difficult for foreigners to come into the US, so other countries have retaliated by making it more difficult for US citizens to enter their countries, even for business or foreign study.

THE FUTURE OF TERRORISM REGARDING THE US

Shortly after 9/11, a plurality of US citizens thought that the war on terror would take a long time. In a 2001 national survey, 33 percent thought that it would take more than five years (Polling Report, 2007, NBC News Poll,

September 20, 2001). But in 2002, 82 percent of the population believed that the most difficult part of the war on terror was yet to come (Polling Report, 2007, ABC News Poll, September 5–8, 2002). Now, surveys show that US citizens expect terrorism to continue to be a threat for many years to come. For instance, 52 percent believes that the US will be no safer from terrorism in 2008 than it was in 2005 (Polling Report, 2007, CBS News/New York Times Poll, January 14–18, 2005). Most (81 percent) acknowledge that the US will most likely always have to live under the threat of terrorism (Polling Report, 2007, CBS/New York Times Poll, August 17–21, 2006). An expert witness on terrorism, testifying before congress, stated that 'Terrorist acts . . . will never be eradicated' (Flynn, 2007).

Affecting the Government

The 2008 US presidential elections will probably make significant changes to US policy just as the 2006 congressional elections have. The politicization of terrorism depends on what is accomplished in Congress and the Executive branch by November 2008, and what happens in Iraq and other trouble spots in the world. The Democrats control Congress now, but it is still unclear whether they will be able to pass new legislation regarding the war on terrorism, or if President Bush's policy will continue to be dominant. If Democrats are able to pass legislation in Congress, then the fate of the 2008 election will depend on the public's approval of the legislation, much as approval ratings for terrorism and homeland security policies played a significant role in the 2004 and 2006 elections.

The Terrorists

There will likely be continued threats of terrorists attacking US interests on American soil, US property overseas, and against US allies and their interests. The latest view of the US State Department is that the main threat to the US is terrorist networks with al Qaeda being the most dominant threat within such networks (State Department, 2008). Unofficially, intelligence leaders have stated that al Qaeda is regrouping and combining with the Taliban in the North Waziristan area of Pakistan (the tribal areas in the mountainous region on the Afghanistan border). It is there that al Qaeda is believed to be building a number of training camps for its operatives and its cohort, the Taliban, that will be the bases for increasing attacks on Coalition Forces in Afghanistan (Oregonian Staff, 2007).

A somewhat different view is that the terrorist networks could be comprised of less formal groups than al Qaeda. The Editor-in-Chief of *U.S. News & World Report* stated that future terrorist attacks on the US would not come from organized international terrorist groups like al Qaeda but rather from homegrown terrorist cells in the US. These are second- and third-generation children of immigrants who have become radicalized by the common grievances and radical Islamic ideology that they share on the Internet, which teaches them not only the radical Islamic doctrine but also how to make bombs and fire missiles (Zuckerman, 2006). Stephen Flynn (2007), an ex-Coast Guard officer who is a prominent homeland-security expert, concurs that much of the terrorist training and organizing comes directly over the Internet (such as instruction on how to build bombs, improvised explosive devices (IEDs) and the recent explosively formed projectiles (EFPs). He believes that Iraq has been and is a prominent training ground for future terrorists to attack other countries, including the US. Related to this a senior FBI official recently stated that the main threat to the US is disenfranchised Muslim youths (whom he termed the 'Pepsi jihad') who report to no particular leader or group but connect haphazardly over the internet. The official indicated that the best hope of alleviating this threat in the future is to deal with these youths who have been radicalized by the common ideology over the Internet; the US being a melting pot of different cultures could be a deterrent to violent Islam (Shrader, 2007). The author of a recent book entitled *American Islam* (New York: Farrar, Straus and Giroux, 2007) also concurs that the tolerance for diversity in the US, including the diversity within Islam, can produce Islamic views that can co-exist with Christian and other views in the US (Barrett, 2007).

Weapons

Although 64 percent of the US population have stated that suicide bombings could easily occur in the US (Polling Report, 2007, Fox News/Opinion Dynamics Poll, March 11–12, 2003), 72 percent believe that chemical and biological weapons are more likely (Polling Report, 2007, Fox News/Opinion Dynamics Poll, February 25–26, 2003). Unfortunately, according to the poll respondents, 70 percent believe that the US is not prepared to deal with such attacks (Polling Report, 2007, CBS News/New York Times Poll, September 2–5, 2002). In addition to the terrorist weapons mentioned earlier, others may arise in the future. Information as a weapon (for example, computer hacking, viruses) could produce great damage at little risk to the terrorist. Laser beams (easily and cheaply obtained) have recently been employed to blind targeted people, such as commercial airline pilots attempting landing approaches (Miller, 2004). Again, uncertainty is the terrorist's main weapon. Therefore, the terrorist will continually seek weapons that have not been considered before.

Likely Terrorist Activities

On US soil

Old and new terrorist targets may emerge. The TSA will continue to monitor airports and airplane travel. US borders – especially with Mexico – will come under increasing scrutiny. In shipping out of US ports, newer devices to discover significant threats should become more available under a Democrat-controlled Congress (Pressman, 2007), and they have stated an intent to require 100 percent of all incoming cargo to be screened *before* it reaches a US port (Marek, 2007). Flynn (2007) also says that there is a desperate need to strengthen the infrastructure (for example, dams and electrical power grids), revamp laws and regulations about placement and use of likely targets (such as oil and chemical refineries near population centers), expand emergency healthcare operations, protect basic sustenance sources (such as food and water supplies), and educate the citizenry regarding surveillance and actions for prevention and response regarding terrorist attacks. Terrorists may target college and university campuses since they are soft targets with easy access and little security. For instance, two of the deadliest bombings in Iraq were at Baghdad universities, where 70 were killed and about 140 wounded in one attack (El-Rashidi, 2007) and where an Iraqi female suicide bomber killed over 40 at another university (Londono, 2007).

Overseas

Iraq is seen by the Bush Administration as at the forefront of areas in the GWOT and will continue to be a problem for the US. Although progress has been made with the first multi-party Iraqi democratic election in 2005, the al Qaeda leadership in Iraq have vowed to fight on until they, not such voters, controlled Iraq (Baker, 2005). As long as thousands of potential foreign terrorist fighters can easily cross the border from Iran or Syria into Iraq within a couple of hours, Iraq will likely be an even greater hotbed of terrorism in the future (Jack, 2004). Another problem for the US is the continual erosion of the supply of military forces to the Iraq coalition forces. The number of countries supplying troops has dropped dramatically, as 17 countries withdrew their troops between 2004 and January 2007, with the UK, South Korea, Denmark, and Poland additionally announcing that they would withdraw all or part of their troops in the near future (Global Security Staff, 2007).

The terrorism that the US will face in the future may be significantly affected by what happens in certain secondary countries in the world. The Institute for National Strategic Studies (2006) of the National Defense University has stated that *Southeast Asia* will likely become more important in the future in the GWOT. The area has recently been undergoing rapid economic changes, geopolitical challenges, disease epidemics, more unified Muslim grievances leading to some jihadist movements, and the shrewd entry of China into more control of the area including sea lanes. The Institute considers Southeast Asia to be the 'second front' in the war on terror and worthy of much greater scrutiny (Ott, 2006). With Fidel Castro transferring power to his brother, Raul, the political leadership and policies of *Cuba*, currently on the State Department's list of countries sponsoring terrorism, may change. Since Cuba is geographically close, those changes could have important repercussions on the US. The US, along with four Asian countries, is negotiating with *North Korea* to terminate its nuclear military development. If not, North Korea could be a threat to the US directly with nuclear missiles and indirectly with terrorism. *Iran* is likewise seeking to increase its nuclear military capability. Allegations of Iranian terrorist activity arising in Iraq have been bandied about but not fully confirmed. *Pakistan* signed a treaty with Pakistani tribes in the mountainous northeast region, stipulating that the government troops would not enter the areas if the peoples of those areas did not attack others in Iran. This mountainous region is purported to be the growing headquarters for Osama bin Laden and an increasing number of al Qaeda training camps (Oregonian Staff, 2007). *Afghanistan* is once again emerging as a significant problem, with the Taliban and al Qaeda increasing their strength and attacking more of the coalition forces (Duffy, 2007). An Islamic terrorist group in *Algeria* is proposing to unite radical Islamic groups throughout North Africa and join al Qaeda; this could pose a serious threat to Europe and thus to the US because travel to the US is visa free for most Europeans (Smith, 2007). Furthermore, there is the increasing power of dictator Hugo Chavez of *Venezuela*. He is an opponent of US foreign policy and has gained unfettered control of Venezuela and other countries in South America and, through oil wealth, other countries in the world who depend upon oil (Terrorism Research Center, 2007).

SUMMARY AND CONCLUSIONS

Terrorism and terrorists are pervasive in the world. Terrorists were around before 11 September 2001 and have continued their attacks since then. However, it is very clear that America was unwillingly thrust into a war on terrorism following the attacks on 9/11. The events of that day produced significant changes in Americans' beliefs and behaviors, in how intelligence is gathered by the US, how the US infrastructure operates (especially airlines and certain utilities), and how companies conduct business. Terrorist events are increasing in number and severity around the world, but the exact form and frequency of terrorism in the future is open to debate. As a recent poll of US citizens stated, terrorism is expected to persist for a long time. Over 80 percent believe that the US will always have to live with some threat of terrorism (Polling Report, 2007, CBS/New York Times Poll, August 17–21, 2006). As long as there are people who have different goals and are willing to harm others to achieve their goals, and objects are available that harm people, there will be terrorists and weapons for terrorists to use. It seems easier to change the people than eliminate anything that could harm people. But even after the great losses from 9/11, Americans still believe that people are basically good and that the world can be a better place . . . but we will have to work to make it that way.

NOTE

1. The authors wish to thank the following for assistance in the preparation of this chapter: Susan Boen, Teal Bohrer, Brenton Clair, Kenneth Ethridge, Steven Greisen, Rebecca Kreger, Pollyana Manning, Janet McElligott, Christy Ohira, Alyssa Oshiro, Andrea Skaar and Sean Wilson.

REFERENCES

Access Control & Security Systems (2002), 'Airport security since 9/11: how far have we come?', *Access Control & Security Systems*, http://www.securitysolutions.com/news/security_airport_security_far/, 12 September 2006.

Appelbaum, S.H. (2005), 'Management of sports facilities: stress and terrorism since 9/11', *Management Research News*, **28**, 69–83, http://proquest.umi.com ABI/INFORM Global, 24 October 2006.

Baker, Luke (2005), 'Al Qaeda vows to fight on in Iraq', http://story.news.yahoo.com/news?tmpl=story&cid=574&ncid=574&e=1&u=/nm/20050131/wl_nm/iraq_dc_201, 31 January 2005.

Barrett, P.M. (2007), 'They're Muslims, and Yankees, too', *U.S. News & World Report*, 15 January, pp. 50–54.

Burke, R.J. (2005), 'Effects of 9/11 on individuals and organizations: down but not out!', *Disaster Prevention and Management*, **14** (5), 629–83.

Chomsky, N. (2006), 'Official definitions of terrorism', 2 February, ZNet Web site: http://blog.zmag.org/node/2502, 1 March 2007.

CNN (2004), 'Purported bin Laden tape calls for Iraqi election boycott', CNN.com, 27 December, http://edition.cnn.com/2004/WORLD/meast/12/27/binladen.tape.ap/.

Deming, W. (2004), 'What is religion?', *Portland Magazine*, **23** (1), 10–11.

Duffy, B. (2007), 'News from the front in that other war', *U.S. News & World Report*, 26 February, p. 9.

El-Rashidi, Y. (2007), 'Deadly blast stuns a Baghdad university already familiar with violence', *The Chronicle of Higher Education*, **53** (21), A42–43.

Flynn, S. (2007), 'Terror's next target?', *U.S. News & World Report*, **142** (6), 19 February 30–35.

Global Security Staff (2007), 'Iraq coalition troops. Globalsecurity.org', http://www.globalsecurity.org/military/ops/iraq_orbat_coalition.htm, 22 February 2007.

Hanley, C.J. (2004), 'Suspect's dirty bomb would have been dud, experts say', *The Oregonian*, 10 June, p. A8.

Hutson, D. (2006), 'Airline industry suffers effects of terrorism', *The Hilltop*, www.thehilltoponline.com, 5 February 2007.

Ifel, G. (2004), 'Struggle for security', NewsHour, PBS, 22 December, www.pbs.org/newshour/bb/middle_east/july-dec04/insurgents_12-22.html, 22 December 2004.

InfoPlease Staff (2007), 'Timeline of history of terrorism in US: a list of major attacks since 1920', *InfoPlease.com*, http://www.com/ipa/A0001454.html, 28 January 2007.

Institute for National Strategic Studies (2006), 'Southeast Asian security challenges: America's response?', National Defense University, 222, October, pp. 1–8.

Jack, J. (2004), 'The Iran connection', *U.S. News & World Report*, 22 November, pp. 32–48.

Johnston, D. and Stout, D. (2004), 'Bin Laden is said to be organizing for a U.S. attack', *New York Times*, 9 July, pp. A1, A16.

Jurist (2007), 'U.S. anti-terrorism laws. Terrorism law and policy, http://jurist.law.pitt.edu/terrorism/terrorism3.htm, 9 March 2007.

Kakutani, M. (2004), 'A dark view of U.S. strategy', *New York Times*, 9 July, pp. 23, 24.

Kaplan, D.E. and Whitelaw, K. (2006), 'Playing defense', *U.S. News and World Report*, **14** (18); http://www.usnews.com/usnews/articles/061103dni. intro.htm, 23 March 2007.

Kellerhals, M.D. Jr. (2008), 'Terrorist group list highlights why terrorism is unacceptable', US Diplomatic Mission to Italy, January 17, 2008, http://www.usembassy.it/viewer/article.asp?article=/file2008_01/alia/a8011701.htm, 18 March 2008.

Kondrasuk, J.N. (2004), 'Effects of 9/11 and terrorism on human resource management: recovery, reconsideration & renewal', *Employee Responsibilities and Rights Journal*, **16** (1), 25–33.

Kondrasuk, J.N. (2005), 'Terrorism hits HRM', *2005 Annual Review of Human Resource Management*, San Jose, CA: Pfeiffer.

Kondrasuk, J.N., Robinson, P., Robles, L. and Niemcziek, W. (2005), 'Leadership in the 21st century: understanding global terrorism', in *Employee Responsibilities and Rights Journal*, **16** (1), 25–33.

Kushner, H.W. (2003), *Encyclopedia of Terrorism*, Thousand Oaks, CA: Sage.

Londono, E. (2007), 'Suicide bomber kills at least 48 at Baghdad college', *The Oregonian*, 26 February, A5.

Magnusson, P., McNamee, M., Arndt, M., Aston, A., Palmeri, C. and Kharif, O. (2004), 'Welcome to security nation', *Business Week*, 14 June, pp. 32–5.

Marek, A.C. (2007), 'Closing the door at the ports', *U.S. News & World Report*, 8 January, pp. 29–30.

Microsoft (2007), *Microsoft Encarta World Dictionary*, www.encarta.msn.com/dictionary_/terrorism.html, 30 January 2007.

Miller, L. (2004), 'Lasers aimed at pilots' eyes: a deadly prank or terrorism?', *The Oregonian*, 31 December, E2.

Nagourney, A. (2005), 'Kerry says bin Laden tape sank his campaign', *The Oregonian*, 31 January, A2.

National Commission on Terrorism (2003), *Countering the Changing Threat of International Terrorism: Report of the National Commission on Terrorism*, Washington, DC: US Government Printing Office.

Neier, A. (2004), 'America's new nationalism', *Social Research*, **27** (4), Winter, 3.

Oregonian Staff (2007), 'U.S. warns of al-Qaida rebound', *The Oregonian*, 19 February, p. A2.

Ott, M.C. (2006), 'Southeast Asian security challenges: America's response?', Strategic Forum, Institute for National Strategic Studies, National Defense University, October, **222**, pp. 1–8.

Phillips, Z. (2006), 'Retail to the rescue', *Government Executive*, **38** (17).

Polling Report (2007), 'War on terrorism', http://www.pollingreport.com/terror.htm, 24 January 2007.

Pound, E.T. (2004), 'Seeds of chaos', *U.S. News & World Report*, 20 December, pp. 20–26.

Pressman, A. (2007), 'Homeland security 2.0', *BusinessWeek*, 22 January, p. 73.

Reuters (2004), 'U.S. closes Bombay Consular Office on terror threat', http://story.news.yahoo.com/news?tmpl=story&u=/nm/20041123/ts_nm/security_india_u.

Rogoway, M. (2004), 'CIA intrigue shadows jet', *The Oregonian*, 29 December, pp. C1, C2.

Rosenberg, E. (2007), 'Budget request soars for defense', *The Oregonian*, 6 February, p. A7.

Rupp, N.G., Holmes, G.M. and DeSimone, J. (2003), 'Airline schedule recovery after airport closures: empirical evidence since September 11th', *NBER Working Papers 9744*, National Bureau of Economic Research.

Schuster, Mark A. (2001), 'A national survey of stress reactions after the September 11, 2001 terrorist attacks', *The New England Journal of Medicine*, **345** (15), http://content.nejm.org/cgi/content/full/345/20/1507?ck=nck, 24 October 2006.

September 11 News (2004), September 11 News.com website/homepage, www.september11news.com.

Shrader, Katherine (2007), 'U.S. melting pot called deterrent to violent Islam', *The Oregonian*, 20 January, p. A3.

Smith, C.S. (2007), 'New terror threat arising in Africa', *The Oregonian*, 23 February, p. A6.

State Department (2007a), 'Foreign Terrorist Organizations (FTOs)', Office of the Coordinator for Counterterrorism, http://www.state.gov/s/ct/rls/fs/37191.htm, 30 January 2007.

State Department (2007b), 'State Sponsors of terrorism', Office of the Coordinator for Counterterrorism, http://www.state.gov/s/ct/c14151.htm, 30 January 2007.

State Department (2007c), 'The terrorist enemy', Office of the Coordinator for Counterterrorism, http//www.state.gov/s/ct/enemy/index.htm, 30 January 2007.

State Department (2007d), 'US Counterterrorism Team', http://www.state.gov/s/ct/team/index.htm, 30 January 2007.

State Department (2008), 'The terrorist enemy', Office of the Coordinator for Counterterrorism, http://www.state.gov/s/ct/enemy, 15 April, 2008.

Stein, B.D. (2004), 'A national longitudinal study of the psychological consequences of the September 11, 2001 terrorist attacks: reactions, impairment, and help-seeking', *Psychiatry: Interpersonal and Biological Processes*, **67** (2), http://www.atyponlink.com/GPI/doi/abs/10.1521/psyc.67.2.105.35964, 24 October 2006.

Stojanovic, D. (2004), '29 die in insurgent ambush of Iraq police', Comcast.net.News, http://www.comcast.net/News/INTERNATIONAL//XML/1107_AP_Online_Regional___Middle_East/9332d8c4-573a-49b4-a246-42f5c35 741ea.html, 24 October 2006.

Terrorism Research Center (2007), Terrorism Research Center website/homepage, www.terrorism.com.

Vick, K. (2004), 'Insurgents shift fight as Fallujah battle ebbs', *The Oregonian*, 15 November, pp. A1, A7.

Ware, M. (2004), 'Meet the new Jihad', *Time*, 5 July, pp. 24–9.

White House (2006), 'National Strategy for Combating Terrorism', September, http://www.state.gov/s/ct/rls/wh/71803.htm, 10 February 2007.

Why Islam? (2005), 'What does Islam say about terrorism?', http://whyislam.org/877/Services/Literature/22.asp, 15 March 2005.

Wikipedia (2004), 'United States', *Wikipedia, the Free Encyclopedia*, http://en.wikipedia.org/wiki/United_States, 30 December 2004.

Wikipedia (2006), 'September 11, 2001 attacks', *Wikipedia, the Free Encyclopedia*, http://en.wikipedia.org/wiki/September_11,_2001_attacks#Long-term_effects, 21 November 2006.

Wikipedia (2007), '7 July 2005 London', *Wikipedia, the Free Encyclopedia*, bombing shttp://en.wikipedia.org/wiki/7_July_2005_London_bombings, 8 March 2007.

Zuckerman, M.B. (2006), 'The threat from within', *U.S. News & World Report*, 18 December, p. 80.

3. Government: target, protector and aggressor

John L. Taylor

> Fighting terrorism is like being a goalkeeper. You can make a hundred brilliant saves but the only shot that people remember is the one that gets past you.
>
> (Paul Wilkinson, Professor of International Relations, University of St Andrews, Daily Telegraph, September 1, 1992)

INTRODUCTION

Governments are responsible for fighting terrorism. They form the defense and the attack: some suggest that governments are responsible for terrorism in the first place.

In any event, the government of the country involved has to organize the rescue services, investigate the incident and bring the perpetrators to court.

Governments are often the direct target of terrorists. Less spectacular than the Twin Towers but significant nonetheless was the attack on the Pentagon on 9/11. And the ultimate destination of the fourth aircraft that crashed into Pennsylvania woodlands is unknown but was probably the White House or Capitol Hill. Had that fourth aircraft reached its destination, the US government would have been even more devastated. Embassies and consulates have been attacked, such as the Australian Embassy, Jakarta and the British Consulate, Istanbul; the Provisional Irish Republican Army (PIRA) launched rocket attacks on 10 Downing Street and MI6 headquarters. There are many examples of governments being the target.

This chapter considers governments' role in terrorism as the potential cause of extremism, the necessary defenses they have to install and how they attack terrorist organizations – in President George W. Bush's words: the 'war on terrorism'. Strategies between governments vary and the chapter explores the different policies and the reasons for those differences.

GOVERNMENTS AS THE CAUSE

> Terrorism is the price of empire. If you do not wish to pay the price, you must give up the empire.
>
> (Pat Buchanan, *Where the Right Went Wrong*, New York: Thomas Dunne Books, 2004)

> Let us never negotiate out of fear. But let us never fear *to negotiate*.
>
> (John F. Kennedy, Inaugural Address, January 20, 1961)

This book is not primarily about the causes of terrorism, but the context of terrorism helps us to understand the targets and the nature of the response. At the heart of virtually all terrorist causes is an anger, frustration or resentment against authority. Peaceful negotiations may or may not precede the turning to violence; they may run alongside; there may be splinter groups, but at the heart is a feeling that the ruling authority – the government – will not listen to argument and recourse to violence is the only alternative.

Thus the Eritreans fought against Ethiopia; the IRA (Irish Republican Army) fought against the British (and occasionally the Irish) governments; the ANC (African National Congress) against the apartheid South African government and the Basques against the Spanish. The Irgun used terrorism in 1946 to promote the state of Israel and Hamas use terrorism to promote the state of Palestine.

Passions run high and arguments about righteous terrorism abound. To those not directly involved it is sometimes hard to follow the logic of those who are on the frontline of terrorism (whichever side) and who are clearly intelligent. Thus we have Benjamin Netanyahu (former Prime Minister of Israel) declaring:

> When I say that terrorism is war against civilization, I may be met by the objection that terrorists are often idealists pursuing worthy ultimate aims – national or regional independence, and so forth. I do not accept this argument. I cannot agree that a terrorist can ever be an idealist, or that the objects sought can ever justify terrorism. The impact of terrorism, not merely on individual nations, but on humanity as a whole, is intrinsically evil, necessarily evil and wholly evil. (Netanyahu, 1981)

And yet he can say the following at the 60th anniversary of the bombing of the King David Hotel in Jerusalem in 1946:

> It's very important to make the distinction between terror groups and freedom fighters, and between terror action and legitimate military action. Imagine that Hamas or Hezbollah would call the military headquarters in Tel Aviv and say, 'We have placed a bomb and we are asking you to evacuate the area'.

Alan Dershowitz (2002, p. 24) is clear about trying to understand the cause of terrorism:

> We must commit ourselves *never to try or eliminate its [terrorism] alleged root causes*, but rather to place it beyond the pale of dialogue and negotiation. Our message must be this: even if you have legitimate grievances, if you resort to terrorism as a means toward eliminating them we will simply not listen to you, we will not try to understand you, and we will certainly never change any of our policies toward you. Instead, we will hunt you down and destroy your capacity to engage in terror. (Original italics)

With such hardline views it is not surprising that a definition of terrorism eludes universal acceptance. Attempts to define terrorism have defeated the United Nations precisely because one man's terrorist is another's freedom fighter.

Governments are at the center of terrorism. Sometimes they have to stand and defend, at other times their actions seem to encourage violence; sometimes there is no alternative but to prepare for the worst and get business back to work as quickly as possible, when the inevitable has happened.

The US and most West European governments now have a standard response to terrorist acts. They are tough and uncompromising. Faced with a hijack, for example, skilled negotiators will engage the hijackers and try to persuade them to release hostages, allow food and drinks, in short to do everything possible to preserve life.

At the same time, the police and special forces will be preparing to attack, but only as a last resort. They will attack only when lives are being lost. So when in 1980 the Iranian embassy in London was taken over by a six-man terrorist team calling itself the 'Democratic Revolutionary Movement for the Liberation of Arabistan' the Special Air Service (SAS) attacked only when the terrorists killed one of their hostages, press attaché Abbas Lavasani. This is now a standard response in the UK. Some governments elsewhere will deploy their special forces earlier and others will give in to terrorist demands in exchange for the release of hostages.

The Italian Premier Romano Prodi allowed negotiation in March 2007 for the release of Taliban militants in exchange for the freedom of an Italian hostage in Afghanistan. Later, in July 2007, Taliban militants in Afghanistan captured 23 South Korean aid workers. Following direct negotiations between the Korean government and the Taliban, the hostages were released and the South Korean government announced the withdrawal of its forces from Afghanistan by the end of the year.

GOVERNMENT AS A CAUSE FOR TERROR

While the original dispute can be the call to arms for the terrorists, there is another fertile ground for the terrorist recruiters. Governments, as we have discussed above, respond to terrorist activity in varying ways, some being perceived as unnecessarily harsh. Ordinary people who may not be militant become incensed; some will take up arms or provide direct help in the form of money or safe havens, or as couriers.

The infamous terrorist Abu Nidal (born Sabri Khalil al-Banna) took up arms as a direct result of seeing Israeli tanks on the streets of his homeland. The British government's policy of internment in Northern Ireland (Monday, 9 August 1971 until Friday, 5 December 1975) might have taken some of the hardline terrorists off the streets of Belfast and mainland Britain, but it proved to be one of the greatest recruitment incentives for PIRA among their otherwise passive supporters and led directly to one of the most deadly phases of bombing by the IRA.

Peace in Northern Ireland came about as the result of a complex mix of reasons, but significant among them was the more sensitive policing and softer military tactics in the province. Talks with what had hitherto been considered as extremists from both the Catholic and Protestant sides were initiated by the British government and resulted in the Joint Declaration on Peace (more commonly known as the Downing Street Declaration) on 15 December 1993. This was signed by John Major, then Prime Minister of the United Kingdom, and Albert Reynolds, then Taoiseach (Prime Minister of the Republic of Ireland), on behalf of the British and Irish governments.

On 10 April 1998, George Mitchell (a Democratic Party politician and US Senator from the state of Maine) was able to announce that 'the two governments and the political parties in Northern Ireland have reached agreement' (public statement, Belfast, April 10). It emerged later that US President Bill Clinton had made a number of telephone calls to party leaders to encourage them to reach a settlement. It became known as the Good Friday agreement and has served as the foundation for the subsequent resolution of critical problems in Northern Ireland.

The Omagh bombing by the Real IRA later that same year resulted in 29 deaths and hundreds of injuries. It was the single worst incident in Northern Ireland during the conflict, but by then the forces for peace and reconciliation were so strong that this terrorist atrocity failed to knock the peace process off course.

The role of governments as potential exacerbator, negotiator and protector is significant in the context of any terrorist incident.

GOVERNMENT AS A TARGET

Defining the target of any terrorist act is fraught with problems. The Omagh bomb in 1998 by the 'Real IRA' caused 29 deaths and seriously injured 220, many of whom were tourists. The stolen car carrying the bomb was parked outside a clothes shop on Omagh's Market Street at 14:00. The driver and another occupant left the car and walked down Campsie Road. Shortly after 14:30 on the day of the bombing, three warnings were issued. Two of them made specific mention of the bomb being close to the court house in Omagh. In fact the bomb went off some 400 yards from the court house at 15:10. It is possible that the target was the court house, a government building, and that the bombers were badly briefed. But the result was that ordinary people were killed and injured and private buildings were badly damaged by the bomb.

The question has to be asked – were the terrorists serious about attacking a British government building or did they just want to terrorize ordinary people?

There were government offices in the Twin Towers but most were occupied by commercial companies. For many terrorists or extremists, the distinction does not appear to have much meaning. Injured people become innocent casualties in any conflict, as we have seen in Afghanistan and Iraq. For the terrorist, loss of innocent life even of their own group is deemed acceptable.

Terrorists do of course target commercial institutions. The HSBC bank was bombed in Istanbul in 2003. The Marriott hotel in Jakarta was bombed in the same year. Terrorists will also focus on rival religions. Synagogues have long been a target for Muslim extremists. More recently, internecine disputes between Sunni and Shia have flared, particularly in Iraq. Finally, transport has been a favourite target – the Madrid train bombings in 2004, following numerous aircraft hijackings in the 1970s and 1980s.

Analysis of 770 terrorist incidents since 1961 shows that innocent civilians and government institutions form the majority of targets. Religious targets are an increasing phenomenon. Purely economic or transport targets form 23 per cent of the target.

Figure 3.1 shows the breakdown of 770 incidents from all over the world between 1961 and 2007. Apart from the first and last two columns, each period represents five years. The last column demonstrates how many incidents there would be if levels of attack in the 2006/07 two-year period were maintained.

The response to an attack on a government building varies from country to country. The UK has had to manage these crises for nearly 35 years, first from Irish extremists and then from Muslim extremist forces. Other states such as Israel and the Lebanon have had more experience, but for the

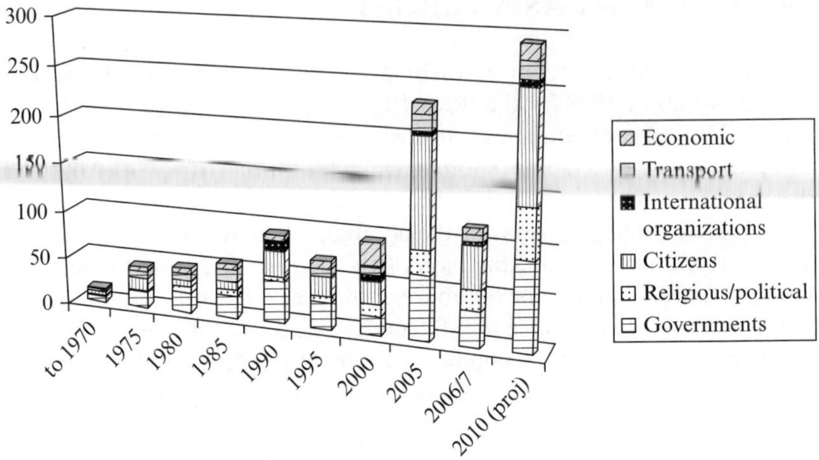

Source: Taken from data held in US State Department, BBC.co.uk, Wikipedia and others, data up to September 2007.

Figure 3.1 Terrorist incidents by target

purposes of this chapter the UK provides a good model and will be described in detail later.

COUNTER-TERRORISM STRATEGIES

United States

Since 9/11, governments have put considerable effort into crafting and promoting counter-terrorist strategies. They vary considerably from country to country, though their overall aim is pretty similar – the protection of their own state from terrorist acts.

The US led the way, and the term 'Homeland Security' quickly became part of our lexicon. Soon after 9/11, George Bush established the White House Office of Homeland Security and the Homeland Security Council to ensure that the 'federal response and protection efforts were coordinated and effective'. The President also directed Homeland Security Advisor Tom Ridge to study the federal government as a whole to determine whether 'the current structure allows us to meet the threats of today while anticipating the unknown threats of tomorrow' (Bush, 2002, p. 2).

The result was the creation of a new department:

> The President proposes to create a new Department of Homeland Security, the most significant transformation of the U.S. government in over a half-century by largely transforming and realigning the current confusing patchwork of government activities into a single department whose primary mission is to protect our homeland. The creation of a Department of Homeland Security is one more key step in the President's national strategy for homeland security. (Ibid.)

The result is a department now employing around 200,000 people and incorporating a number of previous departments. Its aims are described in their website (Box 3.1).

BOX 3.1 DEPARTMENT OF HOMELAND SECURITY: AIMS

Vision

Preserving our freedoms, protecting America . . . we secure our homeland.

Mission

We will lead the unified national effort to secure America. We will prevent and deter terrorist attacks and protect against and respond to threats and hazards to the nation. We will ensure safe and secure borders, welcome lawful immigrants and visitors, and promote the free-flow of commerce.

Strategic Goals

- *Awareness* – Identify and understand threats, assess vulnerabilities, determine potential impacts and disseminate timely information to our homeland security partners and the American public.
- *Prevention* – Detect, deter and mitigate threats to our homeland.
- *Protection* – Safeguard our people and their freedoms, critical infrastructure, property and the economy of our Nation from acts of terrorism, natural disasters, or other emergencies.
- *Response* – Lead, manage and coordinate the national response to acts of terrorism, natural disasters, or other emergencies.
- *Recovery* – Lead national, state, local and private sector efforts to restore services and rebuild communities after acts of terrorism, natural disasters, or other emergencies.

- *Service* – Serve the public effectively by facilitating lawful trade, travel and immigration.
- *Organizational Excellence* – Value our most important resource, our people. Create a culture that promotes a common identity, innovation, mutual respect, accountability and teamwork to achieve efficiencies, effectiveness, and operational synergies.

Source: http://www.dhs.gov/xabout/strategicplan/.

Primarily designed to counter terrorism, it soon became involved in other crises, perhaps most notably the floods in New Orleans as a result of Hurricane Katrina; a crisis which resulted in some criticism for the relatively new Department of Homeland Security (DHS).

The most significant departments now incorporated in the DHS are:

- US Customs Service (Treasury);
- Immigration and Naturalization Service (part) (Justice);
- Federal Emergency Management Agency (FEMA);
- Strategic National Stockpile and the National Disaster Medical System (HHS);
- Nuclear Incident Response Team (Energy);
- Domestic Emergency Support Teams (Justice);
- Chemical, Biological, Radiological and Nuclear (CBRN) Countermeasures Programs (Energy);
- National Communications System (Defense); and
- National Infrastructure Protection Center (Federal Bureau of Investigation: FBI).

The Secret Service and the Coast Guard are also located in the DHS.

The budget request for the DHS for the fiscal year 2008 was $43 billion (www.dhs.gov). The importance and potential power of the DHS are considerable, but policy and strategy reflect not surprisingly the views of the President and those close to him.

In 2006 the President declared his policy on national security (Box 3.2). The document (US Government, 2006) goes on to explore the issues around terrorism and democracy:

> To wage this battle of ideas effectively, we must be clear-eyed about what does and does not give rise to terrorism:
>
> - Terrorism is not the inevitable by-product of poverty. Many of the September 11 hijackers were from middle-class backgrounds, and many terrorist leaders, like bin Laden, are from privileged upbringings.

BOX 3.2 NATIONAL SECURITY STRATEGY

Our national security strategy is founded on two pillars:

The first pillar is promoting freedom, justice and human dignity – working to end tyranny, to promote effective democracies and to extend prosperity through free and fair trade and wise development policies. Free governments are accountable to their people, govern their territory effectively, and pursue economic and political policies that benefit their citizens. Free governments do not oppress their people or attack other free nations. Peace and international stability are mostly reliable built on a foundation of freedom.

The second pillar of our strategy is confronting the challenges of our time by leading a growing community of democracies. Many of the problems we face – from threats of pandemic disease, to proliferation of weapons of mass destruction, to terrorism, to human trafficking, to natural disasters – reach across borders. Effective multinational efforts are essential to solve these problems. Yet history has shown that only when we do our part will others do theirs. America must continue to lead.

Source: From the letter covering the National Security Strategy document written by President George W. Bush, 16 March 2006.

- Terrorism is not simply a result of hostility to U.S. policy in Iraq. The United States was attacked on September 11 and earlier, well before we toppled the Saddam Hussein regime. Moreover, countries that stayed out of the Iraq war have not been spared from terror attack.
- Terrorism is not simply a result of Israeli–Palestinian issues. Al-Qaida plotting for the September 11 attacks began in the 1990s, during an active period in the peace process.
- Terrorism is not simply a response to our efforts to prevent terror attacks. The al-Qaida network targeted the United States long before the United States targeted al-Qaida. Indeed, the terrorists are emboldened more by perceptions of weakness than by demonstrations of resolve. Terrorists lure recruits by telling them that we are decadent and easily intimidated and will retreat if attacked.

The terrorism we confront today springs from:

- Political alienation. Transnational terrorists are recruited from people who have no voice in their own government and see no legitimate way to promote change in their own country. Without a stake in the existing order, they are vulnerable to manipulation by those who advocate a perverse vision based on violence and destruction.

- Grievances that can be blamed on others. The failures the terrorists feel and see are blamed on others, and on perceived injustices from the recent or sometimes distant past. The terrorists' rhetoric keeps wounds associated with this past fresh and raw, a potent motivation for revenge and terror.
- Sub-cultures of conspiracy and misinformation. Terrorists recruit more effectively from populations whose information about the world is contaminated by falsehoods and corrupted by conspiracy theories. The distortions keep alive grievances and filter out facts that would challenge popular prejudices and self-serving propaganda.
- An ideology that justifies murder. Terrorism ultimately depends upon the appeal of an ideology that excuses or even glorifies the deliberate killing of innocents. A proud religion – the religion of Islam – has been twisted and made to serve an evil end, as in other times and places other religions have been similarly abused.

Defeating terrorism in the long run requires that each of these factors be addressed. The genius of democracy is that it provides a counter to each.

- In place of alienation, democracy offers an ownership stake in society, a chance to shape one's own future.
- In place of festering grievances, democracy offers the rule of law, the peaceful resolution of disputes, and the habits of advancing interests through compromise.
- In place of a culture of conspiracy and misinformation, democracy offers freedom of speech, independent media, and the marketplace of ideas, which can expose and discredit falsehoods, prejudices, and dishonest propaganda.
- In place of an ideology that justifies murder, democracy offers a respect for human dignity that abhors the deliberate targeting of innocent civilians.

Solutions are found in the following nine ambitions:

- Champion aspirations for human dignity;
- Strengthen alliances to defeat global terrorism and work to prevent attacks against us and our friends;
- Work with others to defuse regional conflicts;
- Prevent our enemies from threatening us, our allies, and our friends with weapons of mass destruction (WMD);
- Ignite a new era of global economic growth through free markets and free trade;
- Expand the circle of development by opening societies and building the infrastructure of democracy;
- Develop agendas for cooperative action with other main centers of global power;
- Transform America's national security institutions to meet the challenges and opportunities of the 21st century; and
- Engage the opportunities and confront the challenges of globalization.

The focus is clearly on what is perceived to be the external threat, the promotion of democracy, the need for international cooperation and a new era of global economic growth. It is a world seen through the eyes of an America which believes it has the solutions and those are the values and institutions of America.

United Kingdom

The UK did not suffer directly the trauma of 9/11. The pressure therefore of producing a firm and decisive reaction was absent. But the events of 2001 did have their impact and the British government did have to respond with a focused counter-terrorist strategy. The government published their strategy in 2003. It is known as CONTEST and is described in detail later.

There are a number of different emphases between the US and UK strategies. The most recent UK strategy paper says little about creating democracies overseas (the word democratic or democracy appears only twice in a 38-page document compared to 95 references in the 54-page US Strategy paper).

A second difference, which is not dissimilar from the first, is that the UK paper devotes considerable attention to countering radicalization, a subject which caught the attention of Prime Minister Gordon Brown early on in his term of office.

These two differences suggest that the US see the problem of terrorism as an external issue. Non-US citizens or residents are the main perpetrators. It follows that to defeat terrorism they need to tackle it through intervention overseas or through cooperation with allies.

The UK has been tackling terrorism, which has its origins internally, for nearly 40 years. It is equally not surprising, therefore, that their attention is more inward looking. The shift, however, from IRA to Muslim extremism has led to a different emphasis.

The Muslim community in the UK is 1,591,000 or 2.7 per cent of the population. In the US it is 0.6 per cent of a population now over 300 million, some 1.8 million people (2001 census); a figure which doubled between the 1991 and 2001 censuses. The potential threat from inside is proportionally smaller in the US than in the UK, but numerically greater. A group determined to commit a terrorist act does not need to be large nor does it require much support to make a big impact.

The people perpetrating the acts of violence may well be British citizens or residents, but there is considerable evidence that they have received training and much of their inspiration overseas, mostly in Pakistan or Afghanistan.

The bombings in the UK of the London transport systems provide some useful evidence on the nature of the threat facing the UK. The bombers have been identified as Mohammed Siddeque Khan (30), Hasib Hussein (18), Shazad Tanweer (22), and Jermaine Lindsay (19).

All apart from Jermaine Lindsay were British nationals of Pakistani origin, born and brought up in the UK, and at the time of the bombings based in West Yorkshire. Lindsay was a British national of West Indian origin, born in Jamaica and based in Aylesbury prior to the attacks. He was a convert to Islam.

> Siddeque Khan is now known to have visited Pakistan in 2003 and to have spent several months there with Shazad Tanweer between November 2004 and February 2005. It has not yet been established who they met in Pakistan, but it is assessed as likely that they had some contact with Al Qaida figures.
>
> The extent to which the 7 July attacks were externally planned, directed or controlled by contacts in Pakistan or elsewhere remains unclear. The Agencies [security, police and intelligence services] believe that some form of operational training is likely to have taken place while Khan and Tanweer were in Pakistan. Contacts in the run-up to the attacks suggest they may have had advice or direction from individuals there. (Intelligence and Security Committee, 2006)

There is a similar pattern in other recent attacks or attempted attacks. The key suspects in the August 2006 transatlantic aircraft plots are reported to be British-born Muslims, some of Pakistani descent. Those arrested after the attempted London and Glasgow car bombings in 2007 were also living in the UK. In contrast, the 19 hijackers and pilots involved in crashing the four aircraft on 9/11 were all from Saudi or the Yemen and had moved to the US specifically to carry out the attacks.

The US has of course experienced homegrown terrorism (the Black Liberation Army, African–American Muslim Hanafi, the Unabomber), but in the last 20 years there has not been any coordinated series of attacks in the US by any single extremist group. The current US policy of looking outside their shores for an attack is understandable given their recent history. It accounts for the lack of a significant counter-radicalization policy in the United States.

CONTEST

The UK's strategy has developed since 9/11 and is now incorporated under the code-word 'CONTEST'. It has four strands described in the Report presented to Parliament by the then Prime Minister, Tony Blair, summarized in Box 3.3.

BOX 3.3 SUMMARY OF THE UK'S COUNTER-TERRORIST STRATEGY, CONTEST

Preventing Terrorism by Tackling the Radicalisation of Individuals

The PREVENT strand is concerned with tackling the radicalisation of individuals. We seek to do this by:

- Tackling disadvantage and supporting reform – addressing structural problems in the UK and overseas that may contribute to radicalisation, such as inequalities and discrimination;
- Deterring those who facilitate terrorism and those who encourage others [from becoming] terrorists – changing the environment in which the extremists and those radicalising others can operate; and
- Engaging in the battle of ideas – challenging the ideologies that extremists believe can justify the use of violence, primarily by helping Muslims who wish to dispute these ideas to do so.

Pursuing Terrorists and Those that Sponsor Them

The PURSUE strand is concerned with reducing the terrorist threat to the UK and to UK interests overseas by disrupting terrorists and their operations. It has a number of aspects:

- Gathering intelligence – improving our ability to identify and understand the terrorist threat;
- Disrupting terrorist activity – taking action to frustrate terrorist attacks and to bring terrorists to justice through prosecution and other means, including strengthening the legal framework against terrorism, e.g. by introducing legislation to deport those who are judged to be not conducive to the public good; and
- International co-operation – working with partners and allies overseas to strengthen our intelligence effort and achieve disruption of terrorists outside the UK.

Protecting the Public, Key National Services, and UK Interests Overseas

The PROTECT strand is concerned with reducing the vulnerability of the UK and UK interests overseas. This covers a range of issues including:

- Strengthening border security – so that terrorists and those who inspire them can be prevented from travelling here and we can get better intelligence about suspects who travel, including improving our identity management, for example by use of biometrics;
- Protecting key utilities – working with the private sector;
- Transport – reducing the risk and impact of attacks through security and technological advances; and
- Crowded places – protecting people going about their daily lives.

Preparing for the Consequences

The PREPARE strand is concerned with ensuring that the UK is as ready as it can be for the consequences of a terrorist attack. The key elements are:

- Identifying the potential risks the UK faces from terrorism and assessing their impact;
- Building the necessary capabilities to respond to any attacks; and
- Continually evaluating and testing our preparedness – e.g. by frequently exercising to improve our response to incidents and learning lessons from incidents that do take place.

Source: UK Government (2006).

The strategy is commonly referred to as the four Ps: Prevent, Pursue, Protect and Prepare. As with any eye-catching alliterative device such as this, the precise meaning of each word can be obscure. In brief the four strands mean:

- *Prevent*: addressing underlying causes of terrorism in the UK and overseas.

- *Pursue*: gathering intelligence on terrorists and extremist groups, disrupting terrorist activity and cooperating with foreign governments.
- *Protect*: perhaps the easiest strand to understand – improving border controls (largely technological) and the protection of key places including transport and crowded places.
- *Prepare*: identifying the risks and ensuring that institutions are ready for any attack.

When the strategy was first published, the headlines drew attention to the second strand – gathering intelligence – and much was made of the 'Intelligence-led' policy. In truth, all four strands have equal importance. The most innovative is the first – addressing the causes of extremism.

We have already mentioned in this chapter the significance of the change of policy in Northern Ireland. In the early days, the British government resorted to harsh tactics using internment and sometimes brutal policing tactics to try to gain control of the situation. This policy changed to working with the communities concerned, investing money in the province and adopting a lighter touch on the streets of Northern Ireland. But the efforts to acquire intelligence, protect people or manage crisis situations did not diminish.

In 2007, the Department for Communities and Local Government took the lead on much of the frontline counter-radicalization work. Their paper *Preventing Violent Extremism – Winning Hearts and Minds* (2007) describes four approaches to the problem:

- *Promoting shared values* These non-negotiable values are the respect for the rule of law, freedom of speech, equality of opportunity, respect for others and responsibility towards others. The department aims to work with Madrassahs (schools attached to Mosques) and the education system generally to broaden the provision of citizen education to promote faith understanding.
- *Supporting local solutions* The Communities Department consider that approaches driven from Whitehall are not enough. Working with local communities, particularly Muslim communities, to deliver local solutions is key. This is not just about local authorities; it is about wider cultural activities and inter-faith work in our communities.
- *Building civic capacity and leadership* Voices from within Muslim communities and the actions of Muslim organizations can be more powerful than most. The British Government would work with the vast majority of British Muslims who reject violence and who share society's core values. They attempt to fundamentally rebalance their engagement to work more closely with those organizations displaying proactive leadership on these issues.

- *Strengthening the role of faith institutions and leaders* The Department described this approach as follows:

> The overwhelming majority of Muslims condemn terrorism as morally wrong and contrary to the teachings of Islam. But community and faith leaders are not always able to convey their arguments to those vulnerable to violent extremists' messages. It is not for Government to intervene in theological debates. But there is a role for Government in providing support where it is sought or needed. We will support the development of strong faith institutions and leaders capable of engaging effectively with all members of Muslim communities. Many imams still come from overseas; we have introduced new immigration rules for all Ministers of Religion to ensure that those seeking to enter the country meet certain requirements such as basic English skills. But more urgently needs to be done to ensure that imams can connect with all parts of society, particularly young people. (Department for Communities and Local Government, 2007, p. 10)

Other departments in the UK are also working on counter-radicalization, and government ministers have shown considerable interest in this aspect of countering extremism in the UK, including MI5 (www.mi5.gov.uk/output/Page31.html).

A few countries have also pursued policies of counter-radicalization, but as of 2007 it was not universally accepted as a worthwhile or productive aspect of a government's strategy. The European Union produced some proposals as did the governments of Denmark and the Netherlands. But the Australian government received some criticism by their own commentators and journalists for lack of action:

> I find it rather disappointing that we still do not have a national counter-radicalisation strategy in Australia, and I think the first thing we need to do is to get the Council of Australian Governments to address the issue and to embark on the creation of such a plan
>
> (Anthony Bergin, Australian Strategic Policy Institute, 2007)

The Pursue strand of the UK's counter-terrorist strategy caught the public eye as it related to intelligence. Not much is known about the intelligence operations that led to some of the more dramatic anti-terrorist operations. Some information is available from court records and it is clear from these that there has been considerable success. But as the quote from Professor Wilkinson, at the beginning of this chapter describes only too well: 'Fighting terrorism is like being a goalkeeper. You can make a hundred brilliant saves but the only shot that people remember is the one that gets past you'.

The bombings of July 2005 in London and the car bombing in Glasgow on 30 June 2007 show that intelligence is not infallible, but given that there has been radicalization and extremists remain determined to commit terrorist acts, accurate intelligence, which can anticipate violent incidents is the most effective way in which governments can protect their citizens and institutions.

Also under the Pursue strand is the determination to strengthen laws so that suspected terrorists can more easily be brought to justice. This can cause considerable concern. The British government failed to persuade parliament that it was necessary to hold suspected terrorists up to three months before bringing a charge.

There is some irony in these worries because the police and others feel it is necessary to hold prisoners for this length of time because of laws brought in earlier which protect a suspect's rights to have a solicitor and to remain silent during a police interview. The UK holds the dubious position of having one of the least successful confession rates before trial among police forces around the world. The Pursue strand promotes effective legal provisions.

The Protect strand calls for the strengthening of the traditional protection methods to look after the state's significant institutions (public and private) and the people. The new factor is the increasing emphasis on using technology to assist.

Immigration authorities all over the world have taken to this. Iris scanners at airports are now a common sight and the use of other biometric methods such as fingerprinting and digital facial recognition to help identify people as they come into and out of a country is increasing. The UK already has the highest concentration of CCTV cameras in the world.

Scanners are becoming increasingly smart at identifying not just drugs but other harmful substances and hardware. Lorries can be scanned without having to gain access to the inside, car boots can be x-rayed as well as people's baggage. These technologies are already being deployed and can only become more efficient. The Home Office announced in August 2007 that it was spending over a billion pounds on state-of-the-art technology to enhance border controls (www.homeoffice.gov.uk).

That the technology exists is now clear. The challenges come in two forms. First, finding software which can manage the huge amount of data which will be acquired. Governments have not had much success in this field before. The child benefits agency, the Foreign and Commonwealth Office and other government departments have all had their share of IT failures at considerable cost. The second challenge is finding and training people to manage the systems. A human being has in the last resort to make a decision.

The fourth strand, Prepare, is about the provision of procedures and facilities to minimize the effects of a terrorist act. It is about crisis management and ensuring that the businesses affected can get back to work as quickly as possible.

After any major crisis there is almost inevitably a commission or parliamentary report which looks at how well the security services, the emergency services and central government coped. Given that the specific crisis was not anticipated, it will have affected unsuspecting people and institutions. Not surprisingly, the subsequent reports with the benefit of hindsight will find areas of criticism or identify gaps. These have to be taken seriously but sometimes the expense of technology frustrates follow-up action.

Communication is often found wanting. In the US, the 9/11 commission report described a number of failures, both at the national federal Aviation Administration level and at the ground level where police were tragically not able to communicate with firemen who were in the Twin Towers when it was realized that the towers were about to collapse.

Similarly, in London a failure in communication was identified as those above ground were unable to communicate effectively with those below. This was all the more tragic as similar criticisms had been made 18 years earlier when the King's Cross underground fire review had also identified the same problem and recommended urgent action.

Most governments now recognize the need for crisis management centers. The military have had the facility in one form or another for a long time. The civilian equivalents are more recent. Command and control centers also exist with different functions, given that for the latter there is no need for a crisis as such.

In the UK, ministers and other senior leaders are guided by the Cabinet Office Briefing Room known as COBR (often described as COBRA). Details of how and where it operates are treated as confidential and are not therefore publicly accessible. The Cabinet Office website (www.cabinetoffice.gov.uk) describes it as follows:

> COBR(A) is a meeting of the Civil Contingencies Committee, which takes place within the Cabinet Office estate. COBR(A) stands for Cabinet Office Briefing Room (A).
>
> - The Committee deals with the UK's emergency response in a national crisis.
> - It is a meeting of cross-government departments responsible for responding to an emergency.
> - Depending on the nature of the event different representatives will attend the meeting and advise on contingency arrangements.
> - In the past year COBR(A) has met to respond to various emergencies including Avian Flu, alleged terrorist plots and the poisoning of the Russian spy Alexander Litvinenko.

A number of features of COBR are apparent. It is deployed for any crisis, not just those involving counter-terrorist issues.

Various government departments are involved, and by all reports they do exchange information freely. The technology supporting COBR is state of the art and capable of allowing participants to communicate rapidly with their parent departments and with people at the incident itself. This is often the envy of some governments, and for some of them, a mystery.

It is also worth noting that while COBR may well be chaired by the Prime Minister or another senior member of Cabinet, there are cases where COBR can only advise. The UK police have powers which few other forces have. The senior police commander on the spot has absolute control over the incident site and no-one, minister or not, can override his or her decisions.

There are any number of consultants who will offer government departments and businesses advice on how best to establish crisis management centers, particularly all the technology which can go with a crisis center. The technology can make a big difference and provide the essential communications needed both to give the decision makers sufficient information on which they can make a sound decision, but also so that their own recommendations or orders can be received in a timely and accurate fashion.

Experience shows that good crisis management relies on two important human factors:

- The parties coming to the meetings have to be prepared to share all their information with all the participants. Often secret intelligence is involved or for other reasons participants may be reluctant to share their information. The effectiveness of the meeting will be diminished substantially if information is withheld.
- And above all, participants must practice and where necessary learn the new skills needed to make the decision-making process efficient. This means running exercises which replicate as closely as possible the crises that are likely to hit the nation. In the UK, the US and Europe these are done regularly. For some developing countries, the requirement that senior generals, ministers and even heads of state should go through an exercise when they do not know the answers and might therefore be shown up is enormously difficult.

No amount of technical kit will help the decision-making process, if the participants do not know how the system works.

BUSINESS CONTINUITY

Given the number of attacks endured by the British government, it is not surprising that they have well-developed procedures and processes to get business back to normal quickly. This is not just for efficiency but because it minimizes the effects of the terrorist action and shows publicly that the government is defiant.

The bomb against the British Consulate in Istanbul in 2003 stopped work for a few days but it was not long before a consular service was resumed, and a year later Prince Charles went back to Istanbul to open the new building – a very overt act of defiance.

Good business continuity plans are quite complex, largely because they have to account for a range of contingencies from a small letter incendiary device to a large lorry bomb. Typically, they also cover other emergencies such as flooding, riot and fire. Indeed, an early action in developing a business continuity plan is to establish a risk register describing all the most likely risks which might threaten an institution.

Plans have to allow for the removal of personnel to other sites, to ensure that data kept on hardcopy or electronically is retrievable quickly and that communications with staff and the outside world including customers are maintained. Not surprisingly there are many consultancies ready to advise and write business continuity plans. As with crisis management the key is to practice the continuity plan.

CONCLUSIONS

The controversial debate is the extent to which governments are themselves responsible for terrorism in the first place. The US is robust in its arguments against that proposition. They are supported by noted academics such as Alan Dershowitz and nations such as Israel. In the UK, the debate is more finely balanced and the response to terrorism promotes not only an understanding of the causes of terrorism but actively pursues policies which attempt to undermine and reverse extremist views.

The two situations are different. The US sees its predicament as being largely sourced from overseas. The UK has an internal as well as an external threat. There is no doubt that the threat from Muslim extremism is global. The forces of Al Qaeda will attack their enemies anywhere and the closer to the US and her allies the better.

Governments have a multiple role and complex responsibilities. Some such as the UK and the US have developed sophisticated counter-terrorist strategies. There are many other examples of good government practice, but there are still many governments, even those close to the sources of terrorism, who find it impossible to develop coherent and effective counter-terrorist policies. The real losers are their own people and of course their businesses.

Terrorism is the scourge of the 21st century. Its impact on economies, companies and individuals is profound. But it falls to governments to manage the consequences and to lead the fight against terrorism.

BIBLIOGRAPHY

Ashdown, P. (2007), *Swords and Ploughshares*, London: Orion.
Bergin, A., D.M. Jones and C. Ungerer (2007), 'Beyond belief: Islamism, radicalisation and the counter-terrorism response', *Australian Strategic Policy Institute, Strategic Insights*, **37**, September.
Bush, President G.W. (2002), *Department of Homeland Security*, Washington, DC: US Government, June.
Department for Communities and Local Government (2007), *Preventing Violent Extremism: Winning Hearts and Minds*, Norwich: HMSO, April.
Dershowitz, A.M. (2002), *Why Terrorism Works*, New Haven, CT and London: Yale University Press.
Devji, F. (2005), *Landscapes of the Jihad*, London: David Goodwin Associates.
Gambetta, D. (2005), *Making sense of Suicide Missions*, Oxford: Oxford University Press.
Intelligence and Security Committee (2006), *Report into the London Terrorist Attacks on 7 July 2005*, Norwich: HMSO, May.
Netanyahu, B. (1981), *International Terrorism: Challenge and Response*, New Brunswick, NJ: Transaction.
UK Government (2006), *Countering International Terrorism: The United Kingdom's Strategy*, Norwich: HMSO, July.
US Government (2006), *The National Security Strategy of the United States*, Washington, DC, March.

PART II

Impact of terrorism

4. Not business as usual: the psychological impact of terrorism and mass casualty on business and organizational behavior

Kristen H. Walter, Brian J. Hall and Stevan E. Hobfoll

INTRODUCTION

Terrorism and mass casualty have forced businesses and organizations to react to threat and horror that is well beyond the normal range of business activity. Although organizations are collections of individuals, we must understand both individual and group dynamics if we are to understand the special dynamics of organizations exposed to mass casualty. This is the aim of our chapter, where we shall go beyond the mental health impact of terrorism and mass casualty circumstances to the impact on organizational functioning. We apply Conservation of Resources theory (Hobfoll, 1988, 1989, 1998), one of the principal theoretical backdrops for the understanding of stress from burnout to traumatic stress, to organizations and organizational behavior following mass casualty.

CONSERVATION OF RESOURCES (COR) THEORY PRINCIPLES

To understand the influence of mass casualty on organizations, we first review the principles of COR theory. COR theory is perhaps the most widely used theory today for an understanding of mass casualty trauma. It is a motivational stress theory based on the tenet that individuals and groups strive to obtain, retain, and protect the things they most value. These valued items or concepts are termed 'resources'. COR theory predicts that stress will result following the actual loss of resources, threat of loss of resources, or failure to gain resources following considerable resource investment. COR theory differs from appraisal-based stress theories (Lazarus and Folkman, 1984) in that resources are seen as largely observable and objectively quantifiable, rather than subjectively determined by individuals. COR theory further suggests that the subjective component in resource evaluation does exist, but is secondary to the objective elements, and tends to be shared by people in a common culture, including organizational culture.

COR theory posits several key principles. The first principle is that resource loss is disproportionately more salient than resource gain. Studies have repeatedly shown that when considering the effects of loss and gain, loss has considerably greater influence on behavior (Hobfoll and Lilly, 1993; Wells et al., 1997, 1999; Hobfoll et al., 2006). The second principle of COR theory asserts that people and groups must invest resources in order to retain resources, protect against resource loss, recover from losses, and gain further resources. Thus, those individuals and groups with greater resources are less vulnerable to initial resource loss and more capable of resource gain if resources are depleted. Contrariwise, individuals and groups who lack resources are more vulnerable to resource loss and less capable of resource gain (Holahan et al., 1999; Wells et al., 1999; Ennis et al., 2000). This implies not only that people and organizations with resources may be less likely to encounter stressful circumstances that affect psychological and physical well-being (King et al., 1999), but also that those who possess greater resources are likely to be more capable of problem solving in stressful situations and in so doing, be better able to halt resource loss cycles.

In addition, according to COR theory (Hobfoll, 1989; Hobfoll and Lilly, 1993), when individuals and groups experience a traumatic event, such as a mass casualty, they are likely to experience a particularly critical and rapid loss of resources. Hence, it is not uncommon for attempts to respond to these losses by investing resources to fall short, particularly at first.

A primary corollary to COR theory postulates that when stress occurs and resources are lost, those lacking resources are especially susceptible to experience resource loss cycles of increasing strength and speed. These individuals and groups then enter into an escalating spiral of losses termed a 'loss spiral', particularly if they have few initial resources or where the stress is especially severe or chronic. Thus, after a mass casualty event, preventing or at least limiting the accelerating force of loss spirals is critical to post-trauma recovery.

In this chapter, we revisit the key principles of COR theory in relation to organizations following mass casualty and the organizational behavior that might prove likely in response to a mass casualty event. We begin by focusing on the effects that mass casualty and acts of terror have on individuals and organizations. Second, we provide ways to best respond in mass casualty situations. We then examine the organizational behavior that may result in response to mass casualty, as there are individual and organizational reactions that are likely to reveal themselves in situations of chaos and unpredictability under threat. Finally, we explore how management and organizations can help offset problems caused by mass casualty and to foster more optimal environments for their employees and organization.

TERRORISM AND POST-TRAUMATIC STRESS DISORDER (PTSD)

Terrorism is defined by the Code of Federal Regulation as 'the unlawful use of force or violence against persons or property to intimidate or coerce a government, the civilian population, or any segment thereof, in furtherance of political or social objectives' (Code of Federal Regulation, 2007, Title 28, Section 0.85, paragraph L). The goals of terrorism are based on assumptions that political change can be affected by making the public aware of ideological causes for which the terrorists are working, and forced acquiescence on the part of the public to assuage further terrorist threat (Friedland and Merari, 1985), Terrorist activities are designed to intimidate and induce worry and concern disproportionate to the amount of physical damage they cause (ibid.). Terrorism targets business both directly and indirectly, as upset in the economic realm and of the routine of 'business as usual' are two of the primary objectives of terrorism.

Studies conducted in the United States following the terrorist attacks on September 11, 2001, and more recently in Israel during the Al Aqsa Intifada, reported that those who are exposed to terrorist attacks are at higher risk of developing PTSD and depression (Schuster et al. 2001; Galea et al., 2002; Schlenger et al., 2002; Silver et al., 2002; Hobfoll et al., 2006; Hall et al., 2007). PTSD is a disorder that may follow extremely stressful or traumatic events that threaten serious injury or death and are significant enough to invoke concomitant feelings of fear, helplessness and horror. PTSD is marked by a distinct symptom picture involving three diagnostic clusters: re-experiencing the traumatic event in thoughts or dreams, avoidance of thoughts and stimuli that remind people of the trauma, and hyperarousal.

In New York, one month following 9/11, a prevalence rate for PTSD of 7.5 percent was found (Galea et al., 2002), with rates jumping to 20 percent for individuals who lived south of Canal Street, near the World Trade Center. Research also demonstrated that although proximity to the epicenter of terrorist attack exacerbates post-traumatic symptomatology (for example, direct terrorism exposure), individuals living in distant areas from New York City were also affected psychologically (Silver et al., 2002), albeit considerably less severely. Several studies have linked watching television (for example, indirect terrorism exposure) to increases in PTSD symptoms for individuals (Schuster et al., 2001; Silver et al., 2002).

In addition to the development of psychological disorder such as PTSD, trauma has the potential effect of shattering people's assumptions about the world that may lead to profound changes in their thinking. Janoff-Bulman (1992) suggested that trauma is more likely to have an extreme negative impact if it results in shattered assumptions about people's view of the world. According to her model, individuals hold three basic assumptions: the benevolence of the world (that is, the world is a good and just place), the meaningfulness of the world (that is, the world is just; people deserve what they get), and a sense of self-worth (that is, belief of oneself as good, just and decent; positive self-evaluation). Traumatic events have the ability to shatter people's basic assumptions about themselves and their world, thus leaving them vulnerable and more likely to develop post-traumatic distress. Given the overwhelming nature of the disruption to people's assumptive worldviews, the degree of distress that would result would be a function not only of degree and severity of exposure and individual vulnerability factors such as prior psychological distress or a history of prior trauma exposure (Ozer et al., 2003). Rather, the impact will be associated to a significant degree with the extent that the 'new reality of terrorism' results in shattering individuals' view of themselves and their world. This, too, has special implications for business, as people's priorities may naturally turn away from business and toward their families, particularly if businesses and organizations do not make particular efforts to support employees and to see them as people who are nested primarily in families, not primarily in work.

TERROR MANAGEMENT THEORY

Terror management theory (Greenberg et al., 1986) offers another theoretical foundation for explaining post-trauma alterations in thinking and behavior. The theory, not originally proposed in relation to terrorism, posits that humans are endowed intellectually with the ability to identify the inevitability of their own death and annihilation (Becker, 1973; Greenberg et al., 1986). The knowledge of one's finitude has the capacity to produce terror and anxiety. Culture lessens this terror through worldviews that 'consist of humanly constructed beliefs about reality shared by individuals in groups that provide a sense that one is a person of value in a world of meaning' (Solomon et al., 2004, p. 17). Individuals are thought to assuage terror by maintaining allegiance to their cultural worldview and gaining enhanced self-esteem through behaviors consonant with the standards of their worldview (Rosenblatt et al., 1989). Hence, following the terrorist attacks on September 11th, 2001 in the United States, it was clearly evident that people attempted to make meaning following these events by increasing their patriotism and reaffirming their faith in the 'American Way of Life' (Pyszczynski et al., 2003). As symbols of national pride and in-group loyalty were widely displayed (for example, displaying American flags), unfortunate events unfolded in which individuals from other countries or different ethnicities, perceived as out-group members, were threatened or attacked (Jacoby, 2001). National-level trauma such as terrorism that specifically attack another's culture may enhance people's sense of meaning and attachment to in-group members, but may also yield negative consequences in terms of the potential emergence of increased out-group biases.

Research related to terror management theory has demonstrated the hypothesized defense of cultural worldviews when worldviews become threatened with terrorism. Thus, individuals who are reminded of their own mortality evidence an increase in negative reactions toward members of an out-group who threaten cultural worldviews and positive reactions toward in-group members who support and bolster their worldview (Rosenblatt et al., 1989; McGregor et al., 1998; Simon et al., 1998; Halloran and Kashima, 2004; Hall et al., 2007). Moreover, the distrust of others may generalize to yet others, who are not associated with the threat, as we see with increased dialogue about the threat to US security from Mexicans crossing into the US illegally.

INTERVENTIONS IN RESPONSE TO TERRORISM

In addition to the impact that mass casualty events and terror have on society, there are several ways in which traumatic events influence individuals and business and other organizations and disturb routine functioning. First, the physical, social, and psychological demands of situations involving mass casualty may be overwhelming to people (Hobfoll et al., 2007). Concerns regarding pain, injury or loss of life can consume individuals until they see themselves as out of danger. This concern applies to individuals themselves, but also to their family members, friends and co-workers. Furthermore, structures such as homes, places of employment, and places of worship may have been damaged or destroyed by the mass trauma, further contributing to a state of fear and even chaos if destruction is widespread. The worry and anxiety that likely accompanies the aftermath of a traumatic event can incapacitate people on both individual and societal levels. With this increased burden, people may feel that their daily activities are simply more demanding than they are able to handle. In response, their reactions will range from quite adequate coping to rather severe dysfunction and withdrawal. This said, even though a majority of individuals will respond adequately (Bonanno, 2004), higher-order functioning and decision making will be widely impaired, with important implications for business.

It is important to recognize that people's initial reactions should not necessarily be regarded as pathological responses or as indicators of subsequent psychological disorder. Nevertheless, people may experience immense distress and require clinical intervention and major organizational support to decrease psychological symptoms and impairment (Galea et al., 2002).

Second, the devastation of resources can impoverish individuals' and communities' capacities to cope with mass casualty and recover from its consequences. This is especially true where individuals or communities already have depleted psycho-social and economic resources that would assist in recovery efforts (Hobfoll, 1998; Hobfoll et al., 2007). For organizations, this vulnerability may spring in part from a history of economic or workplace instability. This is not so much how the company's stock is doing, but how employees have been treated; so labor unrest, a dissatisfied workforce or infighting, or even a period of rapid growth, may all result in fissures in organizational strength. The ability of companies and individuals alike to be resilient in the face of mass trauma resides, in part, in the amount of various resources that can be devoted to recovery efforts, with organizational support and management and co-worker climate being particularly important.

Third, the loss of territory or sense of safety as a result of mass casualty impacts on individuals and communities as this depleted environment can foster fear and helplessness. Locations within the community that provide structure to daily life, such as places of employment, places of worship and civic centers, may no longer remain, or may be associated with the trauma. This further contributes to the loss of resources and stability that is already being threatened. It is also an aspect of many instances of disaster and mass casualty that ongoing violence, massive failure to provide aid, and secondary losses that follow the initial phase mean that there may be no clear post-trauma phase (ibid.).

Finally, the threat of people's sense of meaning, integrity and order often have particularly stressful effects on the lives of individuals, business organizations, and the collective society (ibid.). People's sense of meaning about themselves and the world are often shattered after experiencing a traumatic event. In addition, their sense of justice may be threatened, especially in situations where people are innocent victims or do not receive the resources and support they need in the aftermath of a traumatic event.

The varied and unpredictable nature of traumatic events, including mass casualty, and their respective aftermath defies any specific guidelines that can be rigidly applied. Rather, we can address a general set of guidelines, while simultaneously highlighting the need for flexibility during intervention and adaptation to the particular situational determinants. These guidelines should be learned and implemented in advance to the extent possible, so that when situations require immediate action, management and business organizations can respond in ways that benefit the individuals who make up their organization and the organization itself. These guidelines can be organized around five principles as follows (ibid.): (i) promote a sense of safety, (ii) promote calming, (iii) promote a sense of self- and collective-efficacy, (iv) promote connectedness and (v) promote hope. We shall explore these intervention guidelines next.

PROMOTE A SENSE OF SAFETY

Following terrorism and mass casualty people must react to the situations that threaten their sense of safety, including their own lives, the lives of family members, friends and co-workers or the items and objects they deeply value (Hobfoll, 1991; Ursano et al., 1994; van der Kolk and McFarlane, 1996; Briere and Elliott, 2000; Basoglu et al., 2005). Interventions that increase both a psychological sense of safety and maximize actual safety have been related to a steady reduction of negative post-trauma reactions (Silver et al., 2002; Ozer et al., 2003). This sense of safety can offer people respite as they begin the recovery process in the post-trauma aftermath. Furthermore, those individuals who are able to re-establish a relative sense of safety have considerably lower risk of developing PTSD in the months following exposure to a traumatic event than those who do not (Grieger et al., 2003), Thus, it is vital that the management of businesses and organizations foster a sense of safety and security among their employees and staff in post-disaster efforts (see Table 4.1). A sense of physical safety can first be implemented in a business setting by utilizing all physical methods of security available, such as use of alarms, security personnel and re-introducing established disaster plans and safety areas.

Businesses should move quickly to safe surroundings at alternative sites if their primary worksite is damaged or otherwise unsafe. Computers may allow this through networking from remote sites, and meetings can also occur at borrowed, rented, or branch offices. If families are unsafe, employees will be unable to function and often will not even come to work, so it is

Table 4.1 Five principles for post-disaster or mass casualty responding

Principle	Management intervention
Promote a sense of safety Can help to reduce biological and cognitive, aspects of post-traumatic stress reactions	Utilize alarms, security personnel and re- introduce established disaster plans and safety areas Discourage negative, unhelpful dialogues that catastrophize the event (e.g., pressure-cooker effect) Provide an accurate and organized voice to help confine threat
Promote calming Calmness can provide a way to restore normalcy and foster improvement of current symptoms	Disseminate accurate information Provide education about common reactions to trauma Provide accessibility to cognitive–behavioral therapy approaches Promote positive emotions through structured activities in the workplace that foster the experience of these emotions Allow initial post-traumatic distress to occur, and resolve naturally Do not engage in psychological debriefing
Promote a sense of self- and collective-efficacy Can provide a belief that actions are likely to result in positive outcomes	Remind employees and staff of their efficacy: remind them of a time when they were successful at work, or dealt with difficult circumstances and coped well Make individuals a part of the decision-making process Aim to provide the skills and resources needed for people to achieve goals and complete tasks through empowerment
Promote connectedness Social connectedness increases opportunities for knowledge essential to mass casualty or terrorism response. For maximum effectiveness, connectedness should be an ongoing process that predates traumatic event exposure	Incorporate social support in reaching a common goal by sharing reactions to the traumatic experiences, solving problems, providing emotional understanding and support, and normalizing reactions and experiences, and offer suggestions regarding coping Change the physical structure of offices, meeting rooms and cafeterias to maximize social interaction Coffee breaks or physical stretching breaks could be implemented at designated times during the day to promote the communication between corporation employees Increase work projects or team-building exercises that provide additional sources to create a socially supportive environment for employees and staff

Table 4.1 (continued)

Principle	Management intervention
Promote hope Instilling hope is critical to ward off despondency, futility, and hopeless resignation	Serve as a unified voice of hope and optimism regarding the future of the corporation Focus on the positive aspects and directions that may need to take place after mass trauma (e.g., rebuilding efforts) Focus on resilience during past traumas or difficult times to serve as a reminder that things can improve

in the best interests of business to support family efforts, including alternative housing, applying pressure to political leaders to provide alternative school sites, and providing safe shelter. For example, for companies with offices abroad in economically undeveloped countries, the office may be the most likely place to be secure from a defense and structural standpoint.

Safety should also be planned in advance through individual and group emergency planning for families. This is especially critical in high-risk areas of the world, but is also quite germane to hospitals and other emergency response organizations. If employees are worried about their families' safety, they will be unable to perform anywhere near capacity at work. Hence, plans should be made and structured in advance for family and even pets, so that employees know how to proceed, contact their families, and if they cannot contact them, that all possible is likely being done. This includes availability of medicines, food, water, and non-electric (wind-up) radio contact, as well as knowledge of safe structures. As phones often become overloaded and cease to work in minutes following disaster and terrorist attack, and roads can become impassable or closed, advance planning is essential and in the best interests of business and organizations. The extent and nature of the planning should follow the rule of reasonable risk, so flood zones require generators and zones of political unrest require proper safe zones and even means of self-defense. The provision to employees of expert advice provided through management encourages reasonable planning that itself lowers anxiety, as over-reaction can be as undermining to safety and perceived safety as overplanning.

PROMOTE CALMING

Traumatic events often produce feelings of intense emotionality as people enter a state of unpredictability and cope with losses that may have resulted. Symptoms of anxiety and depression are commonly experienced following terrorism and mass casualty. In fact, some anxiety is a sign of normal and healthy responses required to maintain a sense of safety in a perceived unpredictable environment. Studies of individual trauma reveal that the majority of individuals show symptoms at the outset that, if continued, would be markers of PTSD. Bryant et al. (2003) contend that slightly elevated levels of anxiety or arousal can act as a psychological buffer during the initial stages of the post-trauma aftermath and as such, should not necessarily be cause for alarm. This initial severe reaction is a normal way of responding to a traumatic event, and most individuals return to more manageable levels of emotions within days or weeks after the event. These symptoms can become problematic when they interfere with daily activities and functioning. Daily activities that may be impaired include such functions as sleeping, eating, concentration, cognitive processing, and decision making. Although anxiety and arousal may be adaptive initially, continued states of intense emotionality may lead to other symptoms, such as depression and somatic problems (Harvey and Bryant, 1998; Shalev and Freedman, 2005). Those who do not return to these lower manageable levels of responding are at considerable risk of eventual development of PTSD (McNally et al., 2003; Shalev and Freedman, 2005).

It is critical that following some initial period of responding the initial levels of heightened emotions be reduced in order for employees and personnel to return to a pre-trauma level of functioning. Business management should keep in mind the importance of using an intervention that contains the critical element of calming in response to a traumatic event.

One type of intervention that has been proposed as a way to ameliorate symptoms and prevent the development of PTSD is psychological debriefing. Psychological debriefing is a way of allowing individuals who have been exposed to trauma to express their emotionality and cognitions to psychological personnel soon after the event (critical incident stress debriefing: CISD; Mitchell, 1983; Raphael, 1986). Unfortunately, debriefing often enhances arousal in the immediate aftermath of trauma exposure, rather than providing a sense of calm that may foster a more positive recovery. There is compelling research evidence that psychological debriefing is not effective in preventing consequent psychological disorders (McNally et al., 2003) and may exacerbate symptoms (Neria and Solomon, 1999; Bisson et al., 2000). Hence, it is not recommended that management provide brief, structured opportunities for debriefing their employees and staff following mass casualty or terrorism, unless they are matched with an ongoing process whereby individuals can be identified or can self-identify for ongoing counseling and support. It is recommended that accurate information, education about common reactions to trauma, and accessibility to cognitive behavioral therapy (CBT) approaches may provide more effective ways to promote a sense of calmness among employees and staff.

After experiencing a traumatic event, some people catastrophize their situation, which refers to thinking that the situation is a worst-case scenario or believing that the circumstances cannot be remedied in any manner. This pattern of thinking is dangerous in that people begin to believe that they have no control over their situation and see no way of improving their condition. When people reveal thoughts that catastrophize their situation, psychological treatments, such as CBT can be implemented to help reduce these negative, unhealthy thoughts or cognitions. CBT is an efficacious treatment that relies on principles of learning to modify faulty thought patterns people may hold regarding themselves, their capabilities and their circumstances. Because such reactions are so common, newsletters, emails, and group meetings can be instituted in which this pattern is normalized and addressed. Just knowing how common such overreactions are, has a calming effect and can then be paired with relaxation techniques and more rational ways of seeing the problems being faced. This is not to minimize real concerns, but to minimize exaggerated extreme fears.

Another way to foster a sense of calm following mass casualty or terror is to incorporate positive emotions into activities and tasks whenever possible (Fredrickson, 2001; Fredrickson et al., 2003). These positive ways of responding and coping can assist employees and staff to recover from post-trauma reactions and return to their previous level of functioning in the workplace. Tragedy may make humor or expectations of joy inappropriate in some circumstances, but the positive emotions of sharing, caring, and faith are virtually always appropriate.

Encouraging participation in yoga, meditation, and aerobic exercise may further assist in providing a calm environment to facilitate recovery. It may be particularly beneficial if business management can provide opportunities for these experiences in the work environment, but at a minimum can promote the participation in these activities outside of work time. Even simply learning some of the breathing techniques offered through yoga can assist in stress reduction, which is often a component used in CBT for stress reduction and anxiety management. Whether through yoga or jogging, people may feel empowered by having a healthy means to reduce stress and anxiety as well as learn a self-soothing behavior. This aspect can be critical during times when chaos is pervasive and people cannot find a source of serenity in their lives.

The Mindfulness Based Stress Reduction program (MBSR; Kabat-Zinn et al., 1992) borrows from yoga concepts in ways that more individuals may find palatable. The MBSR program combines yoga and meditation to help alleviate symptoms of pain and stress by encouraging a sustained focus on the present awareness and does so in a way that those who are more Westernized can feel comfortable with. In this regard, Santorelli (1996) made recommendations regarding how to use mindfulness in the workplace and Davidson et al. (2003) demonstrated that MBSR is effective in corporate work settings. In a randomized clinical trial, the researchers showed that workers who took an MBSR course evidenced improvements in immune functioning, emotional processing under stress, and brain functioning. These results potentially offer promise for the recommendation of incorporating yoga, meditation, or the combination in response to mass disaster. Establishing a calm environment in the post-disaster aftermath can be difficult; incorporating positive emotions into activities and tasks as well as encouraging participation in self-soothing exercises such as yoga, meditation and mindfulness can offer serenity in the critical stages following a traumatic event.

PROMOTE A SENSE OF SELF- AND COLLECTIVE-EFFICACY

One of the best-examined constructs in the field of psychology is the importance of people feeling a sense of control over the events in their lives, and in particular, the positive outcomes they experience (Skinner, 1996). One concept related to personal control is self-efficacy, which is the sense that individuals believe that their actions are likely to result in positive outcomes (Bandura, 1997). It is through the self-regulation of thought, emotions, and behavior that self-efficacy is primarily acquired (Carver and Scheier, 1998), This practice of self-regulation allows people to moderate their behavior so that they can optimize their sense of control in a given situation. In addition to self-efficacy, the construct can be extended on a larger scale to collective-efficacy, which is the sense that one identifies with a group that is likely to experience positive outcomes (Antonovsky, 1979; Benight and Bandura, 2004).

Bandura (1977) contends that there are task-specific self-efficacies that relate to distinct skills. One of these domains is trauma-related self-efficacy, which is the perceived ability to regulate disturbing emotions and to solve problems in the areas of property restoration, relocation, personal relationships, job retraining, resource acquisition and other trauma-related tasks (Benight et al., 1999, 2000). Consistent with this theory, interventions focused on prevention of burnout in the workforce (Freedy and Hobfoll, 1994) have been applied to victims of trauma (Resick et al., 2002) and are based on the idea that people need to feel that they have the necessary skills and abilities to overcome risk and solve problems that arise. In the workforce, employees need to believe that they are capable of attaining goals, and the collective organization can attain positive outcomes. It should also be stressed that typically, most people were living normal and functional lives prior to the disaster or mass trauma, thus management can focus more on reminding employees and staff of their efficacy, rather than building efficacy where there was little or none originally. In addition, management can assign work responsibilities that foster a sense of self- and collective-efficacy to facilitate recovery in the post-trauma aftermath work environment.

It should be recognized that self-efficacy does not occur in isolation, but rather requires successful associates with whom to collaborate and solve large-scale problems that are beyond the reach of any individual (for example, large-scale work projects; Hobfoll et al., 2007). This is particularly important in the work environment, where collaboration is often imperative. Thus, management can aim to provide tasks that are manageable for employees and are likely to have a positive outcome to foster both self- and collective-efficacy. It may be necessary to give more suitable tasks immediately in the post-disaster aftermath so that employees can regain their sense of efficacy before moving on to more complicated work projects. Encouraging group work may also be helpful as individuals can each contribute their strengths and get assistance with their weaknesses.

Hobfoll et al. (1990) emphasized that following mass trauma, people are likely to construe the challenges of disaster and recovery as one overwhelming, unsolvable problem. In these particular structured activities it is important to assist and guide individuals to divide the problem into separate, manageable units. This problem-solving approach will increase a sense of control, provide opportunities to create preliminary gains, and decrease the genuine problems people are facing in the post-disaster recovery (Baum et al., 1993). This approach has been shown to be associated with a positive self-concept, less depression and anxiety, and vocational adjustment, providing an intervention with implications for the business workplace. In the workplace, management can break down larger tasks into smaller units and divide these tasks among groups. Even within groups, assignments can be separated into more concrete tasks. By dividing job responsibilities, people feel productive when they are able to reach a goal successfully and can re-establish worked productivity following disaster.

Research on disasters and trauma has shown the importance of resources in coping with a traumatic event. Unfortunately, those who lose the most personal, social, and economic resources are the most affected by mass trauma (Galea et al., 2002; Neria et al., 2006). Therefore, those individuals who have been exposed to trauma and lost resources not only have to recover from the trauma itself, but also lack the resources they need to aid in this recovery. On the other hand, research shows that individuals who are able to maintain their resources have the best capability to recover (Norris and Kaniasty, 1996; Benight and Bandura, 2004). Hence, another consideration related to self- and collective-efficacy is that resources are needed for empowerment and that advocating control of power over the situation without the requisite resource can actually be disempowering (Rappaport, 1981). Businesses and organizations can help individuals rebuild resources, and make the collective resources of the group more available to employees. This must be wide-ranging and might include such diverse tasks as help with funeral or medical arrangements, help with clearing debris, and short-term loans on the instrumental level. It may also include sharing contacts, networks, and problem-solving ideas, as those most affected will be least likely to be fully functioning on these levels.

When attempting to meet the demands of a post-trauma work environment, management can aim to identify and then provide the skills and resources needed for people to achieve goals and complete tasks. These skills may be taught specifically through workshops or seminars and practiced by those in attendance to foster a sense of self- and collective-efficacy. In this way, management is also modeling self- and collective-efficacy, by responding well to the unusual, troubling, and high demand environment. Due to the unpracticed nature of mass trauma, as well as the unbalanced distribution of resources in society, there is a need to assess the relationship between these factors on the individual, organizational or societal levels when intervening after a traumatic event (Hobfoll et al., 2007).

PROMOTE CONNECTEDNESS

Social support has been well established as an important factor in recovering from stress and trauma (Vaux, 1998; Norris et al., 2002). Social support can include such supportive activities as providing and receiving emotional support and understanding, problem solving, and normalizing experiences (Ruzek, 2006). This social connection facilitates opportunities to share knowledge, which as previously discussed, is important in the aftermath of mass casualty or terrorism (for example, 'What roads are closed?' 'Which buildings are safe?'). Social relationships also facilitate coping situations such as sharing reactions to the traumatic event, solving problems, providing emotional understanding and support, and normalizing reactions. This, in turn, can lead to sense of community-efficacy that can aid in personal and organizational recovery after trauma (Benight and Bandura, 2004).

Social support can offer a positive impact on the psychological health of individuals; however, at times, in the post-trauma aftermath, this typically positive aspect of social support can actually hinder a sense of safety in work environments. When comprehensive information about mass trauma is unavailable, which is a common occurrence following disasters and mass violence, people have a tendency to share anecdotes, gossip or 'horror stories' about the traumatic event. Hobfoll and London (1986) defined this phenomenon as the 'pressure-cooker' effect. While the people who are sharing this information are likely doing so to provide and acquire support in the post-trauma aftermath, research has actually shown that the more people are involved with, or exposed to this type of social 'support', the more they are likely to experience psychological distress (ibid.). What often occurs is that the very individuals whom other people seek out for support are likely to be the most vulnerable to the pressure-cooker effect as people turn to them to cope and in doing so, share their own experiences and interpretations to fill in the gaps in their knowledge. As a result, interventions should limit the number of rumors and anecdotes that are shared regarding the trauma if doing so increases psychological discomfort (Hobfoll et al., 2007). This can be especially important in the workplace as employees seek social support by talking to others about their experiences in the face of a lack of understanding. It is critical to point out that sharing traumatic experiences by itself may not be harmful, but in the context of lacking information it may be detrimental as 'worse-case scenarios' are used to fill in gaps in knowledge. Business leaders can aim to reduce the pressure-cooker effect by providing the most accurate and complete information they are able to provide. In addition, they can also strive to be aware of gossip or rumors that promote distress and concern among employees and personnel. Business leaders must provide, to the best that they are able, an accurate, organized voice to help confine threat, and in so doing increase the perception of safety and security where there is no serious threat at hand (Shalev and Freedman, 2005).

Although too much of this social connectedness can lead to the pressure-cooker effect, social relationships can facilitate recovery as long as individuals do not find them too demanding. While little empirical research on how to translate this to intervention exists (Hobfoll et al., 2007) recommendations can be offered based on what is known about social relationships. In a workplace setting, changing the physical structure of offices, meeting rooms and cafeterias could be done so as to maximize social interaction among those who work in the organization. Coffee breaks or physical stretching breaks could be implemented at designated times during the day to promote communication between corporation employees. Furthermore, increasing work projects or team-building exercises may provide additional sources of creating a socially supportive environment for employees and staff. Social connection is a valuable asset in the post-trauma aftermath. As long as people do not find these relationships particularly demanding, the social support they provide and receive can assist in recovery efforts following a traumatic event. If business management implements a structured time for social interaction aimed at support-building, individuals can foster their own recovery by increasing their social resource reservoir.

Finally, following such traumatic circumstances, the border between family and work becomes blurred physically and psychologically. The human attachments become paramount instead. If business fails to recognize this, employees will lose their commitment to the workplace, even if they may comply in the short run. In contrast, a workplace committed to the employees and their families will achieve a heightened sense of loyalty and commitment to the company mission that will be longlasting.

PROMOTE HOPE

The final principle of providing intervention after mass casualty is supported by research indicating the importance of sustaining hope in the aftermath of a traumatic event. It has been shown that individuals who retain optimism (Carver and Scheier, 1987) are likely to have more favorable outcomes after experiencing mass trauma because they are able to still look forward to their future and feel that it will consist of positive outcomes. A shattered worldview (Janoff-Bulman, 1992) and a sense of a foreshortened future (American Psychiatric Association, 1994) often result after experiencing a traumatic event and can lead people to feel hopeless about themselves and their situation. Furthermore, because mass trauma is not an event that people are usually prepared for, it surpasses their acquired coping skills and can make them feel powerless (Hobfoll et al., 2007). It is here that intervention with a focus on encouraging hope can most benefit individuals, society and corporations.

Similar to the principle of promoting connectedness, instilling hope in the workforce may require creativity on the part of management. One way of promoting a sense of hope would be to establish a unified voice of hope and optimism regarding the future of the corporation. It is important not to instill a sense of false hope, but rather to encourage focusing on the positive aspects and directions that may need to take place after mass trauma. Management can provide specific ways of responding after a traumatic event, so that employees can be aware that a plan is in place and that they can be hopeful about future outcomes. Inspiring hope and optimism as much as possible during the post-trauma aftermath will offer an intervention that will prompt recovery on both an individual and a corporate level.

THE EFFECT OF TRAUMA ON BUSINESS ACTIVITIES AND ORGANIZATIONAL BEHAVIOR

In addition to mental health issues, which themselves affect business culture and activity, mass casualty impacts on people's basic sense of belief in a just world. This shattering of basic assumptions about the world can reverberate throughout an organization. Seeing the world as a threatening place, may lead to defensiveness in business and relations with others. Terrorism disrupts the belief in the continuity of life, thereby attenuating organizations' belief in the probability of continued prosperity. In this sense, fear of repeated terrorist actions and the increase in the perception of threat of such events may cause individuals to enact the very decline in business that they fear. The stock market decline during the three years following the September 11, 2001 terrorist attacks is an example of this perceived threat. It is also to a large extent the result of overreaction on the part of business leaders, such that they created a self-fulfilling prophecy of decline.

Threats to people's belief in the meaningfulness of the world can undermine the very nature of the work for which they are engaged. In some cases, experiencing trauma has the ability to evoke existential questions about the meaning and futility of life (Yalom, 1980). A crisis of purpose can ensue wherein individuals may question their usefulness and engage in a reorientation of their beliefs. If coping following trauma is difficult, and despair ensues, individuals' sense of self-worth may become affected. The confluence of these factors all represents threats to the integrity and purpose of business, and can function to undermine an organization.

Substantial research has demonstrated that as people are reminded of their mortality (called terror management, although it actually was not related to terrorism), they become more supportive of in-group members and aggressive toward out-group members (Rosenblatt et al., 1989; McGregor et al., 1998; Simon et al., 1998; Halloran and Kashima, 2004; Hall et al., 2008). Translated to the business sector, individuals may become less willing to partner with those whom they believe to be unlike themselves. Although relationships with individuals within the workplace may not be strained if they are all from the same country of origin, relationships between individuals from diverse cultures and countries of origin would likely become adversarial. But, this may also extend to people in other departments in the same organization, if business has not fostered true common identity. This can affect business both from an internal standpoint with increased stress in the workplace, and externally as openness to conduct business with those perceived as out-group members could suffer.

COR theory suggests that people strive to protect and maintain resources that they value, and therefore, threats to the loss of valued

Table 4.2 Theoretical predictions of the effect of disaster or mass casualty on managerial decision making

Psychological theory	Effect on decision making
Shattered assumptions	Lack of faith in the probability of continued prosperity may lead to inaction
	Existential questions about the meaning and futility of life may lead to a lack of initiative
	Lack of self-worth can lead to feelings of despair and depression which undermine self-efficacy and goal-directed agency
	Beliefs that the world is a dangerous place can lead to a 'bunker' mentality whereby opportunities are seen as threatening and not acted upon
Terror management	A shift in orientation can lead to supportive decisions that affect in-group members and aggressive impulses toward out-group members
	Lack of willingness to partner and collaborate with individuals who are unlike themselves
	Business relationships between individuals from diverse cultures and countries of origin may be seen as threatening and therefore avoidance of these groups may develop
	Infighting may occur within an organization if business has not fostered true common identity. This can lead to alienation in the workplace, and bullying
	Decisions may neglect important input from team members
Conservation of resources	Defensive reactions to preserve valued resources may limit research and development
	Decisions may be based on risk aversion and not sound business practices
	Individuals may decide to limit business expansion
	Increase in aggressive business approaches in attempts to procure more valuable resources, disregarding potential consequences (e.g., Tragedy of the Commons)

resources may motivate individuals and organizations to respond defensively to mass casualty and terrorism events (see Table 4.2). Defensiveness can take the form of becoming unduly risk averse, and this could limit the potential for business growth. For example, if an organization moves to avoid risk and preserve capital and other resources, lucrative business opportunities may be missed. Contrariwise, businesses may become more aggressive in their approach and may make attempts to procure more valuable resources, disregarding potential consequences. This is illustrated by the parable of the Tragedy of the Commons. Where finite common resources exist under conditions of threat, people will try to garner those resources, depleting the common pool. Since for many organizations, management is relying on this common pool of resources, such overuse of common resources could prove disastrous. In contrast to the defense posture organizations can institute following terrorism exposure, encouraging individuals and organizations to focus on goal-directed behavior, thereby establishing a future orientation, may lessen symptoms of psychological distress (Holman and Silver, 2005).

OVERVIEW

We have outlined the effects of mass casualty and terrorism circumstances on individuals as well as business corporations, provided guidelines for interventions following disaster and offered typical ways that business leaders may respond following terrorism. Although each traumatic event brings with it unique aspects, there are common principles that can be used when implementing interventions following the event. Promoting a sense of safety, calming, self- and collective-efficacy, connectedness, and hope can provide the best possible environment to facilitate recovery and re-establish pre-trauma levels of business functioning. It should also be kept in mind that resources play a vital role following traumatic events. Without the resources to offset losses that result following a traumatic event, little recovery can be expected. In the aftermath of a terrorist attack, people, including business management, may tend to act in particular ways in response to the event. Assumptions of the world and working environment may be threatened for both management and the workforce following a terrorist attack. In addition, as shown by terror management theory, individuals may find others who are different from them to be a source of threat and may be less likely to work collaboratively with co-workers. With awareness of these potential behavioral pitfalls, they can be stopped and more adaptive ways of responding can be employed.

Terrorism and mass casualty events breed chaos and unpredictability in society and business organizations. The devastation created from traumatic events can present extreme challenges to business performance and operations. However, investing available resources in recovery efforts will help foster a durable work environment. Traumatic events will continue to push the boundaries of human coping efforts as they present new situations that require novel ways of responding. As we learn more through scientific research and experience, more specific guidelines can be offered to provide optimal ways of returning to pre-trauma levels of functioning for individuals, society and businesses alike.

REFERENCES

American Psychiatric Association (1994), *Diagnostic and Statistical Manual of Mental Disorders*, 4th edn, Washington, DC: American Psychiatric Association.

Antonovsky, A. (1979), *Health Stress and Coping: New Perspectives on Mental and Physical Well-being*, San Francisco, CA: Jossey-Bass.

Bandura, A. (1977), 'Self-efficacy: toward a unifying theory of behavioral change', *Psychological Review*, **84**, 191–215.

Bandura, A. (1997), *Self-efficacy: The Exercise of Control*, New York: Freeman.

Basoglu, M., Salcioglu, E., Livanou, M., Kalender, D. and Acar, G. (2005), 'Single-session behavioral treatment of earthquake-related posttraumatic stress disorder: a randomized waiting list controlled trial', *Journal of Traumatic Stress*, **18**(1), 1–11.

Baum, A., Cohen, L. and Hall, M. (1993), 'Control and intrusive memories as possible determinants of chronic stress', *Psychosomatic Medicine*, **55**(3), 274–86.

Becker, E. (1973), *The Denial of Death*, New York: Simon and Schuster.

Benight, C.C. and Bandura, A. (2004), 'Social cognitive theory of posttraumatic recovery: the role of perceived self-efficacy', *Behaviour Research and Therapy*, **42**, 1129–48.

Benight, C.C., Freyaldenhoven, R.W., Hughes, J., Ruiz, J.M., Zoschke, T.A. and Lovallo, W.R. (2000), 'Coping self-efficacy and psychological distress following the Oklahoma City bombing', *Journal of Applied Social Psychology*, **30**(7), 1331–44.

Benight, C.C., Swift, E., Sanger, J., Smith, A. and Zeppelin, D. (1999), 'Coping self-efficacy as a mediator of distress following a natural disaster', *Journal of Applied Social Psychology*, **29**(12), 2443–64.

Bisson, J.I., McFarlane, A.C. and Rose S. (2000), 'Psychological debriefing', in E.B. Foa, T.M. Keane and M.J. Friedman (eds), *Effective Treatments for PTSD: Practice Guidelines from the International Society for Traumatic Stress Studies*, New York: Guilford, pp. 39–59.

Bonanno, G.A. (2004), 'Loss, trauma, and human resilience: have we underestimated the human capacity to thrive after extremely aversive events?', *American Psychologist*, **59**, 20–28.

Briere, J. and Elliott, D. (2000), 'Prevalence, characteristics and long-term sequelae of natural disaster exposure in the general population', *Journal of Traumatic Stress*, **13**(4), 661–79.

Bryant, R.A., Harvey, A.G., Guthrie, R.M. and Moulds, M.L. (2003), 'Acute psychophysiological arousal and posttraumatic stress disorder: a two-year prospective study', *Journal of Traumatic Stress*, **16**(5), 439–43.

Carver, C.S. and Scheier, M.F. (1987), 'Dispositional optimism and physical well-being: the influence of generalized outcome expectancies on health', *Journal of Personality*, **55**(2), 169–210.

Carver, C.S. and Scheier, M.F. (1998), *On the Self-regulation of Behavior*, Cambridge and New York: Cambridge University Press.

Code of Federal Regulations (2007), http://frwebgate. access.gpo.gov/cgi-bin/getcfr.cgi?TITLE=28&PART=0&SECTION=85&TYPE=TEXT, retrieved April 9.
Davidson, R.J., Kabat-Zinn, J., Schumacher, J., Rosenkranz, M., Muller, D., Santorelli, S.F. et al. (2003), 'Alterations in brain and immune function produced by mindfulness meditation', *Psychosomatic Medicine*, **65**(4), 564–70.
Ennis, N.E., Hobfoll, S.E. and Schroder, K.E.E. (2000), 'Money doesn't talk, it swears: how economic stress and resistance resources impact inner-city women's depressive mood', *American Journal of Community Psychology*, **28**(2), 149–73.
Fredrickson, B.L. (2001), 'The role of positive emotions in positive psychology: the broaden-and-build theory of positive emotions', *American Psychologist*, **56**(3), 218–26.
Fredrickson, B.L., Tugade, M.M., Waugh, C.E. and Larkin, G.R. (2003), 'What good are positive emotions in crisis? A prospective study of resilience and emotions following the terrorist attacks on the United States on September 11th, 2001', *Journal of Personality and Social Psychology*, **84**(2), 365–76.
Freedy, J.R. and Hobfoll, S.E. (1994), 'Stress inoculation for reduction of burnout: a conservation of resources approach', *Anxiety, Stress, and Coping*, **6**(4), 311–25.
Friedland, N. and Merari, A. (1985), 'The psychological impact of terrorism: A double-edged sword', *Political Psychology*, **6**(4), 591–604.
Galea, S., Ahern, J., Resnick, H., Kilpatrick, D., Bucuvalas, M., Gold, J., et al. (2002), 'Psychological sequelae of the September 11 terrorist attacks in New York City', *New England Journal of Medicine*, **346**(13), 982–7.
Greenberg, J., Pyszczynski, T. and Solomon, S. (1986), 'The causes and consequences of a need for self-esteem: a terror management theory', in R.F. Baumeister (ed.), *Public and Private Self*, New York: Springer-Verlag, pp. 189–212.
Grieger, T.A., Fullerton, C.S. and Ursano, R.J. (2003), 'Posttraumatic stress disorder, alcohol use, and perceived safety after the terrorist attack on the Pentagon', *Psychiatric Services*, **54**, 1380–82.
Hall, B.J., Hobfoll, S.E., Canetti-Nisim, D., Johnson, R. and Galea, S. (2007), 'Posttraumatic growth following terrorism from a terror management perspective', manuscript submitted for publication.
Hall, B.J., Hobfoll, S.E., Palmieri, P., Canetti-Nisim, D., Shapira, O., Johnson, R.J. and Galea, S. (2008), 'The psychological impact of forced settler disengagement in Gaza: trauma and posttraumatic growth', *Journal of Traumatic Stress*, **21**, 22–9.
Halloran, M.J. and Kashima, E.S. (2004), 'Social identity and worldview validation: the effects of in-group identity primes and mortality salience on value endorsement', *Personality and Social Psychology Bulletin*, **30**(7), 915–25.
Harvey, A.G. and Bryant, R.A. (1998), 'Predictors of acute stress following mild traumatic brain injury', *Brain Injury*, **12**(2), 147–54.
Hobfoll, S.E. (1988), *The Ecology of Stress*, New York: Hemisphere.
Hobfoll, S.E. (1989), 'Conservation of resources: a new attempt at conceptualizing stress', *American Psychologist*, **44**, 513–24.
Hobfoll, S.E. (1991), 'Traumatic stress: a theory based on rapid loss of resources', *Anxiety Research*, **4**, 187–97.
Hobfoll, S.E. (1998), *Stress, Culture, and Community: The Psychology and Philosophy of Stress*, New York: Plenum.

Hobfoll, S.E., Canetti-Nisim, D. and Johnson, R. (2006), 'Exposure to terrorism, stress-related mental health symptoms, and defensive coping among Jews and Arabs in Israel', *Journal of Consulting and Clinical Psychology*, **74**(2), 207–18.

Hobfoll, S.E., Freedy, J., Lane, C. and Geller, P. (1990), 'Conservation of social resources: social support resource theory', *Journal of Social and Personal Relationships*, **7**, 465–78.

Hobfoll, S.E. and Lilly, R.S. (1993), 'Resource conservation as a strategy for community psychology', *Journal of Community Psychology*, **21**, 128–48.

Hobfoll, S.E. and London, P. (1986), 'The relationship of self-concept and social support to emotional distress among women during war', *Journal of Social and Clinical Psychology*, **4**, 189–203.

Hobfoll, S.E., Watson, P., Bell, C.C., Bryant, R.A., Brymer, M.J., Friedman, M.J. et al. (2007), 'Five essential elements of immediate and mid-term mass trauma intervention: empirical evidence', *Journal of Psychiatry: Interpersonal and Biological Processes*, **70**, 283–315.

Holahan, C.J., Moos, R.H., Holahan, C.K. and Cronkite, R.C. (1999), 'Resource loss, resource gain, and depressive symptoms: a 10-year model', *Journal of Personality and Social Psychology*, **77**(3), 620–29.

Holman, E.A. and Silver, R.C. (2005), 'Future-oriented thinking and adjustment in a nationwide longitudinal study following the September 11th terrorist attacks', *Motivation and Emotion*, **29**(4), 389–410.

Jacoby, J. (2001), 'Speaking out against terror', *The Boston Globe*, September 23, p. D7.

Janoff-Bulman, R. (1992), *Shattered Assumptions: Towards a New Psychology of Trauma*, New York: Free Press.

Kabat-Zinn, J., Massion, A.O., Kristeller, J., Peterson, L.G., Fletcher, K., Pbert, L. et al. (1992), 'Effectiveness of a meditation-based stress reduction program in the treatment of anxiety disorders', *American Journal of Psychiatry*, **149**, 936–43.

King, D.W., King, L.A., Foy, D.W., Keane, T.M. and Fairbank, J.A. (1999), 'Posttraumatic stress disorder in a national sample of female and male Vietnam veterans: risk factors, war-zone stressors, and resilience-recovery variables', *Journal of Abnormal Psychology*, **108**, 164–70.

Lazarus, R.S. and Folkman, S. (1984), *Stress, Appraisal, and Coping*, New York: Springer.

McGregor, H.A., Lieberman, J.D., Greenberg, J., Solomon, S., Arndt, J., Simon, L. et al. (1998), 'Terror management and aggression: evidence that mortality salience motivates aggression against worldview-threatening others', *Journal of Personality and Social Psychology*, **74**(3), 590–605.

McNally, R.J., Bryant, R.A. and Ehlers, A. (2003), 'Does early psychological intervention promote recovery from posttraumatic stress?', *Psychological Science in the Public Interest*, **4**(2), 45–79.

Mitchell, J.T. (1983), 'When disaster strikes: the critical incident stress debriefing', *Journal of Medical Emergency Services*, **8**, 36–9.

Neria, Y., Gross, R., Olfson, M., Gameroff, M.J., Wickramaratne, P., Das, A. et al. (2006), 'Posttraumatic stress disorder in primary care one year after the 9/11 attacks', *General Hospital Psychiatry*, **28**(3), 213–22.

Neria, Y. and Solomon, Z. (1999), 'Prevention of posttraumatic reactions: debriefing and frontline treatment', in P.A. Saigh and J.D. Bremner (eds), *Posttraumatic Stress Disorder: A Comprehensive Test*, Boston, MA: Allyn & Bacon, pp. 309–26.

Norris, F.H., Friedman, M.J., Watson, P.J., Byrne, C.M., Diaz, E. and Kaniasty, K. (2002), '60,000 disaster victims speak: Part I. An empirical review of the empirical literature, 1981–2001', *Psychiatry: Interpersonal and Biological Processes*, **65**(3), 207–39.

Norris, F. and Kaniasty, K. (1996), 'Received and perceived social support in times of stress: a test of the social support deterioration deterrence model', *Journal of Personality and Social Psychology*, **71**, 498–511.

Ozer, E.J., Best, S.R., Lipsey, T.L. and Weiss, D.S. (2003), 'Predictors of posttraumatic stress disorder and symptoms in adults: a meta-analysis', *Psychological Bulletin*, **129**(1), 52–73.

Pyszczynski, T., Solomon, S. and Greenberg, J. (2003), *In the Wake of 9/11: The Psychology of Terror*, Washington, DC: American Psychological Association.

Raphael, B. (1986), *When Disaster Strikes: How Individuals and Communities Cope with Catastrophe*, New York: Basic Books.

Rappaport, J. (1981), 'In praise of paradox: a social policy of empowerment over prevention', *American Journal of Community Psychology*, **9**(1), 1–25.

Resick, P.A. , Nishith, P., Weaver, T.L., Astin, M.C. and Feuer, C.A. (2002), 'A comparison of cognitive-processing therapy with prolonged exposure and a waiting condition for the treatment of chronic posttraumatic stress disorder in female rape victims', *Journal of Consulting and Clinical Psychology*, **70**(4), 867–79.

Rosenblatt, A., Greenberg, J., Solomon, S., Pyszczynski, T. and Lyon, D. (1989), 'Evidence for terror management theory: the effects of mortality salience on reactions to those who violate or uphold cultural values', *Journal of Personality and Social Psychology*, **57**(4), 681–90.

Ruzek, J.I. (2006), 'Models of early intervention following mass violence and other trauma', in E.C. Ritchie, P.J. Watson and M.J. Friedman (eds), *Interventions Following Mass Violence and Disasters*, New York: Guilford, pp. 16–34.

Santorelli, S.F. (1996), 'Mindfulness and mastery in the workplace: 21 ways to reduce stress during the workday', in A. Kotler (ed.), *Engaged Buddhist Reader*, Berkeley, CA: Parallax, pp. 40–45.

Schlenger, W.E., Caddell, J.M., Ebert, L., Jordan, B.K., Rourke, K.M., Wilson, D. et al. (2002), 'Psychological reactions to terrorist attacks: findings from the national study of Americans' reactions to September 11', *Journal of the American Medical Association*, **288**(5), 581–8.

Schuster, M.A., Stein, B.D., Jaycox, L.H., Collins, R.L., Marshall, G.N., Elliott, M.N. et al. (2001), 'A national survey of stress reactions after the September 11, 2001, terrorist attacks', *New England Journal of Medicine*, **345**(20), 1507–12.

Shalev, A.Y. and Freedman, S. (2005), 'PTSD following terrorist attacks: a prospective evaluation', *American Journal of Psychiatry*, **162**(6), 1188–91.

Silver, R.C., Holman, E.A., McIntosh, D.N., Poulin, M. and Gil-Rivas, V. (2002), 'Nationwide longitudinal study of psychological responses to September 11', *Journal of the American Medical Association*, **288**(10), 1235–44.

Simon, L., Arndt, J., Greenberg, J., Pyszczynski, T. and Solomon, S. (1998), 'Terror management and meaning: evidence that the opportunity to defend the worldview in response to mortality salience increases the meaningfulness of life in the mildly depressed', *Journal of Personality*, **66**(3), 359–82.

Skinner, B.F. (1996), 'The ethics of helping people', in M.A. Mattaini and B.A. Thyer (eds), *Finding Solutions to Social Problems: Behavioral Strategies for Change*, Washington, DC: American Psychological Association, pp. 61–72 (originally published 1975).

Solomon, S., Greenberg, J. and Pyszczynski, T. (2004), 'The cultural animal: twenty years of terror management theory and research', in J. Greenberg, S. Koole and T. Pyszczynski (eds), *The Handbook of Experimental Existential Psychology*, New York: Guilford, pp. 13–34.

Ursano, R.J., McCaughey, B.G. and Fullerton, C.S. (1994), *Individual and Community Responses to Trauma and Disaster*, Cambridge: Cambridge University Press.

van der Kolk, B.A. and McFarlane, A.C. (1996), 'The black hole of trauma', in van der Kolk and McFarlane (eds), *Traumatic Stress: The Effects of Overwhelming Experience on Mind, Body, and Society*, New York: Guilford, pp. 3–23.

Vaux, A. (1998), *Social Support: Theory, Research, and Intervention*, New York: Praeger.

Wells, J.D., Hobfoll, S.E. and Lavin, J. (1997), 'Resource loss, resource gain, and communal coping during pregnancy among women with multiple roles', *Psychology of Women Quarterly*, **21**(4), 645–62.

Wells, J.D., Hobfoll, S.E. and Lavin, J. (1999), 'When it rains it pours: the greater impact of resource loss compared to gain on psychological distress', *Personality and Social Psychology Bulletin*, **25**, 1172–82.

Yalom, I.D. (1980), *Existential Psychotherapy*, New York: Basic Books.

5. Psycho-social impacts of bioterrorism and stress in the wake of 9/11

Ross H. Pastel*

INTRODUCTION

Biological warfare (BW) and bioterrorism (BT) use microorganisms (bacteria or viruses) or toxins (products of living organisms) to induce death and disease. Ultimately, bioterrorism would be used to attack a society by inducing fear and vulnerability in the public. The public perception of bioterrorism changed dramatically in October 2001, following the recognition of casualties, including deaths, due to anthrax in the US mail (CDC, 2001). Mother Nature has also terrorized humans with infectious disease outbreaks, including three plague pandemics (for example, the infamous Black Death during the Middle Ages), the 1918 Spanish influenza pandemic, numerous smallpox epidemics throughout the world, several Ebola hemorrhagic fever outbreaks in Africa, and, most recently, the Severe Acute Respiratory Syndrome (SARS) which emerged from China in 2002 and rapidly spread around the world in a matter of weeks. Currently, many people are concerned (or even fearful), that a new influenza pandemic (perhaps involving avian influenza) may strike the world.

Although terrorism has been around for ages, the recognition of the psycho-social impact of terrorism has been more recent. In 1995, the world was shocked by two major terrorist attacks – the Aum Shinrikyo sarin attack in the Tokyo subway system and the large truck bomb attack of two home-grown American terrorists, Timothy McVeigh and Terry Nichols, on the Murrah Federal Building in Oklahoma City. Since then, Al Qaeda has been responsible for two embassy attacks in Africa, the infamous September 11, 2001 attacks on New York City and Washington, DC, the Madrid train attacks, and the London bombing attacks. In each of these

* Disclaimer: The ideas expressed in this chapter are private views of the author and do not reflect the policies or positions of the US Department of the Army, the US Department of Defense, or the US government.

terrorist attacks, the psycho-social impact and trauma have been tremendous.

In this chapter, we shall introduce and explore the psycho-social impact of bioterrorism. As can be imagined, the impact will be both widespread and complex. We shall begin with the topic of risk, risk perception, and the media. Much of the psycho-social impact will be based on risk perception, which is influenced by risk communication by leaders and information and reporting by the news media. We shall follow that with an introduction to bioterrorism and natural outbreaks of infectious disease. The BW agents will be introduced with various possible bioterrorist scenarios that might occur. Outbreaks of contagious disease, whether bioterrorist induced or natural will also be discussed. We shall then explore who will be affected – community, healthcare workers, and patients and their families.

RISK, RISK PERCEPTION, AND THE MEDIA

How people act and react to a bioterrorist attack changes during the sequence of events that occur. Much of the action and reaction will be based on risk. One approach to risk communication states that risk equals hazard plus outrage (Sandman, 2003). The hazard is a scientifically based risk assessment, whereas the outrage is made up of less-quantifiable factors of the public's concern and perception of the biological exposure event. Thus, the risk continuously changes and evolves, based on the risk perception not only to the initial event, but to each event that ensues, including the actions and risk communication of leaders, media representation of the event and the response, and the success or failure of the various actions taken in response to the event.

In a BT event, the beginning and extent of the danger will not be known immediately. Unless the perpetrator is caught, there may well be a dread of possible multiple attacks spread over time. In a bioterrorist attack or natural contagious disease outbreak, information becomes extremely important. The public is eager for information and needs to know what precautionary measures should be taken to protect themselves. Public health authorities and public officials must calculate the extent of the threat, and inform the media and the public about the event and what precautionary measures to take. Mayor Rudy Giuliani of New York City was extremely effective following the events of September 11 and the ensuing anthrax attacks in October, 2001. He demonstrated the value of daily or twice-daily scheduled briefings with the media and the public.

According to experts, the initial US government response to the 2001 anthrax attack in Florida violated many of the risk communication guidelines for handling crises (Goode, 2001; Chess and Celia, 2002). Key principles include: full disclosure of what is known and not known; avoid speculation; detail what is being done to counter the threat; and give recommendations on how to protect oneself (Goode, 2001). Unfortunately, the early failures in risk communication undermined the government's credibility. For example, the Health and Human Services Secretary provided inappropriately reassuring information when he announced that the anthrax exposure of the first victim was probably due to an exposure while hunting (ibid.). Federal officials also downplayed uncertainty when they initially dismissed the possibility of a threat to postal employees from mail containing anthrax (Lipton, 2001).

In Hong Kong, most respondents to a survey reported actively seeking SARS information on a daily basis, and relied more on mass media (television, newspapers and radio), than on medical professionals, friends, or the internet (Lau et al., 2003). Substantial misinformation and false beliefs persisted in Hong Kong adults even at an advanced stage of the SARS epidemic, despite constant media and public service announcements (Leung et al., 2003). Although information existed on precautionary measures, recommended measures were not practiced uniformly. Many people did not understand transmission routes – only one-third of respondents avoided direct contact by touch with contaminated objects (fomites) and less than half practiced at least five of the seven recommended precautions.

Risk perception is an important driver of the outrage component in risk communication. In the risk literature, a number of factors have been shown to increase the perception of risk, including potentially fatal illness, involuntary exposure (lack of control), catastrophic event, presence of an unknown perpetrator, delayed detection and reaction by authorities, and potential effect on future generations (Covello et al., 2001; Brecher and Flynn, 2002). A number of factors may further amplify risk perception – scapegoating, distrust of governmental and industrial experts, and news media hype and misinformation.

Death is always frightening and some of the BW agents are lethal, including several that have a long history of epidemics and death (for example, smallpox, plague, anthrax, and Ebola). BW agents are invisible and odorless, which leads to uncertainties of exposure, amount of exposure, and the possibility of disease and death. In many cases, the extent and location of an aerosol attack may be unknown until suddenly, throngs of sick and dying converge on hospitals. The BW agents may initially induce non-specific, flu-like symptoms – for example, fatigue, headache, nausea, dyspnea, dizziness, and muscle/joint ache. These ubiquitous symptoms may lead many people to seek treatment, even in the absence of any possible exposure. Thus, a bioterrorist attack would be likely to trigger a multitude of risk perception factors, including involuntariness, uncontrollability, unfamiliarity, unfairness, uncertainty, dread, and distrust of institutions (Covello et al., 2001).

The news media can act as an important risk amplifier depending on how they select and frame risk messages to inform the public. Any ensuing media hype can create waves of fear due to intensive reporting (ibid.). Unfortunately, in the 2001 anthrax attacks, the news media hype and worst-case theorizing may have inflated the actual scope of the crisis and public fears (Goode, 2001). Mass media presentations of terrorist attacks such as the Oklahoma City bombing and the September 11, 2001 attacks have been shown to have a strong effect on post-traumatic stress symptomatology in both children and adults (Pfefferbaum et al., 2001; Schuster et al., 2001; Ahern et al., 2002; Schlenger et al., 2002). Mass media has also played an important role in various outbreaks of multiple unexplained symptoms (OMUS) (Vasterman et al., 2005).

BIOTERRORISM AND NATURAL OUTBREAKS OF INFECTIOUS DISEASE

Although the use of infectious diseases as weapons is very old, sophisticated use of pathogens is relatively recent. Development of offensive biological warfare programs began in earnest during and after the Second World War (Woods et al., 2005). Both the Former Soviet Union and Iraq had known offensive BW programs (Davis, 1999). What seems to be more recent is the use of BW agents in BT acts. Although the anthrax attacks through the mail in the US in 2001 are well known, much less publicized were the attempted use of anthrax by the Aum Shinrikyo cult in Tokyo (Takahashi et al., 2004) and the successful bioterrorist use of a salmonella strain in The Dalles, OR in 1984 (Torok et al., 1997). Of special note is that, as of August 2007, the perpetrator(s) of the US anthrax attacks has not been captured.

Bioterrorists have a wide choice of organisms and toxins to choose from, but some are more frightening and effective than others. The US Centers for Disease Control and Prevention (CDC) convened a meeting of national experts who differentiated biological agents into three categories of concern (Rotz et al., 2002). Category A agents were of most concern because of lethality, public fear, and public health requirements. The Category A agents include botulinum toxin and the organisms responsible for anthrax, smallpox, plague, tularemia, viral hemorrhagic fevers (for example, Ebola and Marburg). Category B agents can also be disseminated widely, but would be less lethal, and Category C agents were emerging threat agents.

In 1970, the World Health Organization (WHO) estimated the effects of 50 kilograms of a dried powder containing various biological agents released as an aerosol over an urban population of approximately 500,000 people (WHO, 1970). Mortality estimates ranged from 150 to 500 for Category B agents (Q Fever, Venezuelan Equine Encephalitis, and Brucellosis) to 30,000 to 95,000 for Category A agents (tularemia, plague, and anthrax). Morbidity and incapacitation ranged from 35,000 to 125,000 for the three Category B agents and from 100,000 to 125,000 for the Category A agents. The US Department of Homeland Security (Homeland Security Council, 2004) has published five biological scenarios: aerosol anthrax, pandemic influenza, pneumonic plague, food contamination with anthrax, and a foreign animal disease (Foot and Mouth Disease). These scenarios were developed to help communities test response capabilities and resources. Fatalities in these scenarios range from 300 in the food contamination scenario to 87,000 for the pandemic influenza.

Many of the BW agents pose problems with diagnosis because they initially present with flu-like symptoms which are non-specific. In the absence of a specific diagnosis, empirical treatments are often attempted, but not all the agents respond well to typical first-line treatment. Most of the bacterial diseases have effective antibiotics, but they may not be the typical empirically prescribed treatment. The only existing treatment for many of the viral diseases and botulism is supportive care. Finally, regardless of whether effective treatment or supportive care exists, most cities do not have adequate resources, drugs, medical equipment or medical personnel, to handle a large outbreak of hundreds to thousands of patients.

Another important differentiation among the different biological agents is whether they are contagious. According to *Stedman's Medical Dictionary* (Pugh, 2006), contagious means 'communicable or transmissible by contact with the sick or their fresh secretions or excretions'. Fortunately, most of the biological agents are not contagious. For non-contagious biological agents (for example, anthrax, tularemia, botulinum toxin), the threat will be from the initial dissemination of the agent. However, smallpox and pneumonic plague are contagious via respiratory droplets, while the viral hemorrhagic fevers (for example, Ebola and Marburg) are transmissible via direct contact with secretions or excretions. Two epidemic diseases found in nature, influenza and SARS, are also transmissible via respiratory droplets. Transmission via respiratory droplets typically requires close, intimate face-to-face contact (3–6 feet), which is why healthcare providers are at risk unless properly protected with personal protective equipment (PPE). Unfortunately, the general public probably will not recognize that most biological agents are not contagious and may stigmatize infected people, regardless of whether they pose a risk.

Contagious diseases present the horror of a continuing epidemic, as infected people may transmit their infection to non-protected people. Protection may be from PPE, such as gloves and masks, and for some diseases, from prophylaxis, such as a vaccine or prophylactic antibiotics. Contagious diseases require an enormous public health effort for medical evaluations, medical surveillance, and possibly quarantine of people exposed to infected people. Given the current managed medical care in many countries, there is very little surge capacity to handle large numbers of sick patients.

During an outbreak of an infectious disease, initial identification of the disease organism may take some time and will probably occur as a result of multiple people converging on hospitals with similar symptom syndromes. Disease outbreaks caused by initially unknown agents are likely to provoke anxiety in the general public. Distinguishing between a natural and a covert intentional outbreak may also take time. However, once an attack has been identified as an intentional bioterrorist attack, the public may well react with additional anxiety and fear. A large-scale aerosol BT attack may be very deadly, and the resulting fear may last for a long time, especially if the perpetrator is not caught.

WHO IS AFFECTED?

Many different kinds of psycho-social effects are likely to be seen following a bioterrorist attack. We shall look at three different groups affected by a bioterrorist attack or natural outbreak of epidemic disease: the community, healthcare workers, and patients and their families.

Community Response

Collective (group) responses

Mass panic Community response can be broken down into collective or group responses and individual responses. For most planners, the most common fear is that there will be a collective response of mass panic. Strictly speaking, mass panic means an acute fear reaction marked by loss of self-control and followed by non-social and unreasoning flight (Quarantelli, 1954). However, flight can be a normal reaction to the presence of a perceived immediate danger. Flight becomes a mass panic only when large numbers of people stampede without regard to others in an attempt to escape danger. Most commonly, mass panic occurs in situations with an immediate danger and limited escape routes, for example, fires in a building with limited escape routes. In 1994, following a pneumonic plague outbreak in Surat, India, an estimated 4–6 hundred thousand people fled the city, including hospital staff, private medical practitioners, and municipal workers (Kumar, 1994; Madan, 1995; Ramalingaswami, 2001). However, there was no report of a mass panic involving non-social behavior. In this case, many people may have lacked confidence in the local government's response to the outbreak and took the precaution of leaving the area. Due to the nature of a bioterrorist attack, by the time the event has been identified as an attack, it will be long after the event. Thus, mass panic will be very unlikely. However, mass flight might be a possibility.

Worried well Another common fear is that hospitals will be swarmed by the 'worried well'. This presupposes that the worried well are not suffering real symptoms and have not been exposed, but are simply worried that they might be ill. However, after a BT event, many people may have an unknown exposure and may or may not be symptomatic. The term 'worried well' is a pejorative term and should not be used as it may interfere with the doctor–patient relationship. Unfortunately, there is no generally acceptable term that has taken its place. Perhaps we should take note of a diagnosis that was used in the First World War – NNYD – Nervous, Not Yet Diagnosed (Salmon and Fenton, 1929).

In the first Persian Gulf War, Israel was attacked with scud missiles. Initially, many Israelis feared that the missiles would contain sarin, a lethal nerve agent. Following the initial scud attack, there were 22 minor injuries, but also 172 stress casualties and 171 people who had injected themselves with atropine, a nerve agent antidote (Bleich et al., 1992).

Depending on the situation, there may or may not be an increase in patients who are 'asymptomatic'. After the anthrax attacks in the US, there were hundreds of prescriptions for ciprofloxacin (used to treat anthrax) doled out to people who had no reason to believe that they had been exposed (Steinhauer, 2001). Hospitals reported that their already busy emergency rooms were filled with people anxious about anthrax, many demanding treatment.

The picture is less clear with SARS, due to a lack of reporting in the literature. One urban emergency department in Taiwan reported that there was a more than 30 percent drop in the number of patients when comparing the 30 days previous to the first reported hospital associated transmission of SARS (period 1) to the 30 days after (period 2) (Chen et al., 2005). There was almost a 4 percent drop in non-trauma patients younger than 18 years, a 2.5 percent increase in patients with a body temperature greater than 38° C, and a 4 percent increase in patients with a chief complaint of fever. In comparing the ratio of patients with fever divided by those complaining of fever, the percentage increased from period 1 (56 percent) to period 2 (80 percent), which suggests that people were more accurate about a suspected fever.

OMUS Another important collective behavior phenomenon has variously been called mass psychogenic illness, mass hysteria, and epidemic hysteria (Colligan et al., 1983; Wessely, 1987; Boss, 1997). Unfortunately, these terms all have a pejorative connotation. For example, 'hysteria' presupposes an overemotional response to an event and implies that mostly females are involved. Use of the term 'mass hysteria' by media, medical personnel, and/or public officials can lead to a negative perception of medical personnel and public officials by the people affected by an event and vice versa. Similarly, mass psychogenic illness implies that symptoms are 'all in the head' and thus, not real. A preferred term is 'OMUS', an acronym for outbreak of multiple unexplained symptoms (Pastel, 2001). Although clumsy, this term is still relatively neutral. OMUS is also descriptive – the symptoms are real and unexplained, not 'all in the head'.

It should also be clear that not all OMUS occurs in young, school-age females. For example, in one incident, 1800 men were evacuated from their barracks due to an epidemic of coughing, dyspnea, and chest pain that broke out at a military recruit training center (Struewing and Gray, 1990). Over 350 men were taken to local hospitals, with eight being admitted. Although recruits and medical personnel suspected an airborne toxin, none was detected.

The cause of OMUS is not an either/or phenomenon that can be deduced by ruling out toxic or infectious agents. Symptoms of psychological origin can co-occur with exposure to an actual agent. In other words, some casualties may have psychological reactions in addition to reactions caused by a toxic agent. However, not all symptomatic casualties may have been exposed to a toxic agent (Singer, 1982). A case in point occurred in 1987, when the radiation accident in Goiania became public. The threat of possible radiation exposure caused an estimated 125,000 people (10 percent of the population) to be screened for contamination (Petterson, 1988). Approximately 5000 (8 percent) of the first 60,000 people to be screened presented with symptoms that mimicked radiation sickness (rash around neck and upper body, vomiting, diarrhea and so on), but none of these individuals was contaminated (ibid.).

In the fall of 2001, the US was shocked by anthrax letters in the mail which led to 22 cases of anthrax (11 inhalational and 11 cutaneous) (CDC, 2001). Less well-known or remembered is that following the airplane attacks of September 11, 2001 and prior to the first case of anthrax, there had already been increased purchases of gas masks and ciprofloxacin.

There were anecdotal reports of people not opening their mail, or only opening it outdoors.

The outbreak of SARS, a new and emerging infection, created much fear and anxiety. In Beijing, schools and universities were closed, hundreds of companies closed their doors, and some surrounding villages shut themselves off (Pomfret, 2003). Rumors of neighborhoods being quarantined led to people stockpiling food (Eckholm, 2003). Although officials asked people to avoid travel, thousands of businessmen, migrant workers, and college students left Beijing.

Individual responses

Risk communication Risk communication will be critical in a bioterrorist event or a natural outbreak of a contagious disease. In one study of multiple focus groups, participants reacted with increasing fear and concern as the scenario of an attack with plague evolved with increasing casualties and deaths (Wray and Jupka, 2004). They wanted to understand: (i) the nature of the threat, (ii) the protective actions they should take if they had not been exposed, (iii) the steps to take if they might have been exposed, (iv) the steps to take if they had been exposed and (v) information related to the actual event. Participants indicated that they would look for information from local civil authorities, emergency responders, medical personnel, and local and national media.

A second study involving focus groups utilized a botulinum toxin bioterrorist scenario (Glik et al., 2004). Emotional responses included fear, anxiety, distress, nervousness, and helplessness and a concern for the safety of children. Three groups of responses were noted for behavioral intents. Most planned to 'stock up' on supplies, while others talked about leaving town, and a third group had a 'wait and see' intent. As the scenario evolved, participants' fear increased and they were eager for information concerning health and safety for themselves and their families. Fears started to reduce as the agent was identified, but there was a continuing hunger for more specific information about the attack.

In a contagious disease outbreak, the public is especially eager for information and needs to know what precautionary measures should be taken. One study in Hong Kong found that most respondents actively sought SARS information on a daily basis, relying on television (90 percent), newspapers (71 percent), and radio (27 percent), but not many (<10 percent) sought information from medical professionals, friends, or the internet (Nickell et al., 2004). However, information seeking does not always translate into knowledge or practice. Another telephone survey done in Hong Kong at an advanced stage of the SARS epidemic revealed substantial misinformation and false beliefs in adults despite constant media and public service announcements (Lau et al., 2003). Lack of understanding of transmission routes led to a deficit in adopting some of the precautionary measures – specifically, only about a third of respondents avoided direct contact by touch with contaminated objects (fomites) and less than half practiced at least five of the seven recommended precautions. Recommended measures were not necessarily practiced uniformly. In Hong Kong, two studies found that the majority of respondents wore face masks and washed their hands with soap frequently, but in Singapore, they practiced hand washing, but only 4 percent wore masks (Lau et al., 2003; Leung et al., 2003; Nickell et al., 2004). All three studies also found age and gender differences associated with taking precautionary measures. Women were more likely than men to take precautions, and older people were more likely than younger people to take precautions. Two studies found that increased level of education also increased the likelihood of taking recommended precautionary measures (Lau et al., 2003; Leung et al., 2003).

Quarantine The application of quarantine during the SARS epidemic became an important issue that had varied psychological effects. In Singapore, the government instituted a 10-day quarantine for persons in contact with a probable SARS patient. The quarantined homes were monitored for compliance and violators were forced to wear electronic bracelets (Grady, 2003). Taiwan did not allow non-citizens to enter from China, Hong Kong, or other affected areas and required citizens returning from affected areas to stay at home for 10 days (Nakashima, 2003). Unfortunately, compliance was low, ranging from 21–42 percent of people who registered with local health authorities. In early May, over 23,000 people were in home quarantine.

In the Greater Toronto Area, public health officials ordered a 10-day quarantine for people exposed to SARS (DiGiovanni et al., 2004). Over 15,000 persons with SARS exposure were instructed to remain in voluntary quarantine during the two SARS outbreaks in Toronto (Hawryluck et al., 2004). In one large Toronto school, 2000 students were quarantined after discovering that a student with symptoms had been attending classes for two days (McNeil, 2003). Other incidents included an infected nurse traveling on two commuter trains and an infected doctor who attended a funeral.

One study measured factors influencing compliance through interviews, telephone polling and focus groups (DiGiovanni et al., 2004). The primary motivation for compliance by nearly all the quarantined healthcare workers and the majority of the general population poll respondents was the recognition of the need to reduce the risk of transmission to others. Fear of breaking the law played little role, except for some of the quarantined high school students who feared their parents would be punished if they broke quarantine. There were a number of obstacles to compliance, including: fear of loss of income, confusion over inconsistencies between jurisdictions in the application of quarantine measurese, logistical support (medical supplies, groceries, and household supplies) and psychological stress resulting from social distancing and stigmatization. Of the quarantined healthcare workers, almost 40 percent were 'pretty stressed' or highly stressed.

A web-based survey of 129 quarantined persons, reported that nearly 30 percent were dissatisfied with the information they received about SARS (Hawryluck et al., 2004). They received their information from the media, public health authorities, occupational health departments, healthcare providers, word-of-mouth, and hospital or other websites. Quarantined persons reported wearing their mask in the presence of household members (85 percent) and remaining inside their home for the duration (58 percent). However, they did not comply as rigorously in self-monitoring their temperature: 33 percent did not monitor as recommended, 26 percent monitored less frequently than recommended, and 7 percent did not measure their temperature. Respondents exhibited a high prevalence of psychological distress; with 29 percent reporting post-traumatic stress symptoms and 31 percent reporting depression symptoms.

According to a descriptive study of 21 quarantined individuals in Toronto, the individuals described four sequential themes: life before quarantine, finding out, being in quarantine, and life after quarantine (Cava et al., 2005). Throughout the phases were three subthemes: uncertainty, isolation, and coping. The primary uncertainties listed were fear for health (of self or loved ones), disruption of daily life, and the uncertainty of an unknown new disease, while isolation included boredom, separation from others, rejection, stigmatization, and public scrutiny. Although isolation was primarily experienced during quarantine, the participants continued to experience stigmatization and scrutiny following the end of quarantine.

Stigma Amoy Gardens was the housing area in Hong Kong which had 329 confirmed cases of SARS and 42 deaths from SARS (Lee et al., 2005). One study, using a self-report questionnaire, found that over 80 percent of respondents from Amoy Gardens reported that SARS had a deep effect on their daily life, 49 percent perceived discrimination by employers, and 48 percent reported problems with colleagues or clients (ibid.). Interestingly, many residents reported persistent symptoms of low mood (73 percent), irritability (57 percent), and insomnia (34 percent). Coping strategies varied – some (38 percent) concealed their residential status, while others (41 percent) moved out, or thought of moving out (36 percent).

Healthcare Workers (HCWs)

In a bioterrorist event or a natural outbreak of disease, the hospital becomes the battleground. Contagious diseases, whether from bioterrorism or nature, pose an additional risk to healthcare providers, in that HCWs can be inadvertently exposed and suffer illness and even death. In Canada, 375 people contracted SARS and 44 died, with most of the infections and deaths occurring in Ontario. Of these 375 SARS cases, 72 percent were infected in the healthcare setting: 45 percent HCWs, 15 percent patients, and 11 percent visitors (SARS Commission, 2006). In Taiwan, 160 doctors and nurses quit work at various hospitals out of fear of SARS and fear that infection control measures were inadequate (McNeil, 2003). Thus, the effects of bioterrorism on the healthcare providers can become very important.

Ability and willingness to work

Several studies have looked at the ability and willingness of HCWs to report for duty during catastrophic disasters. The ability to work refers to the capability of the individual to report for work and includes factors such as proximity of home to work, child- or elder-care responsibilities, and financial concerns (Qureshi et al., 2005). Willingness to work refers to the personal decision to go to work. In one study (ibid.) of healthcare facilities in New York City and the surrounding metropolitan region, HCWs were least willing to work during a SARS outbreak (48 percent), radiological event (57 percent), smallpox epidemic (61 percent) or chemical event (68 percent) compared to snow storm, explosion-related mass casualty incident (MCI), or environmental disaster (all >80 percent). Barriers to willingness were mostly frequently related to fear and concern for family and self. HCWs were least able to report for work during a severe snow storm (49 percent), SARS outbreak or radiological event (64 percent), and smallpox epidemic (69 percent). Also notable was that the highest levels of uncertainty for willingness to report were for SARS (30 percent), radiation (25 percent), and smallpox epidemic (24 percent). The authors stated that the 'most effective methods to allay fears and concerns for personal safety revolve around HCW education, provision of appropriate PPE, and assurance of effective environmental controls' (ibid., p. 386).

Psycho-social stressors

Many HCWs had a perception of personal danger which was heightened by uncertainty, the known lethality of SARS, and the intense media coverage of the outbreak. Social support became a problem in one Toronto hospital due to requirements for wearing a mask while in the hospital, lack of meetings with colleagues outside the hospital, and lack of staff meetings

(Maunder et al., 2003). Often hospital staff avoided identifying themselves on the outside as hospital workers because they feared stigmatization within their communities. One Hong Kong doctor wrote: 'No one wanted to come near me for fear of getting the disease . . . Even in the carpark, people would skirt around me to avoid close contact' (Krauss and Altman, 2003).

Staff with potential contact with SARS were quarantined for 10 days in one Toronto hospital, which led to fears about safety of self and family, stigmatization, and feelings of isolation (Maunder et al., 2003). On the SARS isolation unit, staff had spikes of anxiety in association with changes in isolation procedures, HCWs entering quarantine or treatment, and when a discharged patient was readmitted with fever. Reports of fatigue, insomnia, irritability and decreased appetite were common in SARS staff. A doctor in Hong Kong wrote:

> I was worried about going home in case I would infect my family. When I did get home, I felt physically exhausted and emotionally drained, and didn't really want to talk to anyone. I would not and could not touch my wife or children for fear of giving them the disease. I slept in a separate bedroom; I ate separately. (Krauss and Altman, 2003)

One study measured the psycho-social effects of SARS on hospital staff in a Toronto hospital using questionnaires (Nickell et al., 2004). Almost two-thirds of the respondents reported concerns for their own or their family's health. Factors identified by logistic regression analysis for significant association with increased concerns were perception of a greater risk of death from SARS, living with children, personal or family lifestyle affected by the SARS outbreak, and being treated differently by people because of working in a hospital. Emotional distress (as measured by the General Health Questionnaire–12) was found in almost 30 percent of all responders, but was 45 percent in nurses, who were most at risk from infection. Factors identified for significant association with emotional distress were being a nurse, part-time employment status, lifestyle affected by SARS outbreak, and ability to do one's job affected by precautionary measures.

Patients and Their Families

Acute effects

One hospital in Canada reported that patients, who had to identify recent contacts, often felt guilty and fearful for their friends and family (Maunder et al., 2003). SARS patients often spent hours in isolation between contacts with staff and were deprived of family visits, leading to complaints of sadness, anxiety, boredom, loneliness, and non-specific anger and frustration (Avendano et al., 2003; Maunder et al., 2003). Fear and anxiety often waxed and waned with fever (ibid.).

Referral of 10 SARS patients for psychiatric complications was reported in a Hong Kong hospital (Cheng et al., 2004). Seven patients had 'mild' psychiatric problems (anger, anxiety, suicidal ideas and depressive reaction) which was treated by telephone interview due to the strict infection control and barrier treatment in the hospital. However, three patients with more 'severe' problems – hallucinatory and manic features – required face-to-face interviews. Causes of distress included specific effects of SARS (for example, symptom severity and isolation) and indirect effects of SARS (total social isolation due to barrier nursing). Steroid therapy was implicated in the patients with organic hallucinosis or organic manic disorder.

In a study of 26 survivors of the 1995 Ebola epidemic in Congo (De Roo et al., 1998), almost all reported experiencing fear – of suffering, death, separation from relatives and/or abandonment by family. They reported suffering from the loss of friends and colleagues who died next to them in the ward and of abandonment by hospital staff at the beginning of the epidemic. What caused the most suffering was the isolation, including not being allowed visitors. During convalescence, over one-third reported rejection by society, including family, friends, and neighbors.

Many patients with botulism require ventilation (Cohen and Anderson, 1986). Botulism leads to a descending flaccid paralysis and leaves patients unable to breathe, speak, swallow, or communicate. These patients can deteriorate rapidly, and unless monitored closely, may suffer difficulty breathing, but may be too weak to communicate with hospital staff. In one study of 12 patients, patients reported high levels of fear, anxiety, and helplessness (ibid.). Family members suffered depression in addition to fear, anxiety and helplessness.

Long-term effects

Long-term effects have been reported for various diseases. For example, during the Second World War, American soldiers were exposed to unfamiliar diseases in the Pacific theater – malaria, filariasis, scrub typhus, and schistosomiasis (Frank, 1946). The duration and amount of invalidism was reportedly greater than might be expected from the organic manifestations of illness and greater than similarly affected natives. Q-fever has been associated with a post-infection fatigue syndrome, both in England and in Australia (Marmion et al., 1996; Ayres et al., 1998). Brucellosis has also been associated with chronic fatigue (Harris and Kemple, 1954; Imboden et al., 1959; Alapin, 1976). It is unknown whether these cases of chronic brucellosis are due to subclinical infection, localized infection, or are psychophysiologic.

Both post-traumatic stress disorder (PTSD) and a decreased health-related quality of life (HRQOL) were reported among survivors of an outbreak of Legionnaires' disease (Lettinga et al., 2002). Patients reported symptoms of fatigue (75 percent), neurologic symptoms (66 percent), and neuromuscular symptoms (63 percent), 17 months after diagnosis of illness. HRQOL was impaired in seven of eight dimensions, and 15 percent of patients experienced PTSD. Others have reported decreased HRQOL and PTSD in survivors of acute respiratory distress syndrome (Weinert et al., 1997; Schelling et al., 1998). Most of the Category A biological warfare agents (BWA) cause acute respiratory distress syndrome, so similar long-term effects might well be expected, and may explain the reports in the anthrax survivors.

There have been numerous reports of decreased HRQOL following SARS in Hong Kong and Singapore (Hui et al., 2005; S.T. Lau et al., 2005; Kwek et al., 2006; Tansey et al., 2007). Reported problems include decreases in pulmonary function and exercise capacity (Ong et al., 2004; Hui et al., 2005; S.T. Lau et al., 2005; Tansey et al., 2007) and tachycardia (S.T. Lau et al., 2005). Increased stress, psychological distress, anxiety and depression have also been reported (Mok et al., 2005; Sheng et al., 2005; Wu et al., 2005a; Lee et al., 2007; Tansey et al., 2007). There has also been one report of PTSD from Hong Kong, which found that of 131 patients who responded at both timepoints, 4 percent scored above cutoff on all 3 impact of event scale–revised (IES–R) subscales at 1 month and 5 percent at three months post-discharge (Wu et al., 2005b).

Newspapers reported that survivors of the US anthrax attacks have suffered symptoms of fatigue, shortness of breath, chest pains, memory problems, nightmares, and rage 6–12 months after their illnesses (Broad and Grady, 2002; Stolberg, 2002). Only one was well enough to return to work. Symptoms were similar in both cutaneous and inhalational forms of anthrax. In one published study, survivors reported significant health problems, significantly greater overall psychological distress, and significantly reduced HRQOL one year after the onset of infection when compared with US referent populations (Reissman et al., 2004). The survivors of inhalational anthrax reported worse physical health than the cutaneous anthrax survivors.

A three-year study of 22 botulism patients and 51 family members reported that perceived psycho-social dysfunction persisted longer than perceived physical dysfunction (Cohen et al., 1988). Many of the patients reported fear of death, found the experience of the intensive care unit to be anxiety producing, and felt anxiety about perceived loss of control (Hardin and Cohen, 1988). Anxiety symptoms included a sense of dread, difficulty falling asleep, and difficulty concentrating. Almost half reported panic attacks. Depression and anger were also found in most patients. More recently, 211 people in the Republic of Georgia who had experienced botulism were interviewed regarding long-term consequences (Gottlieb et al., 2007). Most had been hospitalized for one week to one month and 25 percent had required ventilatory assistance. Over two-thirds reported having worse health at the time of the interview compared with17 percent of the case control group. Current health was reported as 'fair' or 'poor' in almost half of patients, but only 25 percent of controls. Fatigue, weakness, dizziness, dry mouth, and difficulty breathing following moderate exertion (for example, walking or lifting heavy items) were reported more often in former botulism patients than in controls. Positive emotional factors (full of pep, calm and peaceful, energetic, happy) were reported less, and negative emotional factors (nervous, unable to sleep) were reported more in former botulism patients.

SUMMARY

This chapter has introduced and reviewed many of the potential psychosocial effects of bioterrorism. The recent outbreak of SARS has been particularly instructive, because many articles on various psychological effects were published. In contrast, publications on the psychological effects of previous epidemics and pandemics are rather sparse.

Although SARS was a natural outbreak, it can serve as a model for likely effects of a BT attack. Such attacks have the potential to cause large-scale illness and death. Measures that may be taken to reduce contagious disease transmission, such as quarantine of exposed persons, isolation of patients, social distancing, and other measures, are likely to increase anxiety and fear in patients, healthcare workers, and the general community. Both infected and exposed persons may well be subject to social stigma. Even nations may be affected by stigma.

Risk communication will be essential and will be a critical variable in the impact of a bioterrorist attack. Although poorly done in the American anthrax attacks in 2001, risk communication was much improved for the SARS epidemic. Lessons from the anthrax attacks and the SARS epidemic are currently being applied in pandemic influenza planning.

Collective behaviors such as OMUS may be common following the attack. Some long-term effects such as syndromes characterized by fatigue and multiple unexplained physical symptoms may persist. Other collective behaviors such as mass panic will be rare. However, many people will take precautionary measures. Effective risk communication will help educate people to take effective precautionary measures, but not all people will follow that guidance.

HCWs will be at the battlefront of any bioterrorist attack. Due to changes in medical care systems, most cities and countries have no surge capacity for a large influx of patients. The influx may well include huge numbers of stress casualties, asymptomatic but worried casualties, symptomatic but non-infected casualties, and symptomatic infected casualties. Hospital staff will be overwhelmed and understaffed. In a contagious outbreak, whether terrorist or natural, necessary protective measures will enhance the fear and stress of the HCWs, as will the fear and risk of death due to patient care.

Despite these fears and stresses, HCWs and hospitals persisted in overcoming a mysterious contagious disease – SARS – which appeared in 2003. The current fears of pandemic influenza are driving increased attention to public health, including the application of lessons learned from SARS. In some areas, funding for public health has increased as well. Bioterrorism has an enormous potential for disaster. However, the increased attention to public health and risk communication may do much to mitigate the psycho-social effects of bioterrorism.

REFERENCES

Ahern, J., Galea, S., Resnick, H., Kilpatrick, D., Bucuvalas, M., Gold, J. et al. (2002), 'Television images and psychological symptoms after the September 11 terrorist attacks', *Psychiatry*, **65**, 289–300.

Alapin, B. (1976), 'Psychosomatic and somato-psychic aspects of brucellosis', *Journal of Psychosomatic Research*, **20**, 339–50.

Avendano, M., Derkach, P. and Swan, S. (2003), 'Clinical course and management of SARS in health care workers in Toronto: a case series', *Canadian Medical Association Journal*, **168**, 1649–60.

Ayres, J.G., Flint, N., Smith, E.G., Tunnicliffe, W.S., Fletcher, T.J., Hammond, K. et al. (1998), 'Post-infection fatigue syndrome following Q fever', *Quarterly Journal of Medicine*, **91**, 105–23.

Bleich, A., Dycian, A., Koslowsky, M., Solomon, Z. and Wiener, M. (1992), 'Psychiatric implications of missile attacks on a civilian population. Israeli lessons from the Persian Gulf War', *Journal of the American Medical Association*, **268**, 613–15.

Boss, L.P. (1997), 'Epidemic hysteria: a review of the published literature', *Epidemiologic Reviews*, **19**, 233–43.

Brecher, R.W. and Flynn, T. (2002), 'Principles of risk communication: building trust and credibility with the public', in W.M. Haschek, C.G. Rousseaux and M.A. Wallig (eds), *Handbook of Toxicologic Pathology*, 2nd edn, Vol. 1, New York: Academic Press, pp. 447–58.

Broad, W.J. and Grady, D. (2002), 'Science slow to ponder ills that linger in anthrax victims', *New York Times*, September 16.

Cava, M.A., Fay, K.E., Beanlands, H.J., McCay, E.A. and Wignall, R. (2005), 'The experience of quarantine for individuals affected by SARS in Toronto', *Public Health Nursing*, **22**, 398–406.

Centers for Disease Control and Prevention (CDC) (2001), 'Update: Investigation of bioterrorism-related inhalational anthrax – Connecticut, 2001', *Morbidity and Mortality Weekly Report*, **50**, 1049–51.

Chen, W.K., Cheng, Y.C., Chung, Y.T. and Lin, C.C. (2005), 'The impact of the SARS outbreak on an urban emergency department in Taiwan', *Medical Care*, **43**, 168–72.

Cheng, S.K., Tsang, J.S., Ku, K.H., Wong, C.W. and Ng, Y.K. (2004), 'Psychiatric complications in patients with severe acute respiratory syndrome (SARS) during the acute treatment phase: a series of 10 cases', *British Journal of Psychiatry*, **184**, 359–60.

Chess, C. and Celia, J. (2002), 'Risk communication is a key to dealing effectively with bioterrorism', *Risk Analysis*, **22**, 1039–40.

Cohen, F.L., Hardin, S.B., Nehring, W., Keough, M.A., Laurenti, S., McNabb, J. et al. (1988), 'Physical and psychosocial health status 3 years after catastrophic illness – botulism', *Issues in Mental Health Nursing*, **9**, 387–98.

Cohen, R.E. and Anderson, D.L. (1986), 'Botulism: emotional impact on patient and family', *Journal of Psychosomatic Research*, **30**, 321–6.

Colligan, M.J., Pennebaker, J.W. and Murphy, L.R. (1983), *Mass Psychogenic Illness*, Hillsdale, NJ: Lawrence Erlbaum.

Covello, V.T., Peters, R.G., Wojtecki, J.G. and Hyde, R.C. (2001), 'Risk communication, the West Nile virus epidemic, and bioterrorism: responding to the communication challenges posed by the intentional or unintentional release of a pathogen in an urban setting', *Journal of Urban Health*, **78**, 382–91.

Davis, C.J. (1999), 'Nuclear blindness: an overview of the biological weapons programs of the former Soviet Union and Iraq', *Emerging Infectious Diseases*, **5**, 509–12.

De Roo, A., Ado, B., Rose, B., Guimard, Y., Fonck, K. and Colebunders, R. (1998), 'Survey among survivors of the 1995 Ebola epidemic in Kikwit, Democratic Republic of Congo: their feelings and experiences', *Tropical Medicine and International Health*, **3**, 883–5.

DiGiovanni, C., Jr, Conley, J., Chiu, D. and Zaborski, J. (2004), 'Factors influencing compliance with quarantine in Toronto during the 2003 SARS outbreak', *Biosecurity and Bioterrorism*, **2**, 265–72.

Eckholm, E. (2003), 'Illness psychological impact in China exceeds its actual numbers', *New York Times*, April 24.

Frank, J.D. (1946), 'Emotional reactions of American soldiers to an unfamiliar disease', *American Journal of Psychiatry*, **102**, 631–40.

Glik, D., Harrison, K., Davoudi, M. and Riopelle, D. (2004), 'Public perceptions and risk communications for botulism', *Biosecurity and Bioterrorism*, **2**, 216–23.

Goode, E. (2001), 'Anthrax offers lessons in how to handle bad news', *New York Times*, October 23.

Gottlieb, S.L., Kretsinger, K., Tarkhashvili, N., Chakvetadze, N., Chokheli, M., Chubinidze, M. et al. (2007), 'Long-term outcomes of 217 botulism cases in the Republic of Georgia', *Clinical Infectious Diseases*, **45**, 174–80.

Grady, D. (2003), 'Fear reigns as dangerous mystery illness spreads', *New York Times*, April 7.

Hardin, S.B. and Cohen, F.L. (1988), 'Psychosocial effects of a catastrophic botulism outbreak', *Archives of Psychiatric Nursing*, **2**, 173–84.

Harris, H.J. and Kemple, C. (1954), 'Chronic brucellosis and psychoneurosis', *Psychosomatic Medicine*, **16**, 414–25.

Hawryluck, L., Gold, W.L., Robinson, S., Pogorski, S., Galea, S. and Styra, R. (2004), 'SARS control and psychological effects of quarantine, Toronto, Canada', *Emerging Infectious Diseases*, **10**, 1206–12.

Homeland Security Council (2004), 'Planning scenarios. Executive summaries', www.globalsecurity.org.

Hui, D.S., Wong, K.T., Ko, F.W., Tam, L.S., Chan, D.P., Woo, J. et al. (2005), 'The 1-year impact of severe acute respiratory syndrome on pulmonary function, exercise capacity, and quality of life in a cohort of survivors', *Chest*, **128**, 2247–61.

Imboden, J.B., Canter, A., Cluff, L.E. and Trever, R.W. (1959), 'Brucellosis. III. Psychologic aspects of delayed convalescence', *Archives of Internal Medicine*, **103**, 406–19.

Krauss, C. and Altman, L.K. (2003), 'Quarantine hits Toronto school in SARS fight', *New York Times*, May 29.

Kumar, S. (1994), 'Plague in India', *Lancet*, **344**, 941–2.

Kwek, S.K., Chew, W.M., Ong, K.C., Ng, A.W., Lee, L.S., Kaw, G. et al. (2006), 'Quality of life and psychological status in survivors of severe acute respiratory syndrome at 3 months postdischarge', *Journal of Psychosomatic Research*, **60**, 513–19.

Lau, H.M., Lee, E.W., Wong, C.N., Ng, G.Y., Jones, A.Y. and Hui, D.S. (2005), 'The impact of severe acute respiratory syndrome on the physical profile and quality of life', *Archives of Physical Medicine and Rehabilitation*, **86**, 1134–40.

Lau, J.T., Yang, X., Tsui, H. and Kim, J.H. (2003), 'Monitoring community responses to the SARS epidemic in Hong Kong: from day 10 to day 62', *Journal of Epidemiology and Community Health*, **57**, 864–70.

Lau, S.T., Yu, W.C., Mok, N.S., Tsui, P.T., Tong, W.L. and Cheng, S.W. (2005), 'Tachycardia amongst subjects recovering from severe acute respiratory syndrome (SARS)', *International Journal of Cardiology*, **100**, 167–9.

Lee, A.M., Wong, J.G., McAlonan, G.M., Cheung, V., Cheung, C., Sham, P.C. et al. (2007), 'Stress and psychological distress among SARS survivors 1 year after the outbreak', *Canadian Journal of Psychiatry*, **52**, 233–40.

Lee, S., Chan, L.Y., Chau, A.M., Kwok, K.P. and Kleinman, A. (2005), 'The experience of SARS-related stigma at Amoy Gardens', *Social Science and Medicine*, **61**, 2038–46.

Lettinga, K.D., Verbon, A., Nieuwkerk, P.T., Jonkers, R.E., Gersons, B.P., Prins, J.M. et al. (2002), 'Health-related quality of life and posttraumatic stress disorder among survivors of an outbreak of Legionnaires disease', *Clinical Infectious Diseases*, **35**, 11–17.

Leung, G.M., Lam, T.H., Ho, L.M., Ho, S.Y., Chan, B.H., Wong, I.O. et al. (2003), 'The impact of community psychological responses on outbreak control for severe acute respiratory syndrome in Hong Kong', *Journal of Epidemiological Community Health*, **57**, 857–63.

Lipton, E. (2001), 'A nation challenged: New Jersey; more questions than answers in contaminated letters from Trenton', *New York Times*, October 31.

Madan, T.N. (1995), 'The plague in India, 1994', *Social Science and Medicine*, **40**, 1167–8.

Marmion, B.P., Shannon, M., Maddocks, I., Storm, P. and Penttila, I. (1996), 'Protracted debility and fatigue after acute Q fever', *Lancet*, **347**, 977–8.

Maunder, R.G., Lancee, W.J., Rourke, S., Hunter, J.J., Goldbloom, D., Balderson, K. et al. (2003), 'The immediate psychological and occupational impact of the

2003 SARS outbreak in a teaching hospital', *Canadian Medical Association Journal*, **168**, 1245–51.
McNeil, D.G. (2003), 'SARS fears shake Taiwan medical staffs', *New York Times*, May 21.
Mok, E., Chung, B.P., Chung, J.W. and Wong, T.K. (2005), 'An exploratory study of nurses suffering from severe acute respiratory syndrome (SARS)', *International Journal of Nursing Practice*, **11**, 150–60.
Nakashima, E. (2003), 'As disease spreads, Singapore life suffers', *Washington Post*, April 14.
Nickell, L.A., Crighton, E.J., Tracy, C.S., Al Enazy, H., Bolaji, Y., Hanjrah, S. et al. (2004), 'Psychosocial effects of SARS on hospital staff: survey of a large tertiary care institution', *Canadian Medical Association Journal*, **170**, 793–8.
Ong, K.C., Ng, A.W., Lee, L.S., Kaw, G., Kwek, S.K., Leow, M.K. et al. (2004), 'Pulmonary function and exercise capacity in survivors of severe acute respiratory syndrome', *European Respiratory Journal*, **24**, 436–42.
Pastel, R.H. (2001), 'Collective behaviors: mass panic and outbreaks of multiple unexplained symptoms', *Military Medicine*, **166**, 44–6.
Petterson, J.S. (1988), 'Perception vs. reality of radiological impact: the Goiania model', *Nuclear News*, **31**, 84–90.
Pfefferbaum, B., Nixon, S.J., Tivis, R.D., Doughty, D.E., Pynoos, R.S., Gurwitch, R.H. et al. (2001), 'Television exposure in children after a terrorist incident', *Psychiatry*, **64**, 202–11.
Pomfret, J. (2003), 'SARS inciting "mass panic" in Beijing', *Washington Post*, May 1.
Pugh, M.B. (2006), *Stedman's Medical Dictionary*, 28th edn, Baltimore, MD: Lippincott Williams & Wilkins.
Quarantelli, E.L. (1954), 'The nature and conditions of panic', *American Journal of Sociology*, **60**, 267–75.
Qureshi, K., Gershon, R.R., Sherman, M.F., Straub, T., Gebbie, E., McCollum, M. et al. (2005), 'Health care workers' ability and willingness to report to duty during catastrophic disasters', *Journal of Urban Health*, **82**, 378–88.
Ramalingaswami, V. (2001), 'Psychosocial effects of the 1994 plague outbreak in Surat, India', *Military Medicine*, **166**, 29–30.
Reissman, D.B., Whitney, E.A., Taylor, T.H., Jr, Hayslett, J.A., Dull, P.M., Arias, I. et al. (2004), 'One-year health assessment of adult survivors of Bacillus anthracis infection', *Journal of the American Medical Association*, **291**, 1994–8.
Rotz, L.D., Khan, A.S., Lillibridge, S.R., Ostroff, S.M. and Hughes, J.M. (2002), 'Public health assessment of potential biological terrorism agents', *Emerging Infectious Diseases*, **8**, 225–30.
Sandman, P.M. (2003), 'Bioterrorism risk communication policy', *Journal of Health Communication*, **8**, 146–7.
Salmon, T.W. and Fenton, N. (1929), 'Neuropsychiatry in the American Expeditionary Force', in T.W. Salmon and N. Fenton (eds), *Neuropsychiatry*, Volume 10, Washington, DC: Government Printing Office.
SARS Commission (2006), *Spring of Fear. Executive Summary*, vol. 1, Toronto, ON.
Schelling, G., Stoll, C., Haller, M., Briegel, J., Manert, W., Hummel, T. et al. (1998), 'Health-related quality of life and posttraumatic stress disorder in survivors of the acute respiratory distress syndrome', *Critical Care Medicine*, **26**, 651–9.
Schlenger, W.E., Caddell, J.M., Ebert, L., Jordan, B.K., Rourke, K.M., Wilson, D. et al. (2002), 'Psychological reactions to terrorist attacks: findings from the

National Study of Americans' Reactions to September 11', *Journal of the American Medical Association*, **288**, 581–8.
Schuster, M.A., Stein, B.D., Jaycox, L., Collins, R.L., Marshall, G.N., Elliott, M.N. et al. (2001), 'A national survey of stress reactions after the September 11, 2001, terrorist attacks', *New England Journal of Medicine*, **345**, 1507–12.
Sheng, B., Cheng, S.K., Lau, K.K., Li, H.L. and Chan, E.L. (2005), 'The effects of disease severity, use of corticosteroids and social factors on neuropsychiatric complaints in severe acute respiratory syndrome (SARS) patients at acute and convalescent phases', *European Psychiatry*, **20**, 236–42.
Singer, J.E. (1982), 'Yes Virginia, there really is a mass psychogenic illness', in M.J. Colligan, J.W. Pennebaker and L.R. Murphy (eds), *Mass Psychogenic Illness*, Hillsdale, NJ: Lawrence Erlbaum, pp. 127–35.
Steinhauer, J. (2001), 'Hysteria can be hazardous', *New York Times*, October 21.
Stolberg, S.G. (2002), 'For anthrax survivors, a halting, painful recovery', *New York Times*, May 7.
Struewing, J.P. and Gray, G.C. (1990), 'An epidemic of respiratory complaints exacerbated by mass psychogenic illness in a military recruit population', *American Journal of Epidemiology*, **132**, 1120–29.
Takahashi, H., Keim, P., Kaufmann, A.F., Keys, C., Smith, K.L., Taniguchi, K. et al. (2004), 'Bacillus anthracis incident, Kameido, Tokyo, 1993', *Emerging Infectious Diseases*, **10**, 117–20.
Tansey, C.M., Louie, M., Loeb, M., Gold, W.L., Muller, M.P., de Jager, J. et al. (2007), 'One-year outcomes and health care utilization in survivors of severe acute respiratory syndrome', *Archives of Internal Medicine*, **167**, 1312–20.
Torok, T.J., Tauxe, R.V., Wise, R.P., Livengood, J.R., Sokolow, R., Mauvais, S. et al. (1997), 'A large community outbreak of salmonellosis caused by intentional contamination of restaurant salad bars', *Journal of the American Medical Association*, **278**, 389–95.
Vasterman, P., Yzermans, C.J. and Dirkzwager, A.J. (2005), 'The role of the media and media hypes in the aftermath of disasters', *Epidemiologic Reviews*, **27**, 107–14.
Weinert, C.R., Gross, C.R., Kangas, J.R., Bury, C.L. and Marinelli, W.A. (1997), 'Health-related quality of life after acute lung injury', *American Journal of Respiratory and Critical Care Medicine*, **156**, 1120–28.
Wessely, S. (1987), 'Mass hysteria: two syndromes?', *Psychological Medicine*, **17**, 109–20.
Woods, J.B., Darling, R.G., Dembek, Z.F., Carr, B.K., Cieslak, T.J., Lawler, J.V. et al. (2005), *USAMRIID's Medical Management of Biological Casualties Handbook*, 6th edn, Fort Detrick, MD: United States Army Medical Research Institute of Infectious Diseases.
World Health Organization Group of Consultants (1970), *Health Aspects of Chemical and Biological Weapons*, Geneva: World Health Organization.
Wray, R. and Jupka, K. (2004), 'What does the public want to know in the event of a terrorist attack using plague?', *Biosecurity and Bioterrorism*, **2**, 208–15.
Wu, K.K., Chan, S.K. and Ma, T.M. (2005a), 'Posttraumatic stress, anxiety, and depression in survivors of severe acute respiratory syndrome (SARS)', *Journal of Trauma Stress*, **18**, 39–42.
Wu, K.K., Chan, S.K. and Ma, T.M. (2005b), 'Posttraumatic stress after SARS', *Emerging Infectious Diseases*, **11**, 1297–300.

6. Fear of terror and health: a study of apparently healthy employees

Arie Shirom, Sharon Toker, Shlomo Berliner, Itzhak Shapira and Samuel Melamed[1]

INTRODUCTION

There is substantial evidence, reviewed below, that mass violence in general, and terrorism in particular, is associated with physical and mental health impairments (Baum and Dougall, 2002; Duffy et al., 2007; Marshall et al., 2007). However, potential effects of fear of terror and objective exposure to terror on biomarkers of physical health and on self-rated health (SRH), as an indicator of general health, have seldom been investigated. As noted in a recent review, the long-term effects of terrorism on physical health, especially in situations when the threat of terrorism continues over time, have hardly been investigated (Yehuda and Hyman, 2005). In general, media coverage of acts of terror focuses on the loss of lives, the injured, and property damage. However, perhaps the major objective of terrorism is ultimately psychological, to create a climate of fear and vulnerability affecting entire populations. For this reason, examining the relationships among objective exposure, fear of terror and mental and physical health provides an important and relevant research focus. In this chapter, we report on the results of two studies. The first study investigated the effects of fear of terror and objective exposure to terror on mental health as indicated by SRH. The second study investigated the effects of fear of terror on C-reactive protein levels, which indicate the presence of low-grade inflammation commonly regarded as carrying a potential risk of cardiovascular disease (Verma et al., 2006).

FEAR OF TERROR AND SUBJECTIVE HEALTH (SRH)

In this study, we assessed the effects of the affective response to terrorism and objective exposure to terror on SRH. We focused on SRH because recent studies have suggested that it could be conceived as a sensitive indicator of physiological health (Goldman et al., 2004; Jylha et al., 2006). Furthermore, several meta-analytic studies supported the assertion that SRH predicted mortality and survival (Idler and Benyamini, 1997; Benyamini and Idler, 1999; Kawada, 2003; DeSalvo et al., 2006a), thus documenting the ability of SRH to reflect a wide spectrum of health conditions.

As posited by Lazarus's cognitive–motivational–relational approach to affective states (Lazarus, 2001), one's affective response to acts of terror reflects one's subjective exposure to and appraisal of them. Studies conducted in the USA, in the wake of the events of September 11, 2001, found that one's proximity to the events' location predicted subsequent incidence of mental health impairments (Schlenger et al., 2002; Silver et al., 2002). However, a single occurrence of several interconnected acts of terror probably impact on health differently relative to ongoing exposure to terrorist attacks (Marshall et al., 2007). For example, follow-up studies found that initially severe symptoms in response to the September 11, 2001 attacks in New York City subsided substantially within several months (Galea et al., 2003; Fraley et al., 2006).

Earlier studies on the health-related effects of objective exposure to continuous terror yielded inconsistent results. Two studies found no significant associations between objective threat or objective exposure and health symptoms (Bleich et al., 2003; Somer et al., 2005); one study found objective exposure to positively predict post-traumatic symptoms (Hobfoll et al., 2006); and yet another study reported that the most severe health symptoms were found among those least directly affected by exposure to terror attacks (Kaplan et al., 2005). This inconsistency in past research may be related to inherent conceptual and methodological limitations. For example, several past studies conceptualized objective exposure based on information provided by the same respondents who also self-reported on their state of health (for example, Bleich et al., 2003; Somer et al., 2005; Cohen and Eid, 2007), thus making the findings prone to common method and other response biases (Podsakoff et al., 2003). In addition, some past studies did not control for crucial third variables that may have influenced SRH, such as depressive symptoms known to influence both subjective stressor appraisals and health symptoms reporting (for example, Somer et al., 2005; Cohen and Eid, 2007). Finally, some studies assessed the health-related consequences of exposure long after it had been experienced (for example, Kaplan et al., 2005).

In the current study, we have addressed these possible reasons for the inconsistency of past research on the health consequences of terrorism. First, we assessed objective exposure independently of the study participants, based on acts of terror that took place in their residential area. Second, the objective exposure clearly anteceded the participants' reports on their affective responses to terror. Third, we controlled for the negative affective states of depressive symptomatology and anxiety. In the literature on terrorism, there is a consensus that a major objective of the perpetrators of acts of terror is to create fear among the populace (Schmid and Jongman, 1988; Baum and Dougall, 2002); therefore, we focused on fear of terror as representing the most relevant affective reaction to terrorism.

In Israel, the civilian population has been subjected to a continuous threat in the form of acts of terror since the onset of the second Intifada (Palestinian uprising) in September 2000. Israeli residents are uncertain as to where and when the next terror attack might strike; this uncertainty, coupled with the uncontrollability of the externally caused threat, could lead to the affective reactions of fear of terror, helplessness, and anxiety (Bleich et al., 2003; Inness and Barling, 2005; Zeidner, 2005). Because of their unpredictability and uncontrollability, acts of terror have had ripple effects on the residents in the area exposed to these acts – the effects magnified and reinforced by the extensive and intensive media coverage of each such act (Keinan et al., 2003). Following the literature documenting the negative impact of these negative effects on people's health (Suls and Bunde, 2005), our first hypothesis (Hypothesis 1) was that fear of terror would negatively predict SRH even after controlling for levels of depressive symptomatology and anxiety.

As noted by Norris et al. (2002) in their integrative review of the disaster literature, it was generally found that as the severity of exposure increased, participants' health-related symptoms increased. This relationship was also found in studies that operationalized exposure as the number of relevant stressors in the respondents' ecological milieu (ibid., p. 225). In a review of 52 studies on the psychological consequences of disasters, it was found that the number of human casualties in a disaster is related to subsequent psychopathology among members of the victimized population (Rubonis and Bickman, 1991). Therefore, we conceptualized objective exposure as referring to the number of acts of terror and number of casualties caused by them, and expected this exposure to negatively predict SRH. In a recent study of a representative sample of Israelis, exposure to terrorism was found to predict stress-related symptoms of impaired mental health even after controlling for subjectively assessed loss of psycho-social resources (Hobfoll et al., 2006). Furthermore, the above review (Norris et al., 2002) noted that ecological measures of exposure, like neighborhood measures, were found in past research to explain variance in outcomes over and above that of individual-level reactions to the stressors. Therefore, we expected (Hypothesis 2) that objective exposure to terrorism would be a negative predictor of SRH even after controlling for the effects of fear of terror on SRH.

In formulating our third hypothesis, expecting objective exposure to interact with fear of terror in predicting SRH, we followed the Person–Environment Fit occupational stress and strain model (French et al., 1982). This model has been extensively used in occupational stress research (Kahn and Byosiere, 1992; Hart and Cooper, 2001). We followed it by making the conceptual distinction between the objective environment, represented in our model by objective exposure, and the subjective environment, represented in our model by fear of terror. More specifically, our hypothesis was based on research on traumatic stressors which suggested that objective exposure might interact with subjective reactions to the stressors in impinging upon individuals' health (Schnurr and Green, 2004, pp. 247–75). Following these past theories and evidence, we expected (Hypothesis 3) that the higher the objective exposure, and the higher the fear of terror, the more pronounced the effects on SRH would be.

We further hypothesized that objective and subjective exposure may affect SRH in a gender-dependent way, because SRH may reflect physiological parameters such as high level of inflammation. Several other studies also found important gender differences in psychological reactions to terror (Solomon et al., 2005; Somer et al., 2005; Zeidner, 2005, 2006; Cohen and Eid, 2007). Based on the above evidence, our hypothesis was that the effects of subjective and objective exposure would be more pronounced among women relative to men (Hypothesis 4).

FEAR OF TERROR AND BIOMARKERS OF INFLAMMATION

Studies of psychological reactions to terrorist attacks in the US and in Israel reported findings of acute post-traumatic stress disorder (PTSD) and distress symptoms (Silver et al., 2002; Galea et al., 2003; Solomon and Lavi, 2005). As we indicated above, the degree of the response could not be predicted by objective measures of exposure to the trauma, or trauma induced (for example, Marshall et al., 2007). It follows that there is a need for studies that directly assess the relationship between fear of terror, as a psychological reaction to terrorist attacks that reflects subjective appraisal of the threat to one's physical well-being and physiological outcomes. There is evidence that repeated episodes of acute psychological stress, chronic psychological stress, and psychological reactions to either condition can lead to a chronic inflammatory process culminating in atherosclerosis and cardiovascular disease (Black, 2002, 2003; Kop, 2003). This body of research has led to the hypothesis that chronic fear of terror may be associated with low-grade inflammation (Hypothesis 5).

We tested this hypothesis in otherwise healthy, employed men and women in Israel. The presence of inflammation was measured by blood levels of C-reactive protein (CRP), found to be a powerful biomarker of systemic inflammation, a consistent predictor of coronary heart disease risk (Pradhan et al., 2001; Libby and Ridker, 2004; Ridker et al., 2004, 2005).

CONTROL VARIABLES

The negative affective states of depressive symptomatology and anxiety may influence fear of terror, by making individuals more vulnerable to the effects of terrorism (Silver et al., 2002), and were shown to be influential predictors of one's state of health (Suls and Bunde, 2005); therefore, we controlled for them in both Study 1 and Study 2. A well-established finding in the scientific literature is the inverse association between socio-economic status (SES) and health: the higher the SES, the lower the prevalence and incidence of health problems, illness, disease and death (Gallo and Matthews, 2003). This major finding was replicated and corroborated in a large-scale study which used SES to predict SRH (Nicholson et al., 2005) and in studies conducted in Israel (for example, Baron-Epel et al., 2005). For example, there is strong evidence supporting SRH–educational attainment linkage (Hill and Needham, 2006). Following this evidence, we also used proxy measures of SES, namely education and financial strain, and also age, as control variables.

In testing our fifth hypothesis, we used an additional set of control variables. As above, to rule out the possibility that fear of terror may be related to CRP levels through their respective associations with general psychological distress, thus fully mediating the effects of fear of terror on CRP, we controlled for general anxiety and depressive symptoms. Heightened general anxiety and depressive symptoms are part of the clinically significant psychological distress symptoms that may be evoked by terror attacks (Silver et al., 2002). Some studies have uncovered a positive association between depressive symptoms and CRP levels (Douglas et al., 2004; Ford and Erlinger, 2004), while other studies have yielded mixed results (Ludwig et al., 2003) or negative results (Penninx et al., 2003). The association between anxiety and CRP levels was less investigated. In testing fear of terror effects on CRP, we also controlled for several other potent confounding variables reported to correlate with CRP levels (Bermudez et al., 2002); these additional control variables are fully described below, in the method section. We analyzed the data separately for men and women, given the important gender differences in elevated CRP (Rifai and Ridker, 2003) as well as the evidence that there are gender-specific paths of influence of psychological stress on the inflammatory process (Steptoe et al., 2002).

METHOD

Sample

Study participants ($n = 5307$, 38 percent women) attended the Center for Periodic Health Examinations of Tel Aviv's Sourasky Medical Center, for a routine health examination some time between 2002 and 2005. They represent 91 percent of the examinees during this period. In assessing bias due to refusal to participate in the study, we systematically compared the participants with the eligible non-participants on gender, age and education, and no non-response bias was found. We excluded from the study 245 men and 124 women who did not work regularly, and 629 men and 498 women who reported that a doctor had informed them that they had at least one of the following diseases or who habitually take medications for treating one or more of them: inflammatory, cardiovascular, rheumatic or peripheral blood disease, previous stroke or cancer, and mental crisis. The decision to exclude participants who self-reported being afflicted with one or more of the above-mentioned diseases or taking medication for treating them was based on the relationship between these diseases and SRH (for example, Franssen et al., 2003). In addition, 489 men and 241 women were excluded because of missing data relating to one of the study parameters. Therefore, the final sample used in testing the first four hypotheses consisted of 2930 males and 1947 females.

For testing our fifth hypothesis, associated with Study 2, we used a subset of the above sample. First, we used only respondents for whom we had results of their blood tests. Second, we excluded 622 participants because of known inflammatory disease (arthritis, inflammatory bowel disease), pregnancy, steroidal or nonsteroidal treatment, acute infection, previous myocardial infarction or stroke, or because of missing data for one of the study parameters. Thus, the final sample consisted of 1153 subjects (see description below).

Measures

The questionnaire covered background, occupational, psychological, and physical morbidity factors. The multi-item indexes constructed for this study, representing affective states, had all been included in previous research in Israel, where they had exhibited high reliabilities and construct validities. They were constructed by combining single items that measured the same variable, as verified by confirmatory factor analyses.[2] The respondent's score on each of the indexes was obtained by computing the mean of his or her responses to the items in the index. Means, standard deviations, and intercorrelations for the study's variables are presented in Table 8.1. The reliability coefficients reported below (Cronbach's internal consistency reliability coefficients or α) refer to the sample actually used in the current study.

Self-rated health (SRH) was measured by a single item, asking respondents to assess their general health with response options of excellent (scale value = 5), very good, good, fair or poor (scale value = 1). Several meta-analytic studies found this measure to have predictive validity relative to the criteria of all-cause morbidity and mortality (for example, DeSalvo et al., 2006a). This measure, extensively used in past studies (for example, Benyamini and Idler, 1999), was found to be a reasonably valid substitute for multi-item measures of SRH (for example, McHorney, 1999; Barofsky et al., 2004), demonstrated good reproducibility and reliability when compared to analogous single-item measures (DeSalvo et al., 2006b), and was found in several qualitative studies (see McCullough and Laurenceau, 2004; Simon et al., 2005) to encompass people's evaluations of many medical, psychological, and social conditions in their lives.

Objective indicators of terror were assessed using two indicators, the total number of casualties (dead and wounded) and the number of acts of terror which had taken place in a respondent's urban area during the 30 days immediately preceding his or her periodic health examination. The data were taken from the Israeli Ministry of Foreign Affairs (Israel Ministry of Foreign Affairs, 2006) official website (www.mfa.gov.il).

Fear of terror was assessed by three items adapted from the studies of Silver et al. (2002) and Gidron et al. (1999). These items were the following: 'Nowadays, I worry about my personal safety; Nowadays, I feel heightened tension when I am in crowded places; I am afraid of a terror strike harming me or my family'. For data on the predictive validity of this scale, see Melamed et al. (2004). For each item, participants checked off one of the following responses: 'not at all', 'a little bit', 'moderately', 'quite a bit', and 'extremely'. They were also asked to indicate the duration (in months) of experiencing such feelings (α = 0.88).

High sensitivity CRP concentrations in serum were determined with the BN II Nephelometer (Dade-Behring, Marburg, Germany) analyzer, as described by Rifai et al. (1999). CRP levels of <1 mg/l were considered indicative of a low risk of a future cardiovascular event, levels of 1–3 mg/dl

Table 6.1 Means, standard deviations, and intercorrelations of all study variables

Measure	1	2	3	4	5	6	7	8	9	M	SD
1. Self-rated health	–	0.01	0.00	*–0.14*	*–0.16*	0.02	*–0.13*	*–0.30*	*–0.22*	4.07*	0.61
2. Casualties, N.[1]	0.00	–	*0.79*	0.02	*–0.06*	–0.02	–0.02	0.01	*–0.03*	14.8*	31.8
3. Incidents, N.[1]	0.00	*0.78*	–	0.03	–0.02	–0.02	0.00	–0.01	0.02	0.30	0.46
4. Fear of terror	*–0.18*	0.02	*0.05*	–	–0.01	*–0.05*	*0.16*	*0.21*	*0.34*	1.88*	0.91
5. Age	*–0.19*	–0.03	–0.01	*0.05*	–	*0.13*	*–0.06*	*0.04*	–0.01	44.8*	10.6
6. Education, in years	*0.07*	0.00	–0.01	*–0.11*	0.03	–	*–0.11*	*–0.06*	0.02	15.6	2.9
7. Financial strain	*–0.20*	0.02	*0.06*	*0.21*	0.00	*–0.17*	–	*0.15*	*0.14*	1.59*	0.64
8. Depression	*–0.30*	0.01	0.01	*0.28*	*0.05*	*–0.06*	*0.25*	–	*0.41*	1.35*	0.37
9. Anxiety	*–0.24*	–0.01	0.02	*0.36*	0.03	*0.05*	*0.16*	*0.51*	–	1.97*	0.85
M	4.21*	17.1*	0.32	1.50*	45.6*	15.2	1.78*	1.19*	1.91*		
SD	0.62	35.4	0.47	0.66	10.1	3.0	0.78	0.27	0.80		

Notes: Entries above the diagonal represent the female respondents' ($n = 1947$) parameters, and those below the diagonal the male respondents' ($n = 2930$) parameters. Italic coefficients and means are significant at the $p < 0.05$ level.

1. Casualties, N. and incidents, N. refer to the number of casualities and incidents, respectively, during the 30 days preceding the respondent's date of completing the questionnaire.

* Represent a significant difference between the means of male and female respondents based on a two-tailed T-test.

of a moderate risk, and levels >3 mg/l of a high risk (Ridker, 2003). For the present study, an elevated CRP level was defined as >3.0 mg/l.

Depression was measured using the Personal Health Questionnaire (PHQ), the depression section of a patient-oriented self-administered instrument derived from the PRIME-MD (Spitzer et al., 1994, 1999). It lists nine potential symptoms of depression (for example, feeling down, depressed or hopeless, little interest or pleasure in doing things), and asks patients to rate the frequency of experiencing each symptom during the past two weeks on a scale ranging from 'never' to 'almost always' ($\alpha = 0.77$). This measure has been used in several past studies (Dietrich et al., 2003; Lubetkin et al., 2003; Nease and Maloin, 2003), and in screening populations (Douglas et al., 2004).

The measure of *anxiety* includes four items (for example, feeling nervous, jittery, fidgety) adapted from questionnaires used in several large-scale studies conducted by the Institute of Social Research, University of Michigan (French et al., 1982). All items are scored on a five-point frequency scale, ranging from 1 = almost never, to 5 = almost always ($\alpha = 0.86$). As we wanted to orient the items toward the measurement of trait anxiety so that we could use it as a control variable gauging a general disposition to feel anxious, they were worded to reflect how the respondent generally felt. *Generalized anxiety* was measured by four items (feeling nervous, jittery, fidgety, and tense) adopted from French et al.

We assessed *financial strain*, representing a proxy variable of economic hardships, by a six-item measure developed by Pearlin and colleagues (Pearlin and Lieberman, 1979; Pearlin et al., 1981; Pearlin, 1999) and used in several past studies (Peirce et al., 1994, 1996). The six items asked respondents to report how often during the past month they had engaged in activities reflecting financial difficulties, like borrowing money to pay bills, delaying elective medical treatment or dental care because of their cost, and cutting or canceling leisure activities. Anchors ranged from 'to a very large extent' (scale value of 5) to 'not at all (scale value of 1) ($\alpha = 0.90$).

Education level was assessed based on the number of years of formal education.

We used various *control variables* to study the effects of fear of terror on CRP. In testing our fifth hypothesis, expecting fear of terror to predict elevated levels of CRP, we controlled for several potential confounding factors, in addition to generalized anxiety and depressive symptoms. We controlled for demographic and biomedical variables that have been found to be associated with CRP concentration, namely, age, sex, race, obesity, smoking, alcohol consumption, physical exercise, lipid levels, glucose levels, blood pressure and hormone replacement therapy (Bermudez et al., 2002). Ethnic origin was documented as self-reported place of birth: Asia, North

Africa, Eastern Europe, Europe, North or South America, Israel. Body mass index (kg/m^2) and years of education were used as continuous variables. Smoking (current, former, never) was documented by self-report, as was alcohol consumption (yes, no), physical exercise (yes, no), and use of hormone replacement therapy (yes, no). Total serum cholesterol and triglycerides were measured with the Roche/Hitachi 747 Analyzer (Roche Diagnostics, Mannheim, Germany) and the Raichem Kit (Reagents Applications, San Diego, California, USA). Low-density lipoprotein was assayed on a Roche/Hitachi 747 Analyzer with the Randox Kit (Randox Laboratories, Crumlin, UK) and was used to compute high-density lipoprotein (HDL) levels. Hypertension was defined as a systolic blood pressure of $\geq$140 mmHg, diastolic blood pressure of $\geq$90 mmHg, or current use of antihypertensive medication. Fasting glucose was determined with the glucose oxidase method using an autoanalyzer (Beckman Instruments, Fullerton, California, USA).

Procedure

Active informed consent procedures were used in this study. The local medical center ethics committee approved the study. Participants were recruited individually by an interviewer, while awaiting their turn for the clinical examination. The interviewer gave each participant an explanation of the survey and asked for her or his voluntary participation. In return, participants were promised detailed feedback of the results. Confidentiality was assured, and each participant signed a written informed consent form. For each respondent, the results of these examinations and his or her responses to the study questionnaire were recorded and computerized. All patients underwent blood sampling (after an overnight fast), anthropometric measurements, physical examination, urinalysis, stress ECG (electrocardiogram), spirometry, and vision and hearing function tests. Respondents also completed a detailed medical questionnaire regarding their recent operations, and based on their responses it was determined that none had been physically injured or hospitalized because of an act of terror. The complete results of the periodic health check-up were mailed to each respondent about two weeks following his/her examination.

Analytic Techniques

The hypotheses were tested by ordinary least squares (OLS) regressions separately for the male and female respondents. For each of the study's variables, we used two-tailed T-tests to check the significance of the mean differences between the male and female respondents. An omnibus test, the

Chow test (Chow, 1960), was conducted in order to test our assumption that the male and female respondents' vectors of regression coefficients differ significantly from each other. The results provided considerable support to our a priori decision to analyze the data separately for men and women with regard to SRH.

In the OLS regressions, we entered the control variables, namely age, education, financial strain, depressive symptomatology (referred to as depression) and anxiety, in the first step in order to weed out their confounding effects, followed by the simple and quadratic effects of the predictors in the next two steps, respectively, including also the quadratic effects of depressions and anxiety. In the fourth step, we entered the multiplicative terms of the two objective indicators of terror with fear of terror (Cortina, 1993). To reduce the possibility of multicollinearity among the interactions and quadratic terms and their component predictors, all component predictors were centered prior to the regression runs (Aiken and West, 1992). The reported OLS results represent those of the last step in the hierarchical regression analyses.

The multivariate association between fear of terror and elevated CRP was examined by means of logistic regression models in which the prediction of elevated CRP was adjusted for potential confounding factors. Anxiety, depressive symptoms, age, BMI (body-mass index), cholesterol, HDL levels, and fasting glucose were entered as continuous variables; whereas hypertension, physical exercise, ethnic origin, smoking status, alcohol consumption, and HRT (hormone replacement therapy) use were included as dichotomous variables (low/high). Odds ratios (ORs) and 95 percent confidence intervals (CIs) were calculated. Finally, we conducted multivariate linear regression analysis to examine whether fear of terror is also associated with CRP levels as a continuous variable. All possible confounding variables listed above were included as covariates. To reduce skewness of CRP values, all values ≥ 13.0 mg/l (prevalent for 3 percent of the study participants) were coded as 13.0 mg/l.

RESULTS

Fear of Terror and Subjective Health

Excepting education and exposure gauged in terms of the number of terror casualties, all male–female mean differences were significant at the $p<0.05$ level (see Table 6.1). Excepting SRH and financial strain, found to be significantly higher among the male respondents, all other significant differences indicated higher levels of the variable among female relative to male respondents. Women had a significantly higher mean score of fear of terror than men, 1.88 and 1.50, respectively ($p < 0.05$). We have reported elsewhere (Melamed et al., 2004) on the complete breakdown of fear of terror by gender: 11 percent of the men and 27 percent of the women had experienced high levels of fear of terror. Most of the high-fear group (83 percent of the men and 97 percent of the women) had experienced these feelings for a year or more, a fact that points to the chronicity of such a feeling. The correlations of fear of terror with the study variables are presented in Table 6.1. As we expected, fear correlated modestly with both anxiety and depressive symptoms. Fear of terror was only weakly related to age, but it was negatively associated with level of education in both males and females. By and large, fear of terror was not associated with health habits. None of the biomedical variables was related to fear of terror in men. In women, fear of terror was positively associated with BMI and fasting glucose levels, and negatively with HRT use. Thus, fear of terror seems to be largely independent of biomedical variables in both sexes (for the complete set of results regarding these correlations, see Melamed et al.).

The correlations of fear of terror with the study variables are presented in Table 6.1. As expected, SRH was negatively correlated with fear of terror for both genders. However, objective exposure was unrelated to SRH for both males and females. Fear of terror was not correlated with either measure of objective exposure for the females and was only marginally correlated ($r = 0.05$) with number of incidents for the male respondents. Fear of terror correlated modestly with both anxiety and depression for the males and females. Fear of terror was not related to age, but in women it was negatively associated with level of education. Because we formulated our hypotheses controlling for the effects of age, education, financial strain and the negative affective states of symptoms of depression and anxiety, we tested them by using regression analyses reported in Table 6.2.

As the results that we report in Table 6.2 indicate, only our first hypothesis, expecting fear of terror to be a negative predictor of SRH, received support for both male and female respondents. Neither measure of objective exposure predicted SRH in the two gender groups, thus disconfirming our second hypothesis. Because objective exposure may vary between respondents and over time, we tested the second hypothesis using different time frames, namely one week, three months, and six months, with the same results of no support found for our second hypothesis. None of the interactive terms used to test the third hypothesis, expecting objective exposure to moderate the prediction of SRH by fear of terror, was found to be significant, and therefore we concluded that there is no support for this hypothesis. The only non-linear effect found, for the male respondents, was

Table 6.2 Summary of multivariate regression of self-rated health on control variables, number of casualties and number of incidents of acts of terror, and fear of terror for female and male respondents

Measure	Female respondents ($n = 1947$)			Male respondents ($n = 2930$)		
	B	(SEB)	β	B	(SEB)	β
Control variables						
Age	–0.01*	(0.00)	–0.18	–0.01*	(0.00)	–0.16
Education, in years	0.01*	(0.00)	0.05	0.00	(0.00)	0.02
Financial strain	–0.09*	(0.02)	–0.12	–0.08*	(0.02)	–0.08
Depression	–0.33*	(0.04)	–0.20	–0.52*	(0.04)	–0.23
Anxiety	–0.07*	(0.02)	–0.10	–0.10*	(0.02)	–0.12
Anxiety^2				0.03*	(0.01)	0.05
Predictors						
Casualties, N.	–0.00	(0.00)	–0.03	0.00	(0.00)	0.00
Incidents, N.	0.05	(0.05)	0.04	0.00	(0.04)	0.00
Fear of terror	–0.04*	(0.02)	–0.05	–0.05*	(0.02)	–0.05
	Adjusted $R^2 = 0.15$			Adjusted $R^2 = 0.13$		

Note: All coefficients were taken from the final step of the OLS regression. B = unstandardized partial regression coefficients; SEB = standard error of the former; β = standardized partial regression coefficients. Number of casualties and number of incidents refer to acts of terror during the month preceding the report on fear of terror.
* $p < 0.05$.

that between anxiety and SRH. Because this non-linear U-shaped relationship was not expected, we did not plot it.

For testing our fourth hypothesis, expecting male–female differences in the above set of predictions, we tested the significance of the difference between the regression coefficients representing the prediction of SRH by fear of terror in the two gender groups. There was no support for the fourth hypothesis. On an explorative basis, we also tested the extent to which depression or anxiety moderated the effects of fear of terror on SRH for either the male or the female respondents. No evidence for such interactive effects was found. We also tested, on an explorative basis, the possibility that the relative lull in the terrorist activities inside Israel, which followed the events in December 2004–January 2005, affected our results. For each gender group, we divided the sample into those examined before and after January 1, 2005 and then conducted the same regression analyses reported in Table 6.2. No significant difference was found for our predictors: for both

Table 6.3 Adjusted[1] OR of elevated CRP associated with fear of terror, anxiety and depressive symptoms

Variable	Men		Women	
	OR 95% CI[2]	p value	OR 95% CI	p value
Fear of terror	0.9 (0.6–1.3)	0.46	1.7 (1.2–2.4)	0.001
Anxiety	0.8 (0.6–1.2)	0.32	0.8 (0.5–1.2)	0.32
Depressive symptoms	3.3 (1.1–9.4)	0.028	1.2 (0.5–3.1)	0.63

Note: [1] This model was adjusted for the following confounding variables: age, ethnic origin, educational level, BMI, smoking status, alcohol consumption, leisure-time physical activity, hypertension, serum cholesterol, HDL cholesterol, triglycerides, fasting glucose, and use of HRT.

gender groups, the results for the pre-2005 period mirrored those obtained for the post-2005 period.

Fear of Terror Associated with Elevated CRP Levels

We next tested our hypothesis (Hypothesis 5) that fear of terror would predict elevated levels of CRP. Elevated CRP levels (>3.0 mg/dl) were found in 24.5 percent of the men and 31.1 percent of the women. These figures correspond to those obtained in earlier studies (Rifai and Ridker, 2003). A preliminary logistic regression analysis (results not shown) was conducted to test the hypothesis of interactive effect of fear of terror x gender of CRP levels. This interactive term was added to the model along with fear of terror, gender and the above-mentioned control variables. As expected, a significant interaction ($p < 0.05$) was uncovered, thus supporting our a priori decision to analyze the data separately for men and women.

The association between fear of terror and risk of elevated CRP was tested through logistic regression analyses (Table 6.3). The control variables entered were anxiety, depressive symptoms, and all correlates of CRP that had been found in the literature or in our univariate analyses to be significantly associated with it (see Table 6.3). Fear of terror was associated with increased risk for elevated CRP in women (adjusted OR = 1.7, 95% CI 1.2–2.4), but not for men (adjusted OR = 0.9, 95% CI 0.6–1.3). No association between anxiety and CRP levels was uncovered. However, in men, depressive symptoms were significantly associated with increased risk of elevated CRP (adjusted OR = 3.3, 95% CI 1.1–9.4). No such association was found for women (adjusted OR = 1.2, 95% CI 0.5–3.1).

Complementary multivariate linear regression analyses were conducted in order to examine the association between fear of terror with CRP treated as a continuous variable. Included in these analyses were the same covariates applied in the above logistic regression models. The results have shown essentially the same trend. The adjusted R-square in men was 0.136. No significant association was found for fear of terror ($\beta=0.04$, $p=0.31$) or anxiety ($\beta=-0.01$, $p=0.80$). However, as before, depressive symptoms were positively and significantly associated with CRP levels ($\beta=0.11$, $p=0.029$). The adjusted R-square in women was 0.293. Again, fear of terror was positively associated with CRP levels ($\beta=0.16$, $p=0.0026$), while no significant association was found for either anxiety ($\beta=-0.07$, $p=0.17$) or depression ($\beta=0.01$, $p=0.80$).

DISCUSSION

Terror can strike anywhere. In the US, this possibility became a painful reality with the vicious attacks on the World Trade Center and on the Pentagon on September 11, 2001. In Israel, the civilian population has been subject to a continuous threat of terror since the onset of the second Intifada in September 2000. Israeli residents have been constantly reminded of the potential for attacks by daily alerts issued by the police and security checks at entrances to practically all public locations. Israeli residents are uncertain as to where and when the next terror attack might strike, and this uncertainty could be regarded as a risk factor for the development of fear, anxiety and chronic stress in individuals.

Fear of Terror and Subjective Health

In accordance with our first hypothesis, fear of terror was a negative predictor of SRH among both male and female respondents. We obtained this result after controlling for age, indicators of SES, depression and anxiety. However, the magnitude of the effect obtained was small, with fear of terror explaining less than 1 percent of SRH variance in the two gender groups. The small effect size could not be explained by the possible moderating effects of objective exposure and levels of anxiety or depression; we systematically tested for these possible moderating effects, with null results. Although it may appear to be a small effect size, Rosenthal et al. (2000) have described several examples of key clinical decisions taken in medicine based on correlations smaller than 0.05. In addition, many other factors not considered here could possibly influence one's SRH, including genetic make-up, physical constitution, personality traits, past accidents and diseases not reported in our questionnaire, health behaviors, environmental factors, and one's available and utilized coping resources. Therefore, the small effect size found for fear of terror as a predictor of SRH is not surprising.

What could possibly account for the lack of support found for our second and third hypotheses, expecting exposure to terror to have main (Hypothesis 2) and interactive (with fear of terror, Hypothesis 3) effects on SRH? From a theoretical viewpoint, our findings provide considerable support to the recently formulated relative risk appraisal perspective (Marshall et al., 2007). The new term 'relative risk appraisal' captures the psychological function that is the missing link between objective exposure and subjective response to acts of terror. This theoretical perspective (ibid.) argues that proximity to the acts of terror is only one of several factors that influence the behavioral and psychological responses, and further proposes that relative risk appraisal is an active, multidimensional process set in action by unpredictable societal threats. From an empirical viewpoint, then, as noted above, other studies which investigated the effects objective exposure to severe and traumatic stressor also found that after controlling for one's affective reaction to the exposure, objective exposure did not have a significant impact on mental health or symptoms of physical health (for example, Norris et al., 2002). Two earlier studies, Bleich et al. (2003) and Somer et al. (2005) arrived at the same conclusion, both assessing objective exposure by asking their study participants to report on their direct (for example, personal involvement in an incident of terror; Bleich et al., 2003) and indirect (for example, providing help to victims; Somer et al., 2005), personal experiences with terrorism.

Our study of fear of terror and SRH replicates and extends past studies in the following ways. First, we used measures of exposure which were assessed independently of the respondents. Second, we used several possible time frames for these measures of exposure, systematically testing whether length of objective exposure modified our findings: it did not. None of the objective exposure measures was found to be predictive of SRH. Third, Bleich et al. (2003, p. 619) found that persons who were directly involved in an act of terror showed no more adverse emotional affective states than those who had not been thus exposed. We found the same lack of effect using exposure measures reflecting the assumption of vicarious traumatization. By vicarious traumatization (Lerias and Byrne, 2003), we refer to the wide-ranging impact of the pervasive terrorism in Israel, engulfing the populace irrespective of direct engagement or involvement in acts of terror. Still, the question remains: what could explain our findings, not supportive of our hypotheses in this regard?

A major component of the current armed struggle of Palestinian terrorist groups against Israel involves the use of suicide bombers (Reuter, 2004), a weapon whose use involves uncertainty and unpredictability. Therefore, we argue that the respondents may have acted like rational decision makers in deciding on their level of personal fear of acts of terror irrespective of recent acts of terror and the recent number of casualties in their immediate vicinity. As noted, fear of terror was not correlated with either measure of objective exposure for the females and was only marginally correlated ($r = 0.05$) with number of incidents for the male respondents. Therefore, the respondents may have acted as rational decision makers because past experience could not be used to forecast the next occurrence of an act of terror. Yet another possibility is that they used alternative indicators of objective exposure in their immediate social milieu in their mental calculation while assessing their levels of fear of terror and SRH: the extent to which these events and casualties were in fact covered by the mass media. It could be that objective exposure impacts on SRH only for those respondents having a daily exposure to the mass media. As a counter-argument, note that media coverage of each act of terror in Israel is without exception immediate, graphic, and extensive, thus making minimal exposure to this media coverage almost unavoidable in the major urban areas.

Solomon et al. (2005) found that terror is not gender-blind, and reported that men and women differed in self-reporting their exposure to terror and in the way they perceived and coped with the threat of terror. Like Solomon et al., we found that women had higher levels of fear of terror and were exposed to a higher number of casualties in their ecological milieu relative to the men in our sample, but unlike the earlier study we did not find any evidence of gender differences in the effects of these predictors on SRH. Terror, unlike war or other types of traumatic events, hit men and women in the same setting and in this respect tends to be gender-blind. Based on a review of 46 meta-analytic studies, Hyde (2005) advanced the thesis that males and females are similar on most, but not all, relationships among psychological variables. We found corroborative evidence for validity of this thesis with respect to the relationship between fear of terror and SRH. We did find considerable gender difference in depression linkage with SRH. Depression, the most powerful predictor of SRH (Table 6.2), played a more powerful role as a predictor of SRH among men relative to women ($\beta = -0.52$, -0.33, respectively). This finding is consistent with our earlier research (see Melamed et al., 2004; Toker et al., 2005) reporting that in comparison with women, depression among men was twice as likely to be associated with elevated levels of inflammation biomarkers. However, as we indicate below, we found considerable gender differences in the associations of fear of terror and the inflammation biomarker investigated in Study 2, CRP.

Fear of Terror and Low-grade Inflammation

There is substantial evidence that mass violence in general, and terrorism in particular, is associated with mental health impairments (Norris et al., 2002). However, the potential association between fear of terror and low-grade inflammation has hardly been explored. We evaluated 1153 otherwise healthy, employed adults who underwent a routine annual physical examination. We found that by and large women scored higher on fear of terror than men ($p < 0.0001$). Some 11 percent of the men and 26.7 percent of the women had experienced high fear of terror, most of them for a year or more. Fear of terror was common to people throughout the country, and was not limited to those living in cities/towns directly affected by terror. Other studies of psychological reactions to terrorist attacks in the US (for example, Silver et al., 2002) and Israel (for example, Bleich et al., 2003) reported similar findings. The major finding of the present study is that in women, chronic fear of terror is associated with low-grade inflammation, expressed by a heightened concentration of CRP, supporting our hypothesis. Fear of terror was associated with a 1.7-fold greater risk of elevated CRP (95% CI 1.2–2.4). This was true even after controlling for general anxiety, depressive symptoms, use of HRT, and several other potent confounding variables found to be associated with CRP levels in earlier studies. In contrast to the results for women, we did not find any significant association between chronic fear of terror and CRP levels in men.

It is noteworthy that although fear of terror has been found to be significantly and positively associated with anxiety levels and depressive symptoms in both sexes, the elevated CRP levels in the women in the present study were directly associated with fear of terror and were not mediated by general psychological distress indexed by the above symptoms. On the other hand, depressive symptoms were found in the multivariate analysis to be positively associated with CRP levels in men (OR = 3.3, 95% CI 1.1–9.4), but not in women (OR = 1.2, 95% CI 0.5–3.1). This differential of depression association with CRP levels for men and women, replicate a similar recent finding in the study by Danner et al. (2003) where a recent history of major depressive episode was used as the predictor variable. These positive findings lend further support to the suggestion that low-grade systemic inflammation could represent a mechanism linking depression to cardiovascular risk. An additional complementary finding here that the associations reported above were also observed with CRP levels taken as a continuous variable, pointed to a linear positive association between fear of terror and CRP levels in women, and between depressive symptoms and CRP levels in men. Our multivariate regression analyses with CRP treated as a continuous variable overcame a possible information loss that may have been the result of using CRP as a dichotomous variable. The significant results indicate that the observed association between fear of terror or depressive symptoms is not an outcome of using a predetermined cut-off point for CRP levels. It was also noteworthy to learn that the percentage of variance in CRP levels explained by predictor variables here, was much higher for women (29.3 percent) than for men (13.6 percent).

The Limitations of the Two Studies

What are some of the limitations of the two studies, conducted to test the possibility that fear of terror predicts self-rated health (first study, Hypotheses 1 to 4) and a biomarker of inflammatory processes in the body (second study, Hypothesis 5)? Our sample of subjects undergoing a periodic health examination may not be representative of the general population. Most of the individuals were highly educated, white-collar workers who exhibited generally good health-behavior patterns: they smoked little, exercised regularly, and drank alcohol only in moderation. Furthermore, only one man and one woman in the sample met the criteria of having symptoms of a major depression (Spitzer et al., 1994). This rate is much lower than that obtained on the average in more representative samples. The point of prevalence of a major depressive disorder in adult community samples varies from 5–9 percent for women and 2–3 percent for men (Aggen et al., 2005). As a result of their superior health habits, our subjects may have been more resilient to the effects of fear of terror on their SRH. However, it is even more likely that the significant findings obtained here with regard to fear of terror–SRH linkage will be replicated in the less resilient general population. Because both fear of terror and SRH were measured concomitantly, our design cannot rule out the possibility of reverse causation – that is, the more vulnerable persons, who appraised their health as poor, would therefore report higher levels of fear of terror. Future researchers may consider investigating this possible causal path in a longitudinal design.

The same limitation is relevant to our second study, testing the fifth hypothesis. Our findings were based on cross-sectional data; therefore, the temporal ordering of the association between fear of terror and CRP levels cannot be definitively established. It is possible that instead of fear of terror leading to inflammation, inflammation leads to subjective appraisal of the situation as posing a threat to one's life and to the lives of one's loved ones. There is compelling evidence that the level of proinflammatory cytokines (IL-1, TNF-α, IL-6) serve as a signal to the brain, inducing adverse effects on affect and cognitive function and thereby contributing to anxiety, depression, learned helplessness and cognitive disturbances (Maier and

Watkins, 1998; Reichenberg et al., 2001; Weaver et al., 2002). Indeed, the presence of one or the other may set off a vicious cycle, ultimately leading to an exacerbation of the threat appraisal, immunosuppresion, and inflammation.

Possible Strengths of the Two Studies

Our two studies have a number of strengths, including the large sample size, using two indicators to assess objective exposure to terrorism and testing several possible time frames in assessing their impact on SRH, using an important objective indicator of micro-inflammatory processes in the body, controlling for several well-known antecedents of SRH CRP, and basing the analyses done on apparently healthy persons. In addition, exposure to terrorism was assessed independently of the respondents and related to time frames predating their reports of their fear of terror and SRH, thus allowing us to test the direction of possible influence from exposure to the criterion of SRH.

Future Research Paths on the Health Effects of Terror

We noted the unexpected U-shaped relationship between anxiety and SRH among the male respondents. Future research may investigate the possibility that this linkage, if replicated, may be indicative of a repressive coping style used by highly anxious males. Future research directions may consider the possibility that the suggested effects of fear of terror on SRH would dissipate following a long-term lull in terrorism, probably reflected also in raised levels of SRH. Future researchers could also examine the possible moderating effects on the relationships investigated in our study by potentially powerful moderators such as racial or ethnic group membership (Read and Gorman, 2006), coping resources such as social support (Olphen et al., 2003), and personality predispositions such as self-efficacy and sense of coherence (Kristenson et al., 2005). Moderating variables such as those mentioned above might lead to a substantial amount of explained variance in SRH in specific subgroups with very low levels of coping resources.

The finding here of an association between fear of terror and elevated CRP levels may have important long-term health implications. Recent studies suggest that inflammation may be associated with risk of atherosclerosis (Libby and Ridker, 2004), myocardial infarction and stroke (Ridker et al., 2004, 2005). Taken together, these data indicate that women who chronically experience high fear of terror and display elevated CRP levels might be at risk of tissue damage and of future cardiovascular events.

In our study, women turned out to be highly vulnerable. Women expressed higher fear of terror than men. This is consistent with the recurring finding (for example, Bleich et al., 2003) that women have a higher prevalence than men of PTSD, depression and other psychological distress symptoms in response to terrorist attacks, and that women are more likely than men to suffer adverse mental health effects in response to all types of disasters (Norris et al., 2002). Second, only among the women was fear of terror associated with a significant increase in CRP level.

CONCLUSIONS

The findings of the two studies, reported in this chapter, demonstrated for the first time, in a sample of apparently healthy workers, that the psychological reaction of chronic fear of terror has a negative impact on indicators of subjective and objective health. We found that for both genders, the effects of fear of terror on SRH were negative. When considering the indicator of micro-inflammation in the body used in our study as an objective indicator of one's physical health, women turned out to be highly vulnerable compared with men. They not only appraised the situation as more threatening and expressed higher fear of terror, but this also appears to have negative implications for their physical health. For our female respondents, fear of terror was associated with a 1.7-fold risk of elevated CRP, a marker of low-grade inflammation and a strong risk factor for diabetes and future cardiovascular events.

Terrorist acts and the threat of terrorism may continue to affect populations worldwide. International studies are needed to explore the extent of fear of terror in other countries, and the relative risk of physical morbidity in women. The replication of our findings in longitudinal studies and in larger and more representative samples may have important health-care policy implications for primary care physicians, nurses and community mental health professionals with regard to stress management interventions.

NOTES

1. We would like to acknowledge with thanks the financial support provided to the study by the Israel Science Foundation (Grant 962/02 – 1).
2. Detailed results of this analysis are available from the authors upon request.

REFERENCES

Aggen, S.H., Neale, C.N. and Kendler, K.S. (2005), 'DSM criteria for major depression: evaluating symptom patterns using latent-trait item response models', *Psychological Medicine*, **35**(4), 475–87.

Aiken, L.A. and West, S.G. (1992), *Multiple Regression: Testing and Interpreting Interactions*, Newbury Park, CA: Sage.

Barofsky, I., Erickson, P. and Eberhardt, M. (2004), 'Comparison of a single item and an index of multi-item health status measure among persons with and without diabetes in the US', *Quality of Life Research*, **13**(6), 1671–81.

Baron-Epel, O., Kaplan, G., Haviv-Messika, A., Tarabeia, J., Green, M.S. and Nitzan Kaluski, D. (2005), 'Self-reported health as a cultural health determinant in Arab and Jewish Israelis: MABAT – National Health and Nutrition Survey 1999–2001', *Social Science and Medicine*, **61**(6), 1256–66.

Baum, A. and Dougall, A.L. (2002), 'Research on terrorism: what have we learned? Understanding and predicting effects of terrorism', *Current Opinion in Psychiatry*, **15**, 617–21.

Benyamini, Y. and Idler, E.L. (1999), 'Community studies reporting association between self-rated health and mortality – additional studies', *Research on Aging*, **21**(3), 392–401.

Bermudez, E.A., Rifai, N., Buring, J., Manson, J.E. and Ridker, P.A. (2002), 'Interrelationships among circulating interleukin-6, C-reactive protein, and traditional cardiovascular risk factors in women', *Arteriosclerosis Thrombosis and Vascular Biology*, **22**(10), 1668–73.

Black, P.H. (2002), 'Stress and the inflammatory response: a review of neurogenic inflammation', *Brain, Behavior, and Immunity*, **16**(6), 622–53.

Black, P.H. (2003), 'The inflammatory response is an integral part of the stress response: implications for atherosclerosis, insulin resistance, type II diabetes and metabolic syndrome X', *Brain, Behavior, and Immunity*, **17**(5), 350–64.

Bleich, A., Gelkopf, M. and Solomon, Z. (2003), 'Exposure to terrorism, stress-related mental health symptoms, and coping behaviors among a nationally representative sample in Israel', *Journal of the American Medical Association*, **290**(5), 612–20.

Chow, G.C. (1960), 'Tests of equality between sets of coefficients in two linear regressions', *Econometrica*, **28**, 591–605.

Cohen, M. and Eid, J. (2007), 'The effect of constant threat of terror on Israeli Jewish and Arab adolescents', *Anxiety, Stress and Coping*, **20**(1), 47–60.

Cortina, J.M. (1993), 'Interaction, nonlinearity, and multicollinearity: implications for multiple regression', *Journal of Management*, **19**(4), 915–22.

Danner, M., Kasl, S.V., Abramson, J.L. and Vaccarino, V. (2003), 'Association between depression and elevated C-reactive protein', *Psychosomatic Medicine*, **65**(3), 347–56.

DeSalvo, K.B., Bloser, N., Reynolds, K., He, J. and Muntner, P. (2006a), 'Mortality prediction with a single general self-rated health question. A meta-analysis', *Journal of General Internal Medicine*, **21**(3), 267–75.

DeSalvo, K.B., Fisher, W.P., Bloser, N., Merrill, W. and Pebody, J. (2006b), 'Assessing measurement properties of two single-item general health measures', *Quality of Life Research*, **15**(2), 191–201.

Dietrich, A.J., Oxman, T.E., Burns, M.R., Winchell, C.W. and Chin, T. (2003), 'Application of a depression management office system in community practice: a demonstration', *Journal of the American Board of Family Practice*, **16**(2), 107–14.

Douglas, K.M., Taylor, A.J. and O'Malley, P.G. (2004), 'Relationship between depression and C-reactive protein in a screening population', *Psychosomatic Medicine*, **66**(5), 679–83.

Duffy, M., Gillespie, K. and Clark, D.M. (2007), 'Post-traumatic stress disorder in the context of terrorism and other civil conflict in Northern Ireland: randomised controlled trial', *British Medical Journal*, **334**(7604), 1147–50.

Ford, D.E. and Erlinger, T.P. (2004), 'Depression and C-reactive protein in US adults: data from the third national health and nutrition examination survey', *Archives of Internal Medicine*, **164**(9), 1010–14.

Fraley, R.C., Fazzari, D.A., Bonanno, G.A. and Dekel, S. (2006), 'Attachment and psychological adaptation in high exposure survivors of the September 11th attack on the World Trade Center', *Personality and Social Psychology Bulletin*, **32**(4), 538–51.

Franssen, P.M.L., Bultmann, U., Kant, I. and van Amelsvoort, L.G.P.M. (2003), 'The association between chronic diseases and fatigue in the working population', *Journal of Psychosomatic Research*, **54**(4), 339–44.

French, J.R.P., Caplan, R.D. and Harrison, R.V. (1982), *The Mechanisms of Job Stress and Strain*, Chichester: John Wiley.

Galea, S., Vlahov, D., Resnick, H., Ahern, J., Susser, E., Gold, J. et al. (2003), 'Trends of probable post-traumatic stress disorder in New York City after the September 11 terrorist attacks', *American Journal of Epidemiology*, **158**(6), 514–24.

Gallo, L.C. and Matthews, K.A. (2003), 'Understanding the association between socioeconomic status and physical health: do negative emotions play a role?', *Psychological Bulletin*, **129**(1), 10–51.

Gidron, Y., Gal, R. and Zahavi, S. (1999), 'Bus commuters' coping strategies and anxiety from terrorism: an example of the Israeli experience', *Journal of Traumatic Stress*, **12**(1), 187–97.

Goldman, N., Glei, D.A. and Chang, M.-C. (2004), 'The role of clinical risk factors in understanding self-rated health', *Annals of Epidemiology*, **14**(1), 49–57.

Hart, P.M. and Cooper, C.L. (2001), 'Occupational stress: toward a more integrated framework', in N.R. Anderson, D.S. Ones, H.K. Sinangil and C. Viswesvaran (eds), *Handbook of Industrial Work and Organizational Psychology*, vol. 2, London: Sage, pp. 93–115.

Hill, T.D. and Needham, B.L. (2006), 'Gender-specific trends in educational attainment and self-rated health, 1972–2002', *American Journal of Public Health*, **96**(7), 1288–92.

Hobfoll, S.E., Canetti-Nisim, D. and Johnson, R.J. (2006), 'Exposure to terrorism, stress-related mental health symptoms, and defensive coping among Jews and Arabs in Israel', *Journal of Consulting and Clinical Psychology*, **74**(2), 207–18.

Hyde, J.S. (2005), 'The gender similarities hypothesis', *American Psychologist*, **60**(6), 581–92.

Idler, E.L. and Benyamini, Y. (1997), 'Self-rated health and morbidity: a review of twenty-seven community studies', *Journal of Health and Social Behavior*, **38**(1), 21–37.

Inness, M. and Barling, J. (2005), 'Terrorism', in J. Barling, E.K. Kelloway and M.R. Frone (eds), *Handbook of Work Stress*, Thousand Oaks, CA: Sage, pp. 375–99.

Israel Ministry of Foreign Affairs (2006), *Terror since 2000*, http://www.mfa.gov.il/mfa/terrorism/ [Context Link], June 18, 2006.

Jylha, M., Volpato, S. and Guralnik, J.M. (2006), 'Self-rated health showed a graded association with frequently used biomarkers in a large population sample', *Journal of Clinical Epidemiology*, **59**(5), 465–71.

Kahn, R.L. and Byosiere, P. (1992), 'Stress in organizations', in M.D. Dunnette and L.M. Hough (eds), *Handbook of Industrial and Organizational Psychology*, second edition, vol. 3, Palo Alto, CA: Consulting Psychologists Press, pp. 571–651.

Kaplan, Z., Matar, M.A., Kamin, R., Sadan, T. and Cohen, H. (2005), 'Stress-related responses after 3 years of exposure to terror in Israel: are ideological–religious factors associated with resilience?', *Journal of Clinical Psychiatry*, **66**(9), 1146–54.

Kawada, T. (2003), 'Self-rated health and life prognosis', *Archives of Medical Research*, **34**(4), 343–7.

Keinan, G., Sadeh, A. and Rosen, S. (2003), 'Attitudes and reactions to media coverage of terrorist acts', *Journal of Community Psychology*, **31**(2), 149–65.

Kop, W.J. (2003), 'The integration of cardiovascular behavioral medicine and psychoneuroimmunology: new developments based on converging research fields', *Brain, Behavior, and Immunity*, **17**(4), 233–7.

Kristenson, M., Olsson, A.G. and Kucinskiene, Z. (2005), 'Good self-rated health is related to psychosocial resources and a strong cortisol response to acute stress: the LiVicordia Study of Middle-Aged Men', *International Journal of Behavioral Medicine*, **12**(3), 153–60.

Lazarus, R.S. (2001), 'Relational meaning and discrete emotions', in K.R. Scherer, A.A. Schorr and T. Johnston (eds), *Appraisal Processes in Emotions: Theory, Research, Methods*, Oxford and New York: Oxford University Press, pp. 37–67.

Lerias, D. and Byrne, M.K. (2003), 'Vicarious traumatization: symptoms and predictors', *Stress and Health*, **19**(1), 129–38.

Libby, P. and Ridker, P.M. (2004), 'Inflammation and atherosclerosis: role of C-reactive protein in risk assessment', *American Journal of Medicine*, **116**(6, Supplement 1), 9–16.

Lubetkin, E.I., Jia, H. and Gold, M.R. (2003), 'Depression, anxiety, and associated health status in low-income Chinese patients', *American Journal of Preventive Medicine*, **24**(4), 354–60.

Ludwig, K.H., Marten-Mittag, B., Lowel, H., Doring, A. and Koenig, W. (2003), 'Influence of depressive mood on the association of CRP and obesity in 3205 middle aged healthy men', *Brain, Behavior, and Immunity*, **17**, 268–75.

Maier, S.F. and Watkins, L.R. (1998), 'Cytokines for psychologists: implications of bidirectional immune-to-brain communication for understanding behavior, mood, and cognition', *Psychological Review*, **105**(1), 83–107.

Marshall, R.D., Bryant, R.A., Amsel, L., Suh, E.J., Cook, J.M. and Neria, Y. (2007), 'The psychology of ongoing threat: relative risk appraisal, the September 11 attacks, and terrorism-related fear', *American Psychologist*, **62**(4), 304–16.

McCullough, M.E. and Laurenceau, J.P. (2004), 'Gender and the natural history of self-rated health: a 59-year longitudinal study', *Health Psychology*, **23**(6), 651–5.

McHorney, C.A. (1999), 'Health status assessment methods for adults: past accomplishments and future challenges', *Annual Review of Public Health*, **20**, 309–35.

Melamed, S., Shirom, A., Toker, S., Berliner, S. and Shapira, I. (2004), 'Association of fear of terror with low-grade inflammation among apparently healthy employed adults', *Psychosomatic Medicine*, **66**(4), 484–91.

Nease, D.E., Jr and Maloin, J.M. (2003), 'Depression screening: a practical strategy', *Journal of Family Practice*, **52**(2), 118–24.

Nicholson, A., Bobak, M., Murphy, M., Rose, R. and Marmot, M. (2005), 'Socio-economic influences on self-rated health in Russian men and women – a life course approach', *Social Science and Medicine*, **61**(11), 2345–54.

Norris, F.H., Friedman, M.J., Watson, P.J., Byrne, C.M., Diaz, E. and Kaniasty, K. (2002), '60,000 disaster victims speak: Part I. An empirical review of the empirical literature, 1981–2001', *Psychiatry*, **65**(3), 207–39.

Olphen, J., Schulz, A.J., Israel, B.A., Chatters, L., Klem, L., Parker, E.A. et al. (2003), 'Religious involvement, social support, and health among African–American women on the East Side of Detroit', *Journal of General Internal Medicine*, **18**(7), 549–57.

Pearlin, L.I. (1999), 'Stress and mental health: a conceptual overview', in T.L. Scheid and A.V. Horwitz (eds), *A Handbook for the Study of Mental Health: Social Contexts, Theories, and Systems*, Cambridge and New York: Cambridge University Press, pp. 161–76.

Pearlin, L.I. and Lieberman, M.A. (1979), 'Social sources of emotional distress', in R. Simmons (ed.), *Research in Community and Mental Health*, Vol. 1, Greenwich, CT: JAI Press, pp. 217–48.

Pearlin, L.I., Lieberman, M.A., Menaghan, E.G. and Mullan, J.T. (1981), 'The stress process', *Journal of Health and Social Behavior*, **22**(4), 337–56.

Peirce, R.S., Frone, M.R., Cooper, M.L. and Russell, M. (1996), 'Financial stress, social support, and alcohol involvement: a longitudinal test of the buffering hypothesis in a general population survey', *Health Psychology*, **15**(1), 38–47.

Peirce, R.S., Frone, M.R., Russell, M. and Cooper, M.L. (1994), 'Relationship of financial strain and psychosocial resources to alcohol use and abuse: the mediating role of negative affect and drinking motives', *Journal of Health and Social Behavior*, **35**, 291–308.

Penninx, B.W., Kritchevsky, S.B., Yaffe, K., Newman, A.B., Simonsick, E.M., Rubin, S. et al. (2003), 'Inflammatory markers and depressed mood in older persons: results from the Health, Aging and Body Composition study', *Biological Psychiatry*, **54**, 566–72.

Podsakoff, P.M., MacKenzie, S.B., Podsakoff, N.P. and Lee, J.Y. (2003), 'Common method biases in behavioral research: a critical review of the literature and recommended remedies', *Journal of Applied Psychology*, **88**(5), 879–903.

Pradhan, A.D., Manson, J.E., Rifai, N., Buring, J.E. and Ridker, P.M. (2001), 'C-reactive protein, interleukin 6, and the risk of developing type 2 diabetes mellitus', *Journal of the American Medical Association*, **286**, 327–34.

Read, J.G. and Gorman, B.K. (2006), 'Gender inequalities in US adult health: the interplay of race and ethnicity', *Social Science and Medicine*, **62**(5), 1045–65.

Reichenberg, A., Yirmiya, R., Schuld, A., Kraus, T., Haack, M., Morag, A. et al. (2001), 'Cytokine-associated emotional and cognitive disturbances in humans', *Archives of General Psychiatry*, **58**(5), 445–52.

Reuter, C. (2004), *My Life Is My Weapon: A Modern History of Suicide Bombing*, Princeton, NJ: Princeton University Press.

Ridker, P.M. (2003), 'Clinical application of C-reactive protein for cardiovascular disease detection and prevention', *Circulation*, **107**, 363–9.

Ridker, P.M., Rifai, N., Cook, N.R., Bradwin, G. and Buring, J.E. (2005), 'Non-HDL cholesterol, apolipoproteins A-I and B100, standard lipid measures, lipid ratios, and CRP as risk factors for cardiovascular disease in women', *Journal of the American Medical Association*, **294**(3), 326–33.
Ridker, P.M., Wilson, P.W.F. and Grundy, S.M. (2004), 'Should C-reactive protein be added to metabolic syndrome and to assessment of global cardiovascular risk?', *Circulation*, **109**(23), 2818–25.
Rifai, N. and Ridker, P.M. (2003), 'Population distributions of C-reactive protein in apparently healthy men and women in the United States: implication for clinical interpretation', *Clinical Chemistry*, **49**(4), 666–9.
Rifai, N., Tracy, R.P. and Ridker, P.M. (1999), 'Clinical efficacy of an automated high-sensitivity C-reactive protein assay', *Clinical Chemistry*, **45**, 2136–41.
Rosenthal, R., Rosnow, R.L. and Rubin, D.B. (2000), *Contrasts and Effect Sizes in Behavioral Research*, Cambridge: Cambridge University Press.
Rubonis, A.V. and Bickman, L. (1991), 'Psychological impairment in the wake of disaster: the disaster–psychopathology relationship', *Psychological Bulletin*, **109**(2), 384–99.
Schlenger, W.F., Caddell, J.M., Ebert, L., Jordan, B.K., Rourke, K.M., Wilson, D. et al. (2002), 'Psychological reactions to terrorist attacks: findings from the National Study of Americans' Reactions to September 11', *Journal of the American Medical Association*, **288**(5), 581–8.
Schmid, A.P. and Jongman, A.J. (1988), *Political Terrorism: A New Guide to Actors, Authors, Concepts, Data Base, Theories, and Literature*, New Brunswick, NJ: Transaction Books.
Schnurr, P and Green, B. (2004), *Trauma and Health: Physical Health Consequences of Exposure to Extreme Stress*, Washington, DC: American Psychological Association.
Silver, R.C., Holman, E.A., McIntosh, D.N., Poulin, M. and Gil-Rivas, V. (2002), 'National longitudinal study of psychological responses to September 11', *Journal of the American Medical Association*, **288**(10), 1235–45.
Simon, J.G., De Boer, J.B., Joung, I.M.A., Bosma, H. and Mackenbach, J.P. (2005), 'How is your health in general? A qualitative study on self-assessed health', *European Journal of Public Health*, **15**(2), 200–208.
Solomon, Z., Gelkopf, M. and Bleich, A. (2005), 'Is terror gender-blind? Gender differences in reaction to terror events', *Social Psychiatry and Psychiatric Epidemiology*, **40**(12), 947–54.
Solomon, Z. and Lavi, T. (2005), 'Israeli youth in the second Intifada: PTSD and future orientation', *Journal of the American Academy of Child and Adolescent Psychiatry*, **44**, 1167–75.
Somer, E., Ruvio, A., Soref, E. and Sever, I. (2005), 'Terrorism, distress and coping: high versus low impact regions and direct versus indirect civilian exposure', *Anxiety, Stress, and Coping*, **18**(3), 165–82.
Spitzer, R.L., Kroenke, K., Williams, J.B. and and the Patient Health Questionnaire Primary Care Study Group (1999), 'Validation and utility of a self-report version of PRIME-MD: the PHQ primary care study', *Journal of the American Medical Association*, **282**(18), 1737–44.
Spitzer, R.L., Williams, J.B., Kroenke, K., Linzer, M., deGruy, F.V., 3rd, Hahn, S.R. et al. (1994), 'Utility of a new procedure for diagnosing mental disorders in primary care. The PRIME-MD 1000 study', *Journal of the American Medical Association*, **272**(22), 1749–56.

Steptoe, A., Owen, N., Kunz-Ebrecht, S.R. and Mohamed-Ali, V. (2002), 'Inflammatory cytokines, socioeconomic status, and acute stress responsivity', *Brain Behavior and Immunity*, **16**, 774–89.

Suls, J. and Bunde, J. (2005), 'Anger, anxiety, and depression as risk factors for cardiovascular disease: the problems and implications of overlapping affective dispositions', *Psychological Bulletin*, **131**(2), 260–300.

Toker, S., Shirom, A., Shapira, I., Berliner, S. and Melamed, S. (2005), 'The association between burnout, depression, anxiety, and inflammation biomarkers: C-reactive protein and fibrinogen, in men and women', *Journal of Occupational Health Psychology*, **10**(4), 344–63.

Verma, S., Devaraj, S. and Jialal, I. (2006), 'C-reactive protein promotes atherothrombosis', *Circulation*, **113**(17), 2135–51.

Weaver, J.D., Huang, M.H., Albert, M., Harris, T., Rowe, J.W. and Seeman, T.E. (2002), Interleukin-6 and risk of cognitive decline: MacArthur studies of successful aging', *Neurology*, **59**(3), 371–8.

Yehuda, R. and Hyman, S.E. (2005), 'The impact of terrorism on brain, and behavior: what we know and what we need to know', *Neuropsychopharmacology*, **30**(10), 1773–80.

Zeidner, M. (2005), 'Contextual and personal predictors of adaptive outcomes under terror attack: the case of Israeli adolescents', *Journal of Youth and Adolescence*, **34**(5), 459–70.

Zeidner, M. (2006), 'Individual differences in psychological reactions to terror attack', *Personality and Individual Differences*, **40**(6), 771–81.

7. March 11, 2004, terrorist attacks in Madrid, Spain: psychopathological aftermath and comparisons with September 11, 2001, NYC

Hector González-Ordi, Antonio Cano-Vindel, Iciar Iruarrizaga and Juan Jose Miguel-Tobal

INTRODUCTION

Traumatic events are mainly conceptualized by their capacity to evoke terror, fear, helplessness, or horror in the face of a threat to life or serious injury, as defined by the American Psychiatric Association (1994). Among them, terrorist acts or attacks are human-made events that cause intentional interpersonal violence with the aim of provoking disruption to the experience of safety in communities (Fullerton et al., 2003). In addition, clinical and epidemiological literature have shown that often human-made disasters, such as terrorist acts, can be more disturbing and disruptive than natural disasters, such as hurricanes, tornadoes, floods and so on (see Norris et al., 2002; Galea et al., 2005).

Rather than merely an act of war in any traditional sense of destroying the material, technical or personal resources of an enemy, 'terrorism is about taking strategic actions that incite terror and fright in civilian populations' (McDermott and Zimbardo, 2007, p. 358). In this sense, as it is pointed out elsewhere, terrorism is fundamentally about psychology (Bongar, 2007).

MADRID, MARCH 11, 2004: PERI-TRAUMATIC REACTIONS

In the morning of March 11, 2004, Madrid, Spain, suffered the biggest terror attack in its history, which can be compared to the London terror attacks in 2005. Both were the biggest terror attacks that Western Europe has ever experienced. The Al Qaeda group placed 10 bombs which exploded in four commuter trains during the rush hour when most people were heading for work. There were 191 fatalities and 1800 people were injured, causing a state of horror and uncertainty among the population.

In the days immediately after the attack, the attacks dominated national and international headline news, up to 2 million Madrid citizens participated in protest demonstrations throughout the country and a previously scheduled national election resulted in an upset election of the opposition party to power.

The first data obtained on the March 11 psychological aftermath related to the two-weeks-after-disaster surveys that focused mainly on acute stress disorder symptoms and general psycho-social functioning in the general population of the affected zones.

Strong evidence indicates that most people who have been recently exposed to a traumatic experience show a wide range of peri-traumatic reactions within the initial weeks after the trauma such as emotional numbing, reduced awareness of one's environment, derealization, depersonalization, dissociative amnesia, intrusive thoughts, avoidance behaviors, insomnia, concentration deficits, irritability and autonomic arousal (see Bryant, 2004, for a review). However, the majority of these reactions are transient and, after the disruptive psychological and physical effects of the trauma impact, people adapt naturally to their experience in the following months.

Following the March 11 terrorist attacks in Madrid, Muñoz et al. (2004a and b) carried out a survey of 1179 adult Madrid citizens (at least 18 years old) in order to explore the degree of peri-traumatic reactions in the general population soon after the attacks, and their impact on daily life psychosocial functioning during the second week after the terrorist attack (from March 18 to 24). Data showed that almost half of the sample reported acute stress symptoms (46.7 percent); the re-experiencing and dissociative symptoms were the more pervasive (presented in the 72.5 percent and the 71.8 percent of the sample, respectively), while the avoidance and arousal symptoms were less pervasive (48.6 and 44.6 percent, respectively). However, 18.1 percent of the sample had general functioning problems only, and 12.8 percent presented functioning problems for two or more days that interfered, at least moderately, with their daily life activity. According to Muñoz et al. (2004b), such data are comparable to percentages found after the September 11, 2001 attacks (Schuster et al., 2001; Silver et al., 2002), and would reflect an overreporting effect due to the critical situation in which people are immersed. In this sense, although acute stress prevalences were really high, they did not have a critical impact on or interference in people's general daily life functioning, in terms of rates of percentage of people affected. One must keep in mind that the people surveyed were members of the general public, not victims or relatives. It is likely that, after the initial impact of the disaster, they tried to return to their normal lives as soon as possible. Rather than indicating a pathological aftermath, these data reflect that acute symptoms would be part of a normal response in people who are faced with an abnormal situation (Silver et al., 2002).

MADRID, MARCH 11, 2004: POST-TRAUMATIC REACTIONS

A few hours after the attacks on the commuter trains, when the great magnitude of the mass disaster had become apparent, the authors of this chapter enrolled in the first-aid units, set up by the Colegio Oficial de Psicólogos de Madrid (Madrid Psychological Association), in order to give psychological support to survivors, victims' relatives, and emergency personnel.

During these turbulent hours and the days following, we planned to carry out an epidemiological study in order to explore the psychopathological aftermath in the general population, as well as in victims, relatives and emergency personnel.

About a week after the terrorist attacks, we contacted Sandro Galea, Associate Director of the Center for Urban Epidemiologic Studies at the New York Academy of Medicine (currently at the University of Michigan School of Public Health), and his team. He and his colleagues had carried out an outstanding amount of epidemiological research on the psychopathological aftermath of the September 11, 2001, World Trade Center attacks. The empirical research carried out by Galea and his colleagues on September 11 was considered an authoritative scientific reference to be followed in our research (see Galea et al., 2002, 2003), because regardless of the injuries, fatalities, and urban disruption effects, March 11 and September 11 were the biggest terrorist attacks for both Madrid and New York City, respectively.

The September 11 multiple attacks in New York City (NYC) and at the Pentagon increased exponentially the interest in the psychological impact of terrorism. In fact, several studies carried out on the World Trade Center disaster have noted significant consequences both in those people directly affected or exposed – those who directly witnessed the events (Boscarino et al., 2002; Galea et al., 2002, 2003; Piotrkowski and Brannen, 2002) – and to a lesser extent, in the US general population (Schuster et al., 2001; Silver et al., 2002).

One month after the Manhattan attacks, Galea et al. (2002) found prevalence rates of 7.5 percent for post-traumatic stress disorder (PTSD) (20 percent in the 'Zero Zone') and 9.7 percent for depression in a NYC sample population (n = 1008), although this prevalence decreased considerably in time, at least in those not directly affected (Galea et al., 2003). However, the residual psychological impact in the American population is still remarkable. It is estimated that two months after the NYC and Washington disasters, 12 percent of the US population was still experiencing clinically significant psychological distress (Schlenger et al., 2002), and between 25 and 30 percent of Americans were still experiencing anxiety and avoidance of situations that reminded them of September 11 (Silver et al., 2002). Six months later, it was estimated that approximately two-thirds of Americans reported fear of future attacks, and that 41 percent of the general population still feared harm to their families as a result of terrorism (ibid.). Even two years after the event, two-thirds of New Yorkers were still concerned about another attack, and about one-third of respondents indicated that they had not succeeded in returning to normal (Kleinfield and Connelly, 2003).

Sandro Galea readily agreed to provide us with the instruments and methodological framework to carry out the research immediately, and he was delighted to be a part of the research team in Madrid. As well, we also asked for the expertise and support of Stevan E. Hobfoll, Distinguished Professor, Director of the Department of Psychology, Applied Psychology Center at Kent State University and Past-President of the Stress and Anxiety Research Society (STAR) as scientific consultant.

This project consisted of three longitudinal studies which began approximately 5–15 weeks, 6–7 months and 1 year, respectively, after the March 11 terror attacks in Madrid. The results that appear in this chapter relate mainly to the first study that was carried out between 13 April and 28 June 2004 (one month after the events). The Ethics Committee of the Faculty of Psychology of the Complutense University of Madrid approved the study, and oral informed consent was obtained from the study subjects.

The scope of this research was to assess the prevalence of psychopathology, its evolution, and the establishment of vulnerability and protection or resistance predictors.

The instrument used is a Spanish adaptation and modification of the WTCD (WTC disaster) telephone structured interview used by Galea et al. (2002) in the study of the psychological sequelae of the September 11 terrorist attacks. The WTCD interview was modified and adapted to the type of terrorist attacks (train bombings) that occurred on March 11 in Madrid. The current instrument, PE-11M, included some variables not previously considered, and others were adapted for the Madrid city population. The final Spanish structured interview contains more than 300 variables in its 31 pages, including the following modules: demographic data, experiences related to the attack, addictive behaviors, health, interests and concerns, social support, panic attack, professional counseling and aid, life-span stressful events, consequences, depression, PTSD in children and adolescents, children's experiences and events, emotionality, and need for and propensity to seek psychological help. The protocol time of administration varied from 15 to 35 minutes.

It is important to emphasize that this instrument was designed with the aim not only of gathering data for the investigation but also of assessing the needs of the individual being interviewed. In this sense, the last module of the interview evaluated the actual state of the respondent in case he/she needed assistance or support.

Thirty phone lines were installed in the Faculty of Psychology of the Complutense University of Madrid in order to gather the necessary data for the different studies. During the first phase of the study, 80 volunteers were trained to be interviewers; they collected data from 10:00 a.m. to 9:00 p.m., on a non-stop basis from Monday to Saturday for 11 weeks. The great majority of interviewers were psychologists who had studied or were in the process of studying for the *Master en Intervención en la Ansiedad y el Estrés* (Master of Anxiety and Stress Treatment), PhD students and psychologists from the training courses on urgency and emergency assistance from the Madrid Psychological Association.

Psychological Aftermath in the General Population

The sampling frame consisted of adults living in Madrid in households with a telephone. Specifically, the general population ($n = 1265$; 18 years plus), was randomly selected and we strictly maintained the demographic and gender (1:1.1 male/female) distribution of the population of Madrid City along its 21 districts (see Table 7.1), estimated as 2,454,682 adults (≥ 18 years), following the Madrid City 2001 Census. Ages ranged from 18 to 96, with an average age of 47. In addition, an oversample of 324 people was drawn from the three sites most affected by the explosions (Atocha, El Pozo and Santa Eugenia) (see Figure 7.1). Overall, 4412 persons were contacted, and 1589 interviews were completed for an effective survey response rate of 38.7 percent.

The percentage of people living in the affected sites who personally witnessed some relevant aspect of the attacks was 29.7 percent whereas in the 21 city districts of Madrid it was 5.6 percent. Of the latter, 8.8 percent knew somebody who was injured, 7.5 percent knew somebody who had died and

Table 7.1 *Comparison of the studied sample and the demographic data of Madrid's inhabitants (Madrid Census, 2001)*

	Population (%) (≥ 18 years old) Madrid Census 2001 ($n = 2{,}494{,}709$)	Sample (%) (≥ 18 years old) weighted percent ($n = 1589$)	Chi-square	*p*-value
Sex			0.001	0.975
Male	45.9	47.1		
Female	54.1	52.1		
Age			0.036	0.998
18–29	21.7	26.3		
30–44	28.7	22.5		
45–59	20.8	25.5		
≥ 60	28.8	25.6		
Residence by districts			0.763	1.000
1	4.5	4.0		
2	4.5	4.4		
3	4.2	3.5		
4	4.9	4.4		
5	4.5	4.9		
6	4.8	4.8		
7	5.0	5.4		
8	6.8	6.8		
9	3.7	3.6		
10	8.7	8.3		
11	7.5	6.2		
12	4.0	3.9		
13	7.4	7.6		
14	3.6	4.2		
15	7.5	7.0		
16	4.8	6.0		
17	4.2	4.0		
18	1.9	1.9		
19	1.7	2.8		
20	4.6	5.1		
21	1.2	1.3		

Source: Miguel-Tobal et al. (2006a).

3 percent knew somebody who had died and somebody who had been injured. In the case of the former, these percentages rise to 17.3, 16 and 12.3 percent, respectively. The facts prove that the proximity of the residence to the terrorist attack increases the impact considerably.

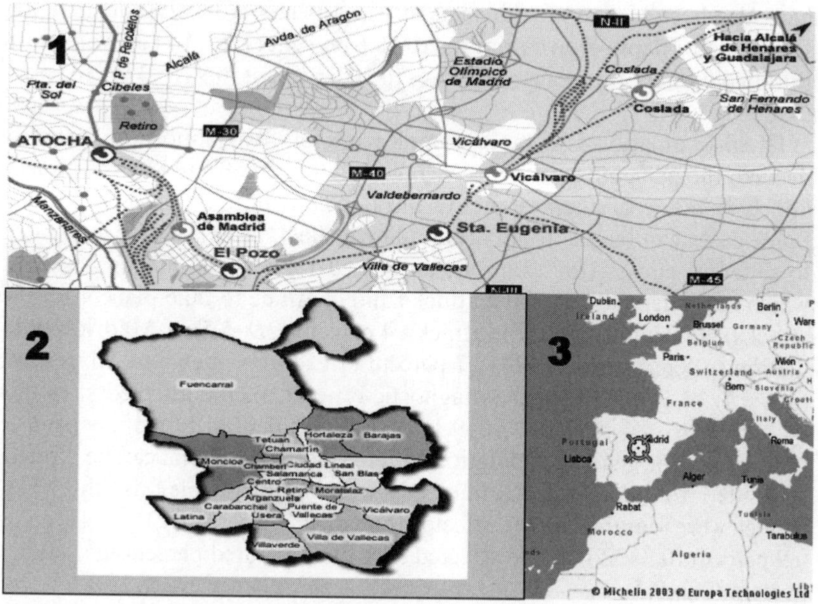

Figure 7.1 *Map of Madrid City including: (1) site of the train bombing explosions, (2) the 21 districts and (3) general overview*

The amount of time that passes before people know the fate of their loved ones after they receive the news of the attacks is of special importance. This period is a time of high levels of negative emotions and acts as a breeding ground for psychopathological reactions such as panic attacks. In addition to this, telephone lines were dead (a common feature during catastrophes of this kind), increasing levels of frustration and anticipation of negative consequences among the population. After becoming aware of the attacks, the average time that elapsed before Madrid's residents knew that their families were safe was 3.5 hours; in the case of friends and acquaintances it increased to 5.8 hours.

Some 10.9 percent (95 percent confidence interval (CI) = 8.8–13.3) of the general population experienced symptoms congruent with the diagnostic criteria of panic attacks (about 270,000 people). Women were more than twice as likely to be affected as men: 14.6 percent (11.4–18.5) versus 6.8 percent (4.7–9.7). In the case of the most directly affected sites, the prevalence of panic attacks was 17.1 percent (12.4–23.1).

In terms of depression prevalence after March 11, 8 percent (6.3–10.1) of Madrid's population has shown symptoms consistent with the diagnostic criteria of major depression (about 200,000 people). In this case, women have also been twice as troubled as men: 5.1 percent (3.4–7.6) of men presented major depression compared to 10.6 percent (7.9–14.1) of women. In the affected sites, this percentage rises to 10.4 percent (6.9–15.3) of the population.

PTSD evaluation is a more complex matter. Two important facts are needed: the percentage of people who were directly affected by the attacks and how many of them attribute the PTSD symptoms to such attacks. Some cases can be due to an accumulative effect, others can be attributed to different causes (for example, an armed robbery or serious accident, which are independent of the attacks and continue to take place after the attacks). One month after the attacks, 4 percent (2.9–5.5) of Madrid's population suffered from PTSD; 2.7 percent (1.6–4.8) of men and 5.1 percent (3.4–7.4) of women met the diagnostic criteria. In the affected sites this amounted to 5.4 percent (3.2–9.1) of the population. Some 2.3 percent (1.6–3.4) of the general population suffered from PTSD that can be directly attributed to the terrorist attacks (about 57,000 people). In this case, women were four times more affected than men, 3.6 percent (2.3–5.4) versus 0.9 percent (0.3–2.5). In the affected sites the estimated percentage was 4.3 percent (2.2–8.3).

According to the results, after March 11, 9.5 percent (7.7–11.6) of Madrid's population suffered from depression or PTSD; 8.4 percent (6.7–10.5) can be directly attributed to the terrorist attacks. Only 2 percent (1.2–3.2) of the population suffered from both disorders concurrently; 1.4 percent (0.8–2.3) is directly attributed to the attacks. Table 7.2 presents a summary of these results.

The fact that we used comparable methodology and instruments between Galea et al.'s (2002) September 11 research and our March 11 research provided us with a unique opportunity to compare the prevalence of PTSD and depression in two different populations after different terrorist attacks, while minimizing possible differences caused by the study design.

Comparing data from both cities, the NYC attack's psychopathological indicators present higher prevalences than the Madrid attacks but with similar tendencies (see table 7.2). Current PTSD related to the terrorist attacks was 2.3 percent (1.6–3.4) in Madrid City and almost three and a half times more in NYC: 7.5 percent (5.7–9.3). Current PTSD not related to the attacks was 4 percent (2.9–5.5) in Madrid and this was duplicated in NYC: 8.9 percent (7.0–10.8). Madrid City lifetime prevalence of PTSD is lower than that in NYC: 12.2 percent (10.3–14.5) and 19.3 percent (16.6–22.0), respectively.

Regarding depression, although the prevalence values found in NYC are higher than those in Madrid, differences are less marked than in the case of

Table 7.2 Comparison of prevalence of PTSD and depression between March 11, 2004, in Madrid and September 11, 2001, in Manhattan, NYC

Disorders	Madrid City % (95% CI)	New York City % (95% CI)
PTSD		
Ever	12.2 (10.3–14.5)	19.3 (16.6–22.0)
Since events	4.0 (2.9–5.5)	8.9 (7.0–10.8)
Terrorist-attacks related	2.3 (1.6–3.4)	7.5 (5.7–9.3)
Depression		
Ever	26.8 (24.1–29.6)	27.0 (23.9–30.1)
Since events	8.0 (6.3–10.1)	9.7 (7.6–11.8)
PTSD and Depression		
Since events	2.0 (1.2–3.2)	4.4 (3.0–5.8)
Terrorist-attacks related	1.4 (0.8–2.3)	3.7 (2.4–5.0)
PTSD or Depression		
Since events	9.5 (7.7–11.6)	14.3 (11.9–16.7)
Terrorist-attacks related	8.4 (6.7–10.5)	13.6 (11.2–16.0)

Source: Miguel-Tobal et al. (2006b).

PTSD. The prevalence of current depression not related to the attacks was 8.0 percent (6.3–10.1) in Madrid and 9.7 percent (7.6–11.8) in NYC. Lifetime prevalence of depression in Madrid and NYC is almost the same: 26.8 percent (24.1–29.6) and 27.0 percent (23.9–30.1), respectively.

Prevalence values of comorbidity show a similar pattern: Madrid prevalence is lower than that found in NYC, both for PTSD and depression and PTSD or depression categories (Table 7.2).

One of the central issues in studying the impact of exposure to a traumatic event is to explore whether this event, that is, a terrorist attack, is able to predict who will develop PTSD, depression or another psychopathological outcome after the exposure, and who will not.

Taking into account the complexity of the trauma situation in terms of the possible number of variables that might play a role in how an individual copes with trauma stressors, there are a number of variables that have been studied more extensively. In a comprehensive meta-analysis study, Brewin et al. (2000) examined 14 variables that may contribute to trauma vulnerability: sex, age, socioeconomic status, education, intelligence, race, psychiatric history, childhood abuse, other previous trauma, other adverse childhood factors, family history of psychiatric disorder, trauma severity, social support, and life stress. In another meta-analytic study carried out three years later, Ozer et al. (2003) focused mainly on seven variables more proximal to the event and more subjective than objective: prior trauma, prior adjustment, family history of psychopathology, perceived life threat, perceived support, peri-traumatic emotions, and peri-traumatic dissociation.

The main results of these comprehensive meta-analysis studies suggest that psychological factors are more important than demographic or intellectual variables in predicting who develops PTSD, and because of the complexity and diversity of the variables addressed, it is difficult to establish a causal model for developing PTSD (Weiss and Ozer, 2006).

Returning to March 11 data, we carried out multivariate analyses based on general population data suggesting that variables such as sex, age, social support, number of stressors before the attacks, presence of panic attacks during the traumatic events, level of exposure and the fear of being wounded or killed have shown a strong predictive capacity (Miguel-Tobal et al., 2004, 2006b).

Specifically, significant predictors of PTSD were the number of life stressors suffered by the individual 12 months before March 11, 2004 as compared to no significant life stressors (one or two life stressors increases the probability of suffering from PTSD by 2.4, three stressors multiplies the probability by 7.2); the presence of symptoms of panic attack during or soon after the event multiplies the probability for developing PTSD by 8.2 as compared with the absence of such symptoms; perceived low social support is another significant predictor of PTSD: 2.2 times more than medium or high social support; women were 2.6 times more likely to suffer from PTSD than males; people who directly witnessed the events were 2.9 times more likely to develop PTSD than those who did not; and finally, the elderly (more than 60 years old) were the most resistant to the trauma impact during the March 11 attacks (see Miguel-Tobal et al., 2006b).

We found that significant predictors of depression were quite similar to the case of PTSD, except that people who were afraid of being wounded or killed during the attacks were 2.1 times more likely to develop major depression than people who had no such fears. Other significant predictors were the number of life stressors 12 months before the event (odds ratio (OR), 7.1 for three or more stressors), having symptoms of panic attack during or soon after March 11 (OR, 3.4), perceived low social support (OR, 1.8), and females as compared to males (OR, 1.7) (see Miguel-Tobal et al., 2006b).

Looking at September 11 multivariate data (Galea et al., 2002), we found some common significant predictors of PTSD such as low social support, number of life stressors a year before the events, having a panic attack during or soon after the events, and personally witnessing the events; similarly, common significant predictors of depression are the number of life stressors before the events and low social support. Although both events were the largest attacks of their kind in Spain and the US, respectively, the September 11 attacks resulted in some 3000 fatalities, compared to the 191 fatalities in Madrid. In addition, the World Trade Center attacks and its collapse were more highly visible to the population and also destroyed surrounding buildings and stores. This serves as a reminder that, in addition to common significant predictors of PTSD and depression, other predictors were found for September 11 but not for March 11: loss of possessions due to the attacks initiated PTSD symptoms; the death of a friend or relative in the attacks and loss of job because of the attacks initiated depression symptoms. Finally, ethnicity was a significant predictor both of PTSD and depression during the September 11 attacks but this was not addressed in our research on March 11 mass trauma.

Another exception concerns the sex and age variables. Sex was clearly a significant predictor both for PTSD and depression, while age was significant for PTSD only with regard to the March 11 data. Neither was a significant predictor in Galea et al.'s (2002) research. Our data coincide basically with the assumption from the literature that females are at greater risk of developing PTSD and other internalizing symptoms following traumatic events (Craske, 2003).

Taking both 'attacks' as a whole, data from multivariate analyses highlight the importance of psychological and emotional factors in the risk of developing psychopathological disorders. First, the role of recent life stressors suggests that people who experience them are particularly vulnerable to the consequences of terrorism (Miguel-Tobal et al., 2006b). Second, peri-traumatic symptomatology, such as panic attacks, seems to play an important role as another vulnerability factor in the development of subsequent psychopathology after the experience of mass trauma (Freedman et al., 1999; Ozer et al., 2003). Finally, perceived social support is important in the way that people cope with trauma (see Kaniasty and Norris, 2004; Kaniasty, 2005).

Psychological Aftermath in Victims and Relatives

The sample of victims and relatives consisted of 117 people, of at least 18 years of age, who had been direct victims and/or whose families or close friends had died and/or been injured during the attacks. Those included in this sample are not representative but incidental; they were contacted during our clinical intervention and kindly answered the survey. The average age of this sample was 39.8 years and comprised 59.5 percent women and 40.5 percent men.

Table 7.3 Summarized data of March 11, 2004, panic attack, PTSD, and depression prevalence in the general population, victims and relatives, and emergency personnel samples

Disorders (since events)	General population ($n = 1{,}589$)	Victims and relatives ($n = 117$)	Emergency personnel ($n = 165$)
Panic attack (%)	10.9	45.3	13.9
PTSD (%)	4.0	35.9	1.2
Depression (%)	8.0	31.3	2.0

Some 66.1 percent of the sample had lost at least one family member in the attacks and 87.9 percent had a friend or acquaintance who had died; 7.7 percent of those polled had been injured to some extent.

Due to the intensity of the impact, this group has the highest psychopathological prevalence. When receiving the news or a short time afterwards, 45.3 percent had a panic attack. Women were almost three times more affected than men: 73.6 and 26.4 percent, respectively. Some 31.3 percent of this group suffered from depression; women, too, were more vulnerable than men: 77.1 versus 22.9 percent, respectively.

As far as PTSD is concerned, the statistics of this group are also considerably higher than the general population's (see Table 7.3). One month after the attacks, 35.9 percent of this group met the diagnostic criteria. In this case, women were four times more troubled than men: 81 and 19 percent, respectively. The fact that more than three out of four women in this group have developed PTSD is especially significant. This outcome is in accordance with other studies in the sense that direct victims reported higher PTSD prevalence than the general population after disasters (see Galea et al., 2005 for a review).

The comorbidity between PTSD and depression is also much higher than that of the general population: 18.8 percent of this group suffered from both disorders concurrently.

Psychological Aftermath in Emergency Personnel

This group of 165 people who actively participated in emergency rescue and assistance comprised 30.1 percent men and 69.9 percent women, with an average age of 34.9 years. This sample was made up of many different professionals, such as psychologists, police officers, social workers, pedagogues, educators and others. Again, this group must be addressed as an incidental sample but is not representative of all the professionals and rescue workers who had been involved during the first-aid interventions after the March 11 attacks in Madrid.

Some 20.6 percent of the sample was involved in rescue operations inside the trains. Some of them, together with the rest of the sample, offered direct assistance to the survivors, families and friends in places like IFEMA,[1] hospitals, mortuaries, and so on.

Some 13.9 percent of people in this group suffered from panic attacks, 3 percentage points higher than the general population (see Table 7.3). However, only 2 percent of the people suffered from depression, much less than that found in the general population. Similarly, only 1.2 percent suffered from PTSD. Even though the majority of the emergency personnel were trained and had experience in this type of operation, they were unable to avoid the adverse peri-traumatic psychological reactions due to the high level of exposure to the stressor. In spite of this, they were in fact capable of handling their emotions in order to prevent chronic reactions, giving rise to disorders such as depression or PTSD.

Although the general population, victims and relatives, and emergency personnel samples are not truly comparable, since victims and emergency personnel are not strictly representative samples, an overview of the general tendencies shows us the differences found regarding psychopathological aftermath due to the trauma impact (see Table 7.3). The victims' sample shows a much higher prevalence of panic attack, PTSD and depression than either the general population or the emergency personnel due to the severe consequences of the attacks, and it is similar to systematic research published in the scientific literature. However, we found divergent results regarding data on the emergency personnel sample. Emergency personnel showed a lower incidence of PTSD and depression than the general population, whereas the literature indicates the contrary. These results are likely to be due to the fact that this incidental sample of emergency personnel can return to their normal lives rapidly after their first-aid interventions, and they did not suffer personal or physical loss that might significantly affect their subsequent daily life activities.

We have drawn an overview of the main results of the psychopathological outcomes caused by the March 11, 2004, terrorist attacks in Madrid. From peri- to post-traumatic aftermath, from general population to victims and relatives and emergency personnel, the terrorist legacy has produced a considerable shock in the lives of many people. Although the initial trauma impact can be managed, pain, grief and bereavement remain in the population, and much more so in the survivors, for a long period. Nothing will be the same for Madrid, and adapting John M. Oldham's (2003) quotation on New York and September 11, 2001, 'the toll taken by this massive disaster, produced by malevolent human intent, has been enormous. But in many ways, the current New York [Madrid] is tougher and stronger than ever' (Oldham, 2003, p. 30).

NOTE

1. The biggest trade fair institution in Madrid, which was the place where corpses were taken for family identification after terrorist attacks.

REFERENCES

American Psychiatric Association (1994), *Diagnostic and Statistical Manual of Mental Disorders*, 4th edn, Washington, DC: American Psychiatric Press.

Bongar, B. (2007), 'The psychology of terrorism. Defining the need and describing the goals', in B. Bongar, L.M. Brown, L.E. Beutler, J.N. Breckenridge and P.G. Zimbardo (eds), *Psychology of Terrorism*, Oxford and New York: Oxford University Press, pp. 3–12.

Boscarino, J.A., Galea, S., Ahern, J., Resnick, H. and Vlahov, D. (2002), 'Access to psychiatric services in New York City following the September 11th terrorist attacks', *Annals of Epidemiology*, **12** (7), 514.

Brewin, C.R., Andrews, B. and Valentine, J.D. (2000), 'Meta-analysis of risk factors for posttraumatic stress disorder in trauma-exposed adults', *Journal of Consulting and Clinical Psychology*, **68**, 748–66.

Bryant, R.A. (2004), 'Assessing acute stress disorder', in J.P. Wilson and T.M. Keane (eds), *Assessing Psychological Trauma and PTSD*, 2nd edn, New York: Guilford Press, pp. 45–60.

Craske, M.G. (2003), *Origins of Phobias and Anxiety Disorders: Why More Women than Men?*, Amsterdam: Elsevier.

Freedman, S.A., Brandes, D., Peri, T. and Shalev, A. (1999), 'Predictors of chronic post-traumatic stress disorder: a prospective study', *British Journal of Psychiatry*, **174**, 353–9.

Fullerton, C.S., Ursano, R.J., Norwood, A.E. and Holloway, H.H. (2003), 'Trauma, terrorism, and disaster', in R.J. Ursano, C.S. Fullerton and A.E. Norwood (eds), *Terrorism and Disaster: Individual and Community Mental Health Interventions*, Cambridge and New York: Cambridge University Press, pp. 1–20.

Galea, S., Ahern, J., Resnick, H., Kilpatrick, D., Bucuvalas, M., Gold, J. and Vlahov, D. (2002), 'Psychological sequelae of the September 11 terrorist attacks in New York City', *New England Journal of Medicine*, **346**, 982–7.

Galea, S., Nandi, A. and Vlahov, D. (2005), 'The epidemiology of post-traumatic stress disorder after disasters', *Epidemiologic Reviews*, **27**, 78–91.

Galea, S., Vlahov, D., Resnick, H., Ahern, J., Susser, E., Gold, J., Bucuvalas, M. and Kilpatrick, D. (2003), 'Trends of probable post-traumatic stress disorder in New York City after the September 11th terrorist attacks', *American Journal of Epidemiology*, **158**, 514–24.

Kaniasty, K. (2005), 'Social support and traumatic stress', *PTSD Research Quarterly*, **16**, 1–3.

Kaniasty, K. and Norris, F.H. (2004), 'Social support in the aftermath of disasters, catastrophes, and acts of terrorism: altruistic, overwhelmed, uncertain, antagonistic, and patriotic communities', in R.J. Ursano, A.E. Norwood and C.S. Fullerton (eds), *Bioterrorism: Psychological and Public Health Interventions*, Cambridge and New York: Cambridge University Press, pp. 200–229.

Kleinfield, N.R. and Connelly, M. (2003), '9/11 still strains New York's psyche, poll finds', *New York Times*, September 8, http://www.nytimes.com/2003/09/08/nyregion/08POLL.html, September 8, 2003.

McDermott, R. and Zimbardo, P.G. (2007), 'The psychological consequences of terrorist alerts', in B. Bongar, L.M. Brown, L.E. Beutler, J.N. Breckenridge and P.G. Zimbardo (eds), *Psychology of Terrorism*, Oxford and New York: Oxford University Press, pp. 357–70.

Miguel-Tobal, J.J., Cano-Vindel, A., González-Ordi, H. and Iruarrizaga, I. (2006a), 'Consequences of terror attacks in Madrid alter 3/11', in P. Buchwald (ed.), *Stress and Anxiety. Application to Health, Work Place, Community, and Education*, Newcastle, UK: Cambridge Scholars Press, pp. 251–62.

Miguel-Tobal, J.J., Cano-Vindel, A., González-Ordi, H., Iruarrizaga, I., Rudenstine, S., Valhov, D. and Galea, S. (2006b), 'PTSD and depression after the Madrid March 11 train bombings', *Journal of Traumatic Stress*, **19**, 69–80.

Miguel-Tobal, J.J., Cano-Vindel, A., Iruarrizaga, I., González-Ordi, H. and Galea, S. (2004), 'Consecuencias psicológicas de los atentados terroristas del 11-M en Madrid. Planteamiento general de los estudios y resultados en la población general', *Ansiedad y Estrés*, **10**, 163–79.

Muñoz, M., Crespo, M., Pérez-Santos, E. and Vázquez, J.J. (2004a), 'We were all wounded on March 11 in Madrid: immediate psychological affects and interventions', *European Psychologist*, **9**, 278–81.

Muñoz, M., Crespo, M., Pérez-Santos, E. and Vázquez, J.J. (2004b), 'Presencia de síntomas de estrés agudo en la población general de Madrid en la segunda semana tras el atentado terrorista del 11 de Marzo de 2004', *Ansiedad y Estrés*, **10**, 147–61.

Norris, F.H., Friedman, M.J., Watson, P.J., Byrne, C., Diaz, E. and Kaniasty, K. (2002), '60,000 disaster victims speak: Part 1. An empirical review of the empirical literature, 1981–2001', *Psychiatry*, **65**, 207–39.

Oldham, J.M. (2003), 'September 11, 2001 and its aftermath, in New York City', in R.J. Ursano, C.S. Fullerton and A.E. Norwood (eds), *Terrorism and Disaster. Individual and Community Mental Health Interventions*, Cambridge and New York: Cambridge University Press, pp. 23–30.

Ozer, E.J., Best, S.R., Lipsey, T.L. and Weiss, D.S. (2003), 'Predictors of posttraumatic stress disorder symptoms in adults: a meta-analysis', *Psychological Bulletin*, **129**, 52–73.

Piotrkowski, C.S. and Brannen, S.J. (2002), 'Exposure, threat appraisal, and lost confidence as predictors of PTSD symptoms following September 11, 2001', *American Journal of Orthopsychiatry*, **72**, 476–85.

Schlenger, W., Cadell, J., Ebert, L., Jordan, B., Rourke, K., Wilson, D., Thalji, L., Dennis, J., Fairbank, J. and Kulka, R. (2002), 'Psychological reactions to terrorist attacks: findings from the National Study of Americans' reactions to September 11', *Journal of the American Medical Association*, **288**, 581–8.

Schuster, M.A., Stein, B.D., Jaycox, L.H., Collins, R.L., Marshall, G.N., Elliott, M.N. et al. (2001), 'A national study of stress reactions after the September 11, 2001 terrorist attacks', *New England Journal of Medicine*, **345**, 1507–12.

Silver, R.C., Holman, E.A., McIntosh, D.N., Poulin, M. and Gil-Rivas, V. (2002), 'Nationwide longitudinal study of psychological responses to September 11', *Journal of the American Medical Association*, **288**, 1235–44.
Weiss, D.S. and Ozer, E.J. (2006), 'Predicting who will develop posttraumatic stress disorder', in G. Young, A.W. Kane and K. Nicholson (eds), *Psychological Knowledge in Court: PTSD, Pain, and TBI*, New York: Springer, pp. 85–96.

PART III

Building individual and organizational resilience

8. The war against psychological warfare

Michelle Slone and Anat Shoshani

INTRODUCTION

19 October, 1994; 8:45 a.m.
It was a regular morning in downtown Tel Aviv, people commuting to work, stores opening, sidewalk cafes filling up for breakfast. Suddenly, an ear-splitting blast seared through the air, resounding for kilometers. A suicide bomber from the Hamas movement exploded on bus number 5 near Dizengoff Circle in downtown Tel Aviv, Israel, killing 22 and injuring 46 people. The two main television channels rushed reporters to the site and, in an impressive journalistic campaign, began broadcasting live coverage almost immediately. As the events unfolded, the cameras of the two channels competed for the most dramatic broadcast. Television screens all over the country were flooded with unedited real-time images of the charred skeleton of the bus, bodies of the victims and evacuation of the injured, even prior to official release of the identities of the dead and wounded and notification to the families. In one case, a mother identified the body of her daughter on television, constituting her first encounter with her personal tragedy.

This disturbing case description exemplifies the complex ethical issues surrounding both the rights and functions of the media and protection of the public in this current era of terrorism. Although the case description relates to the intricate relationship that has developed between the media and the public in Israel over recent years, no less important are the lessons to be learned from this particular local relationship for international concerns with media coverage of terrorism. With the recent wave of terrorism against the background of the current Israeli–Palestinian conflict, the media in Israel have gradually developed exceptional expertise in covering terrorist events with immediate distribution of reporters to the site and real-time live coverage of events as they develop. In a small country like Israel, characterized by its strong community ethos, every attack is almost immediately communicated to the public, who follow the events closely. However, this immediacy ushers in a multitude of predicaments for media agencies, including issues of regulation, censorship of distressing material, rights to privacy, and decisions regarding portrayal of sensitive content. The case description of bus number 5 raised a fierce debate about the function and commission of the media that eventually led to the drafting of ethical guidelines for censorship of terrorist-related material. However, in this period of global terrorism and immediate worldwide coverage, the assignment of responsibility solely to the media may be insufficient, prompting questions about the sharing of responsibility for regulation between the media networks and viewers. Further, sharing of responsibility must be translated into practical strategies both for the media and for viewers.

This chapter addresses the interrelated function of the media and the viewing public in dealing with the psychological warfare of terrorism. In order to examine these complex issues, the chapter proceeds as follows. The next section deals with implications of media viewing of terrorism that have emerged in the research literature, followed by a section that presents a case description of a laboratory study of the efficacy of a preparatory intervention in moderating negative emotional and attitudinal effects of media exposure to terrorism. The final section suggests a set of principles and guidelines for media agencies and for viewers for coping with the psychological warfare of terrorism.

IMPLICATIONS OF MEDIA VIEWING OF TERRORISM

Fusion of Geographical and Time Dimensions in Media Coverage of Terrorism

The increase in terrorism worldwide has become an issue of global concern. Whereas in the past conventional warfare was fought along frontlines by the military, modern warfare occurs in the very midst of civilian populations. This change has necessitated deconstruction of traditional concepts of warfare and composition of new perceptions of the enemy and strategies for civilian defense.

One of the major constituents of this new frontier of modern warfare is terrorism that is projected into individuals' lives either directly by victimization, loss, and acquaintance with victims, or vicariously via the media. Thus, the new enemy is not solely the terrorist group but its penetration and extension into the very fabric of society and the individual home.

Foremost among the primary mechanisms by which this extension and magnification occurs is that of psychological warfare, which constitutes the planned use of communications to influence the opinions, emotions, attitudes, and behaviors of target groups usually in order to attain some national objective (Department of Defense, 1987). The nature and intensity of psychological warfare has been facilitated by the rapid developments occurring in media technology. This progress has created a channel for rapid dispersal of scenes of war, terror, and violence as they occur in real time into almost every living room. Modern global networks enable almost immediate delivery of information and, in this way, contribute to increased awareness of and involvement in the conflicts of distant others (Carruthers, 2000).

This fusion of geographical distance between the frontline and the civilian population and of physical distance between actual events and their delivery to the civilian population has generated extensive exposure to violence that is penetrated both overtly and covertly into civilian life (Signorielli and Gerbner, 1995). Within this arena of the psychological warfare of terrorism, the two players compelled to participate are the media and their audience. Media viewers who have not been directly exposed to the terrorist event frequently become 'victims-by-proxy' by bearing psychological effects derived from witnessing personal pain and frightening scenes of carnage and destruction. However, despite increasing acknowledgement of the power of psychological warfare inherent in media coverage of terrorism (Schleifer, 2006), the need to construct coping interventions for the increasing number of indirect psychological victims has been largely overlooked.

Most research has been directed to the negative emotional and attitudinal influences of media exposure to terrorism, generating instructive evidence of pervasive effects. Although valuable in itself, this evidence could be used as a springboard for identification of viewers' needs and application into coping strategies that enhance resilience.

Emotional and Attitudinal Effects of Media Exposure to Terrorism

The corpus of research on the power of the media to exact emotional and attitudinal effects has intensified with the advent of global terrorism. Research findings have demonstrated that media exposure to terrorism produces increased anxiety (Gunter, 1994; Slone, 2000), anger (Newhagen, 1998; Yukawa et al., 2001), pity and fury (Sontag, 2005), and symptoms of nightmares and fear (Joshi and Kaschak, 1998). In addition, studies of the effects of extended media exposure to the events of September 11th revealed associated high levels of post-traumatic symptoms and an assortment of severe negative emotions from acute psychological distress to more ingrained feelings of vulnerability and insecurity (Cho et al., 2003; Aber et al., 2004; Blanchard et al., 2004; Gil-Rivas et al., 2004; Marshall and Galea, 2004).

In addition to emotional effects, media coverage of political violence and terrorism makes a significant contribution to changing attitudes, enemy perceptions and stereotypes (Rivenburgh, 2000). A study of the effects of terrorist attacks in Israel on enemy perception showed escalation of negative stereotypes toward Palestinians (Bar-Tal and Labin, 2001). Media broadcasts of terrorism may act as a catalyst for radicalization of these perceptions. After the September 11 terrorist attacks, presidential address broadcasts included greater stereotypic rhetoric and negative images of Arabs (Merskin, 2004). In addition, qualitative analyses of central television channels after the attacks found a predominance of Islamophobic, evangelical, and ethnocentric worldviews toward Arabs (Domke et al., 2003; Sheridan, 2006).

Beyond disparate emotional and attitudinal consequences of media exposure to terrorism, these two dimensions may co-exist within a mutually influential interrelation. Terrorism, specifically, represents a domain that demonstrates this combined relationship with its creation of anxiety and insecurity and defensive activation of radicalized attitudes. Credible explanation for this relationship may be found in recent conceptualizations of terror management theory (Greenberg et al., 2003). This theory posits that awareness of death induces a universal need to cope with inevitable mortality by defensively engaging in certain social actions and preserving particular sets of beliefs and values. Social and cultural group affiliation enhances security and self-esteem and enables some denial of mortality by supporting a sense of existential meaning. By its very nature, exposure to terrorism, both direct and indirect, is anxiety provoking since it compels confrontation with existential threat. Rehabilitation of a sense of security and intact self-esteem necessitates defensive psychological response in the form of increased advocacy of cultural worldviews, frequently manifesting as depreciation of the party perceived to be threatening and hostile.

The significant changes in the media during this current era of multichannel, immediate and live coverage of terrorism have not been paralleled with concomitant adjustments in viewing conventions on the part of the public. The passive viewer may not be sufficiently aware of the risks and demands encumbered by the new media era, and thus may continue with old habits of attraction to the abundant and dramatic information presented in coverage of terrorism. In contrast, an active viewer would be apprised of these risks and would instigate regulatory psychological mechanisms of information processing, screening and restriction of consumption, enabling greater personal control over indirect exposure and its effects. Active viewing necessitates provision of psychological tools for enhancing resilience by moderating heightened negative emotional and extremist attitudinal responses.

Although there may be a possible claim for the adaptive function of increased emotions, such as anxiety and anger, and the justification for increased negative attitudes toward the enemy, after viewing news broadcasts of terrorist acts, the legitimacy of this claim should be dependent on the intensity and protracted nature of the reaction. In terms of mental health indicators, reactions of anxiety, anger and enemy derogation can exceed the adaptive range, ceasing to be personally or nationally adaptive. Such cases could include post-traumatic stress disorder (PTSD) symptoms (Blanchard et al., 2004) and depression (Knudsen et al., 2005) as documented in some instances of prolonged media exposure to the events of September 11. Another case could be found in conflicted countries where protracted media exposure to terrorism increased the significant emotional load already borne by the society (Slone, 2000).

In such circumstances of protracted threat and heightened distress, extension of therapeutic services to prevent negative consequences or, at the least, to constrain deleterious outcomes in emotional and attitudinal responses, could aid community health and welfare concerns. However, notwithstanding an acknowledged need for construction of community-based programs for dealing with concerns of terrorism and its wide-ranging population effects, research into design and evaluation of clinical programs has been scant. In illustration of potential avenues for empirically based development and evaluation of clinical community-targeted programs, the next section describes a laboratory study that examined a preparatory intervention for moderating common negative emotional and attitudinal responses to media coverage of terrorism.

ENHANCING RESILIENCE: A LABORATORY STUDY

In extreme trauma circumstances, the major therapeutic efforts have been concerned with mitigation of psychological symptoms and prevention of full-blown disorder. Although traditionally therapeutic efforts have been directed almost exclusively toward post-event intervention, the clinical field has recently begun to recognize the potential embodied in preparatory processes antecedent to possible exposure to trauma for prevention of full-blown pathology. This type of preparatory intervention signifies a modification in therapeutic approach from a focus on individual post-exposure treatment to preparation and prevention for high-risk community sectors. The communal nature of threat of terrorism in regions of conflict and the target of healthy at risk rather than clearly defined pathological populations make group format preparatory intervention particularly appropriate.

The efficacy of preparatory intervention is reinforced by studies conducted during the cultural revolution in China (1968), during the Vietnam War (1968) and in Croatia (1991), which showed that in areas that instituted preparatory processes for coping with conflict in the schools, media and community institutions, there were lower levels of distress as reflected in the number of psychiatric and psychological referrals and in self-report measures than in areas in which no preparatory measures were taken (Minkowski, 1993). Comparable effectiveness was found for a preparatory program instituted in Israel for the threat of scud missile attacks during *Desert Storm* (Somer et al., 2005).

Despite increased concern with preparatory intervention, almost no studies have assessed its effectiveness in dealing with terrorism in general, and indirect media exposure to terrorism, in particular. In the field of terrorism, almost all interest has been directed toward secondary intervention after acute incidents or toward supportive therapies (Kaplan et al., 2001). However, results for these post-event therapeutic interventions have not always been persuasive. For example, one of the most popular and widely used post-trauma therapies is debriefing, constituting a technique in which individuals are helped to reproduce the feelings associated with their experienced traumatic event in a controlled environment in order to facilitate its processing, reformulation, and containment (Brewin, 2001). Although some success has been reported with the use of this technique, the approach is controversial due to its danger for re-emergence of latent frightening material that could disrupt natural processes of rehabilitation and healing (Rose and Bisson, 1998). For media exposure, no intervention strategies have been studied and this, together with the disappointing results for debriefing techniques, suggests a significant need for new intervention patterns, including preparatory interventions that can be subjected to empirical study.

The relevance of examining the use of clinical interventions to enhance resilience in the face of indirect exposure to terrorist activity has escalated during this present era of global terrorism and its widespread media portrayal. Thus, although this experimental study of preparatory clinical intervention was conducted in Israel, which has recently been exposed to a protracted period of severe terrorism, the research findings could hold wide significance for public welfare concerns raised by this new type of warfare.

Terrorism in Israel

Terrorism in Israel over the past few years has mainly been related to a protracted conflict with the Palestinians that peaked recently into what has been termed the second Palestinian Intifada. The long and historically conflicted relations between the Israelis and Palestinians deteriorated severely since the 6 Day War in 1967 during which Israel conquered Gaza from Egypt and the West Bank from Jordan. The conquered sections of land were inhabited by Palestinians who, deprived of other citizenship, were forced to live under Israeli occupation. This situation eventually led to the rise of fundamental organizations that promoted armed force in the struggle for independent nationhood (Smith, 1996). Against this background of unrest, two violent uprisings of the Palestinians occurred, the second of which, Al Aqsa Intifada, broke out in October 2000. The eruption of these hostilities was characterized by increasing cycles of violence that disrupted civilian life on both sides and led to an extremely high number of casualties (Wallach, 2000). The Palestinians lost the autonomy they had achieved in peace agreements with re-occupation by the Israeli army, and suffered massive destruction of homes and property, losses, casualties and economic hardship (Fielding, 2004). Concurrently, the Intifada led to a wave of terrorist attacks and suicide bombing within Israel, unprecedented in its intensity, that severely damaged every sense of personal and social security.

Study Design

Participants in the study were 300 young adults aged 21–30 who were exposed to a broadcast either of terrorist attacks or of political news material not related to terrorism. These broadcasts comprised two 7-minute movie clips, produced from items extracted from news material that had been broadcast on local channels. The terrorist clip contained items of terrorist attacks that had occurred in Israel, including suicide bombings on buses, a bombing at a bus stop outside a rural train station, and one of the many attacks occurring in the pedestrian mall in Jerusalem. The non-terrorist clip contained items from political news broadcasts on local channels unrelated to terrorism, including a political debate between two leaders, visits of political leaders to public places, and reporting about contentious governmental coalition issues.

At the beginning of the experiment, participants completed a battery of questionnaires that assessed emotional variables of state anxiety and state anger and the attitudinal variables of attribution of stereotypes to the enemy and enemy perception. The enemy perception scale included the four aspects of trust in the enemy, empathy toward the enemy, willingness to negotiate, and perceptions of enemy hostility. Participants were then randomly assigned to a preparatory intervention or to a control condition and, thereafter, to the terrorist or non-terrorist media exposure. Immediately after the exposure, participants were administered the post-test battery, which included the same pre-test measures of state anxiety, state anger, negative enemy perception and stereotype attribution.

The preparatory intervention session was based on the rational restructuring approach developed by Ellis (1962). This cognitive approach asserts that emotions are mediated by cognitive processes that attribute meaning to emotional arousal. Thus, cognitive interventions are based on identification of cognitions that appraise a stimulus as threatening, on equipping the individual with more adaptive cognitive skills (Meichenbaum, 1985; DeRubeis and Beck, 1988) and on substituting negative automatic cognitions with learned positive ones (Ellis, 1962).

In the preparatory intervention, after general introductions, the facilitator introduced the topic for discussion as media coverage of terrorism in Israel. After opening comments, participants noted on a piece of paper their reactions when they watch television coverage of terrorist attacks. Participants were asked to share their experiences, personal meanings of their feelings, and ways in which these feelings affect their personal lives. The facilitator emphasized similarities and differences in the feelings expressed, all of which were legitimized. For example, many of the participants spoke of their fear of traveling on public transport and taking part in crowded activities that contain a threat of terrorist attacks. Further, some reported a sense of despair about the future and frustration at the constant bombardment of media exposure to negative events. Others spoke about hopelessness of peace negotiations and expressed feelings of anxiety, anger, and lack of control. Thereafter, participants were asked to distinguish between and discuss thoughts that aided or threatened their sense of control and well-being during viewing. Many of the threatening cognitions they described were related to particular motifs running through the coverage of the attacks such as the sight of blood and charred personal belongings of victims, sounds of hysteria of the bystanders, and emotionally charged interviews with the wounded. The facilitator provided regulatory principles for more adaptive consumption of this threatening material. Cognitive principles included the suggestion for selective and critical consumption of the material, screening of critical information from the dramatic background noise, distraction of attention from unbearable elements and self-control over amount of exposure. Emotional principles included the suggestion for multiparty viewing, sharing of emotions and relaxation techniques.

Participants shared with one another their own strategies. Finally, the facilitator summarized the group process from identification and acknowledgment of negative feelings and associated cognitions through to the adoption and specification of personally helpful cognitive and emotional strategies for coping and, thereafter, the group session was terminated.

In the control condition, after introductions, the facilitator distributed a booklet containing a collection of short stories that participants were requested to read. After completion of this task, the group session was terminated by the facilitator. Both conditions consisted of one group session of one hour duration with 5–7 randomly assigned participants in each group.

Study Results

Results of the study showed that exposure to the terrorism clip produced extremely high levels of state anxiety, state anger, stereotypes toward the enemy and negative enemy perception. For the subscales of the enemy perception measure, participants in the terrorism exposure group showed lower levels of trust in the enemy, empathy toward the enemy, and willingness to negotiate with the enemy and higher levels of perception of enemy hostility than participants in the non-terrorism exposure group.

The therapeutic preparatory intervention was highly effective in producing lower levels of post-exposure state anxiety and perception of enemy hostility than the control condition. In alignment, the preparatory intervention produced higher post-exposure levels of trust in the enemy and willingness to negotiate with the enemy. The preparatory intervention did not succeed in moderating state anger, empathy toward the enemy or stereotypic attributions.

It is possible to detect a certain consistency in this pattern of findings. Anger, low empathy, and stereotypes were more resistant to regulation, possibly since the defensive function of these variables in conditions of threat ensures a sense of preservation of supremacy against the enemy (Rogers, 1983). In this sense, anger and its derivatives of low empathy and derogative stereotypes may be adaptive and their relinquishment threatening. In opposition, the preparatory intervention reinforced those variables relating to a desire for conflict resolution, even though anger and negative stereotypes toward the enemy continue to exist. At first glance, this pattern may appear paradoxical since common assumptions presume that conflict resolution necessitates fundamental decreases in hatred toward and derogation of the enemy. Thus, the findings of this case study illustrate ways in which desire for conflict resolution may be enhanced without depending on basic changes in stable characteristics, such as protective anger and firm stereotypes, that are resistant to relinquishment. Further the findings reinforce the need for public health services to institute preparatory processes for high-risk populations.

TOWARD SHARED MEDIA–VIEWER RESPONSIBILITY

Regulation of Exposure to Terrorism in the Media: The Role of Viewers

In an environment of violent conflict that perpetuates a sense of powerlessness and lack of control, mechanisms of active mastery may easily become latent. The present era necessitates applied principles of active, regulated consumption of media coverage of terrorism for the viewing public. The coping skills provided in the reported study do not necessarily demand lengthy professional clinical intervention. Appropriate configuration and promotion of the tested strategies could be used by the media, educational and mental health systems with the potential to effect significant changes in viewing conventions. In this era of civilian-focused violence and conflict, basic media studies could make a contribution by inclusion in the educational system curricula, with a particular focus on viewer–media interactions.

Several principles emerge from the reported study that could contribute to regulation of media exposure to terrorism. Probably primarily, indirect media exposure to terrorism should be recognized at the public level as a significant stressor worthy of attention, intervention, and supportive services, particularly during crisis periods. In line with this, encouragement should be publicized for seeking social and professional support both during and subsequent to media exposure. The public should be alerted to specific elements in media coverage of terrorism that have been demonstrated to evoke severe negative emotional reactions, such as repeated presentation of harsh scenes, personal interviews with victims and families that increase identification, child victims, scenes and sounds of chaos, and presentation of bodies and dismembered bodies.

Given these detrimental effects of media exposure to terrorism, the public should be equipped with regulatory skills including critical consumption, selective attention and replacement of automatic negative thoughts with controlled cognition. Personal control of dosage and avoidance of personally unbearable elements and professional service seeking should be legitimized by public authorities. Options for selective sourcing of material should be encouraged both between the different information modalities, such as television, radio, newspapers and the internet, and between multiple alternatives within each.

Regulation of Exposure to Terrorism in the Media: The Role of Media Agencies

Parallel to direction of consumption guidelines to the public, an onus falls on the media to confine terrorist coverage within predetermined principles. At the very least practical level, the media holds some responsibility to alert viewers to the severity of imminent exposure with signing of sensitive and disturbing material, similar to that incurred in many cases of other news items.

Moreover, current media capabilities allow for varied typologies of presentation of the same material. Objectivity of coverage is untenable with almost unlimited possibilities for photographic capturing of material, editing, sound, and narrative, all of which create a particular prism through which reality is presented. Comparative analysis of coverage style of the same terrorist attack across different channels demonstrates wide differences in images, messages, and perceptions selected for presentation. For example, a comparative analysis of the coverage of the events of September 11th by CNN and by the central Israeli channel showed wide differences in presentation (Ben-Shaul, 2005). Contrary to expectations for a more dramatic presentation from the local channel closest to the events, the CNN coverage was restrained with reserved reporting focusing strongly on the devoted work of the rescue teams and selection of rhetorically proficient interviewees. In contrast, the distant reporting of the main Israeli channel was emotionally laden and included a collection of clips demonstrating the disorganization, shock and hysteria (ibid.).

Despite reflecting the same reality, this comparative research demonstrates the sovereignty of media networks in determining their presentation of material, even in the case of live coverage. This flexibility and freedom necessitates construction of binding ethical guidelines by media networks aimed at securing appropriate perimeters.

Notwithstanding these general principles, it is untenable that intra-media agency guidelines will resolve all the complex ethical dilemmas in the coverage of terrorism. In view of this, personal responsibility must be allocated to discretion of photographers, reporters and editors in the field whose sensitivity to viewer effects is a critical element in regulating material.

Drawing a distinction between censorship of content and control over populist coverage trends could contribute to the predicament of information regulation facing media agencies. Presentation of information can assume almost infinite forms, leading to the need to distinguish between coverage of terrorism as an issue in itself and between dilemmas surrounding problematic motifs and items. For example, research has shown

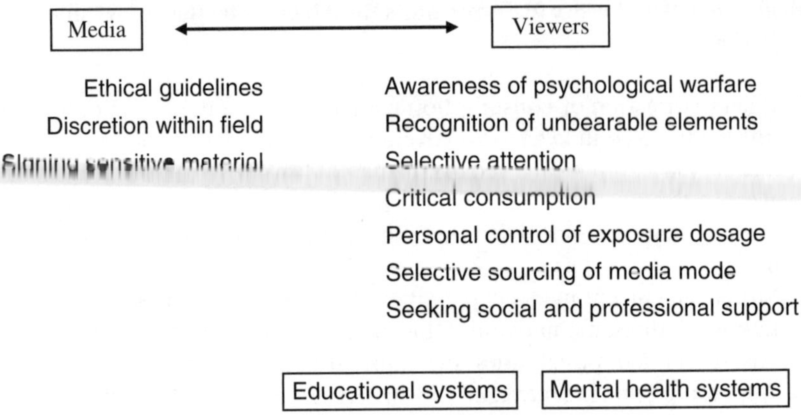

Figure 8.1 Dual media–viewer regulatory principles

that repeated viewing of the single motif of people jumping out of the World Trade Center in the September 11th terrorist attacks accounted for the higher post-traumatic symptoms and depression than viewers unexposed to this element (Ahern et al., 2002). Thus, the media cannot free itself of responsibility for the particular form selected for coverage of terrorist events. In this way, the media companies assume a significant and critical role during violent political conflict.

The above proposals for dual media and viewer regulatory principles can be conceptualized in the model depicted in Figure 8.1.

Mobilizing toward Conflict Resolution

The enormous power of the media in molding public discourse and influencing emotional climate regarding a conflict can be mobilized for the creation of positive social process. Thus, this empowerment could be perceived not as burdensome but as a potent force in this modern era. In the context of violent political conflict, the media can fertilize fear, hopelessness and ethnocentricity or it can create avenues for conflict resolution. Together with this, even the most stringent media regulation cannot mollify the cruel reality of terrorism. As in every battlefield, the public must be equipped, prepared and proactive for its defense.

Decisions surrounding the design of preventive strategies for coping with media exposure to terrorism are complex and multifaceted. Beyond the direct victims of terrorism, multiple resources are required for large at-risk populations. However, despite the multifarious nature of these challenges, regulation of exposure is a vital element of present realities if mental health, values, tolerance, pluralism, and desire for peace are to be enhanced.

REFERENCES

Aber, J.L., Gershoff, E.T., Ware, A. and Kotler, J.A. (2004), 'Estimating the effects of September 11th and other forms of violence on the mental health and social development of New York City's youth: a matter of context', *Applied Developmental Science*, **8**(3), 111–29.

Ahern, J., Galea, S., Resnick, H., Kilpatrick, D., Bucuvalas, M., Gold, J. and Vlahov, D. (2002), 'Television images and psychological symptoms after the September 11 terrorist attacks', *Psychiatry: Interpersonal and Biological Processes*, **65**(4), 289–300.

Bar-Tal, D. and Labin, D. (2001), 'The effect of major event on stereotyping: terrorist attacks in Israel and Israeli adolescents' perceptions of Palestinians, Jordanians and Arabs', *European Journal of Social Psychology*, **31**, 265–80.

Ben-Shaul, N. (2005), 'TV news: visual ideological frames and the coverage of terror events', *Third Text*, **19**(2), 145–55.

Blanchard, E.B., Kuhn, E., Rowell, D.L., Hickling, E.J., Wittrock, D., Rogers, R.L., Johnson, M.R. and Steckler, D.C. (2004), 'Studies of the vicarious traumatization of college students by the September 11th attacks: effects of proximity, exposure and connectedness', *Behaviour Research and Therapy*, **42**(2), 191–205.

Brewin, C.R. (2001), 'Cognitive and emotional reactions to traumatic events: implications for short-term intervention', *Advances in Mind and Body Medicine*, **17**(3), 163–9.

Carruthers, S.L. (2000), *The Media at War: Communication and Conflict in the Twentieth Century*, New York: St. Martin's Press.

Cho, J., Boyle, M.P., Keum, H., Shevy, M.D., McLeod, D.M., Shah, D.V. and Pan, Z. (2003), 'Media, terrorism and emotionality: emotional differences in media content and public reactions to the September 11th terrorist attacks', *Journal of Broadcasting and Electronic Media*, **47**(3), 309–27.

Department of Defense (1987), *Glossary: Military and Associated Terms*, Washington, DC: Joint Chiefs of Staff, Pub 1.

DeRubeis, R.J. and Beck, A.T. (1988), 'Cognitive therapy', in K.S. Dobson (ed.), *Handbook of Cognitive Behavioral Therapies*, New York: Pergamon, pp. 273–306.

Domke, D., Garland, P., Billeaudeaux, A. and Hutcheson, J. (2003), 'Insights into U.S. racial hierarchy: racial profiling, news sources, and September 11', *Journal of Communication*, **53**(4), 606–25.

Ellis, A. (1962), *Reason and Emotion in Psychotherapy*, New York: Lyle Stuart.

Fielding, D. (2004), 'How does violent conflict affect investment location decisions? Evidence from Israel during the Intifada', *Journal of Peace Research*, **41**, 465–84.

Gil-Rivas, V., Holman, E.A. and Silver, R.C. (2004), 'Adolescent vulnerability following the September 11th terrorist attacks: a study of parents and their children', *Applied Developmental Science*, **8**(3), 130–42.

Greenberg, J., Solomon, S. and Pyszczynski, T. (2003), *In the Wake of 9/11: The Psychology of Terror*, Washington, DC: American Psychological Association.

Gunter, B. (1994), 'The question of media violence', in J.B. Bryant and D. Zillman (eds), *Media Effects: Advances in Theory and Research*, Hillsdale, NJ: Lawrence Erlbaum, pp. 17–41.

Joshi, P.T. and Kaschak, D.G. (1998), 'Exposure to violence and trauma: questionnaire for adolescents', *International Review of Psychiatry*, **10**(29), 208–15.

Kaplan, Z., Iancu, I. and Bodner, E. (2001), 'A review of psychological debriefing after extreme stress', *Psychiatric Services*, **52**(6), 824–7.

Knudsen, H.K., Roman, P.M., Johnson, J.A. and Ducharme, L.J. (2005), 'A changed America? The effects of September 11th on depressive symptoms and alcohol consumption', *Journal of Health and Social Behavior*, **46**(3), 260–73.

Marshall, R.D. and Galea, S. (2004), 'Science for the community: assessing mental health after 9/11', *Journal of Clinical Psychiatry*, **65**(11), 37–43.

Meichenbaum, D. (1985), *Stress Inoculation Training*, New York: Pergamon.

Merskin, D. (2004), 'The construction of Arabs as enemies: post-September 11 discourse of George W. Bush', *Mass Communication and Society*, **7**(2), 157–75.

Minkowski, A. (1993), 'Mental stress on children exposed to war and natural catastrophes', *Infant Mental Health Journal*, **14**(4), 273–82.

Newhagen, J.E. (1998), 'TV news images that induce anger, fear and disgust', *Journal of Broadcasting and Electronic Media*, **42**(2), 265–76.

Rivenburgh, N.K. (2000), 'Social identity theory and news portrayals of citizens involved in international affairs', *Media Psychology*, **2**, 303–29.

Rogers, R.W. (1983), 'Cognitive and physiological processes in fear appeals and attitude change: a revised theory of protection motivation', in J.T. Cacioppo and R.E. Petty (eds), *Social Psychophysiology: A Sourcebook*, New York: Guilford, pp. 153–76.

Rose, S. and Bisson, J.I. (1998), 'Brief early psychological interventions following trauma: a systematic review of the literature', *Journal of Traumatic Stress*, **11**, 697–710.

Schleifer, R. (2006), 'Psychological operations: a new variation on an age old art: Hezbollah versus Israel', *Studies in Conflict and Terrorism*, **29**(1), 1–19.

Sheridan, L.P. (2006), 'Islamophobia pre- and post-September 11th, 2001', *Journal of Interpersonal Violence*, **21**(3), 317–36.

Signorielli, N. and Gerbner, G. (1995), 'Violence on television: the cultural indicators project', *Journal of Broadcasting and Electronic Media*, **39**(2), 278–83.

Slone, M. (2000), 'Responses to media coverage of terrorism', *Journal of Conflict Resolution*, **44**(4), 508–22.

Smith, C.D. (1996), *Palestine and the Arab–Israeli Conflict*, New York: St. Martin's Press.

Somer, E., Tamir, E., Maquen, S. and Litz, B.T. (2005), 'Brief cognitive–behavioral phone-based intervention targeting anxiety about the threat of attack: a pilot study', *Behaviour Research and Therapy*, **43**(5), 669–79.

Sontag, S. (2005), *Regarding the Pain of Others*, Ben-Shemen, Israel: Modan.

Wallach, J. (2000), *Not on a Silver Platter. A History of Israel 1900–2000*, Jerusalem: Carta.

Yukawa, S., Endo, K. and Yoshida, F. (2001), 'The effects of media violence on aggression: focus on the role of anger evoked by provocation', *Japanese Journal of Psychology*, **72**(1), 1–9.

9. Integrating organizational and behavioral health principles to promote resilience in extreme events

Brian W. Flynn and C. Frederick Lane

INTRODUCTION

The goal of this chapter is to provide information that will lead to the effective integration of organizational and behavioral health knowledge and principles in the service of protecting, promoting, and sustaining health of individuals and organizations facing extraordinary circumstances. In this case, those circumstances involve a wide variety of emergencies and disasters. These may be naturally occurring (for example, floods, hurricanes) or technological (both intentional and unintentional).

To reach that goal, the chapter begins by exploring key factors required to sustain and enhance organizations both in normal and in extraordinary circumstances. Exploration of behavioral health factors in the context of organizations follows. Once a basis for understanding and applying these principles and factors has been established, the chapter then continues with suggestions intended to enhance readiness, response, and recovery.

This chapter attempts to unify organizational and behavioral health factors, elements, and considerations. In so doing, it attempts to provide practical information and suggestions as well as model this important integrated perspective.

WHY IS THIS IMPORTANT?

I've just survived a tornado. The only thing I care about is keeping my family together and getting on with our lives. What's my job is the last thing on my mind.

If a disaster wipes out my plant, my job is to get the plant up and going as quick as I can. If employees have emotional problems I would send

> them to the EAP [Employee Assistance Program]. That's what it is for. If I don't get the plant back on line, nobody has jobs. Then they've got real problems.

Both of the above statements reflect truth. Yet both fail to recognize or reflect the complex relationships of both individual and organization realities and needs. Both also fail to reflect that, for most people, work and personal lives are intertwined and that nothing happens in organizations without people.

People face stress and adapt to that stress continuously. The same can be said of organizations (this will be expanded upon in later sections). Both individuals and organizations respond and react based upon a complex set of factors that include: history/experience, core beliefs and values, skills/tools, resources, and adaptability/flexibility

In most cases, both individuals and organizations manage to survive routine stresses and strains, some more efficiently and effectively than others. Some just manage to muddle through. But what happens when faced with extraordinary stressors such as a flood that destroys homes and workplaces? Are people and the places where they work prepared? Do each share the common goal of protecting, sustaining, and nurturing the other? If both share that goal, do both have the mechanisms to make it happen?

Much has been written in the expanding field of disaster behavioral health. Several important publications include Ursano et al. (1994, 2003); Danieli et al. (2005); Halpern and Tramontin (2007); Shultz et al. (2007); Ursano et al. (2007). There is increasing understanding and acceptance of the potential adverse physical, behavioral, cognitive, emotional, and social consequence of exposure to disasters (Norris, 2002; Norris et al., 2002; IOM, 2003; Shultz et al., 2007). At the same time, there is a growing understanding and acceptance of the presence and nature of individual and community resilience (Shultz et al., 2007).

There is also an established, if recent, field of specialization in developing and assuring business continuity. Much of the focus of this expanding field has been on establishing business continuity plans (BCPs) or continuity of operations plans (COOPs). Typically, the content of these plans is heavily weighted toward protecting and preserving data while human factors remain a secondary focus if addressed at all. Most disturbing, less than 50 percent of US corporations have any disaster plans at all in place (Mankin and Perry, 2004).

Some attention has been paid to the integration and alignment of human factor and organizational issues in disasters. Flynn et al. (2005, p. 187) commented:

> [We] begin with the assumption that any community's recovery from mass disaster, or any critical incident, is not complete until economic recovery has been achieved. We further propose that economic recovery is predicated upon psychological resistance, resilience, and recovery of the workforce itself. This notion we refer to as 'human continuity'.

They further define human continuity as 'the promotion and practice of psychological infrastructure in the workplace' (p. 187).

There is no question that disasters adversely impact on organizations in serious ways. As many as 40 percent of businesses that are directly affected by disasters are out of business within five years (Owens, 2003). Virgo (2001) reports weakened consumer confidence, rising unemployment levels, fears of deflation, and low world economic growth following the attacks on 9/11 2001.

Similarly, there is ample evidence that exposure to disasters impacts on the people within organizations in negative ways. For example, in a large study following the 9/11 attacks in New York (CDC, 2002), 75 percent of the respondents had problems attributed to the attack; 27 percent missed work; 21 percent smoked more; 1 percent started smoking; and 3 percent drank more. Stein et al. (2003) following up two months after the 9/11 attacks found that distress disrupted work, with 65 percent reporting accomplishing less.

ORGANIZATIONS AS SYSTEMS

The focus of this chapter is how to protect and sustain organizations, and the people critical to those organizations, in the face of extreme events such as those caused by weather, widespread illnesses or terrorism. It is useful, therefore, to understand *how* organizations work, or better said, how they work best. Thinking of organizations as *systems* recognizes that there are many elements to consider in understanding how organizations perform and what the dynamics of these various components are. Understanding these dynamics of organizations can be useful in managing and dealing with challenges or dangers presented to the organization.

All organizations fundamentally exist to serve a variety of constituents, including clients, customers, employees, stakeholders and shareholders. The organization itself is created to perform tasks in a logical manner, utilizing resources of various types to produce an output or product which individuals or groups perceive to be of value. For the organization to perform logically, alignment across intent, execution and result must be continually maintained. Maintaining this alignment is critical for the organization to sustain and perpetuate itself. Early in the 20th century, one of the largest influences on organizational thinking was the work of Frederick Taylor

(Nadler et al., 1992, p. 112). Taylor's 'scientific' approach, developed to serve a factory environment, was a fairly rigid approach with narrow job descriptions, task repetitiveness and limited freedom for individual variation in the performing of tasks. In the last half of the century, social and industrial scientists began to explore and create a much broader concept for how to make organizations more effective by utilizing better the variability of human performance. That variability, however, creates the classic challenge for the organization in maintaining its desired alignment. Creativity and self-direction run hard against the need for a degree of orderliness and compliance. If the organization is threatened by one of the circumstances cited above, can it be assured that the alignment it requires for fulfilling its mission will withstand the crisis? The secret, therefore, is to understand how organizations create such alignment, not as fixed and rigid processes, but as a shared understanding of a culture which permits performance to occur within more broadly defined parameters established by and grounded in common values.

David Nadler, a leading professor and consultant in the field of leadership and organizational change, and his colleagues have done considerable work towards understanding and predicting organizational performance. They have developed what they refer to as the 'Congruence Model' for understanding organizational performance (Nadler and Nadler, 1997, p. 41). While the model has several elements, one, in particular, is very relevant to our discussion because it may have the largest impact on the behavior of individuals in an organization. Specifically, the impact and role of the 'informal organization', which they define to include:

- the organizational culture, values, beliefs and behavioral norms;
- the informal rules and work practices;
- the pattern of communication and influencers; and
- the actual behavior of leaders, rather than their prescribed roles.

No organization operates in a perfectly logical and orderly way. Mistakes are made, systems fail, and confusion sometimes abounds. Particularly in the private sector, competing companies constantly question how the competitor is winning. Better marketing campaigns, better salespeople and top-notch research resources are often cited as reasons. And, they all may be true. The question is *why* are those functions, in fact, performing better. Looking at the functional areas for answers often masks the problem because it is an examination of *what* you do, rather than *how* you do what you do. Examining how organizations do what they do to maintain that critical alignment to achieve their goals provides a basis for understanding what they must do when their organization is challenged, not by normal market circumstances, but by extreme events.

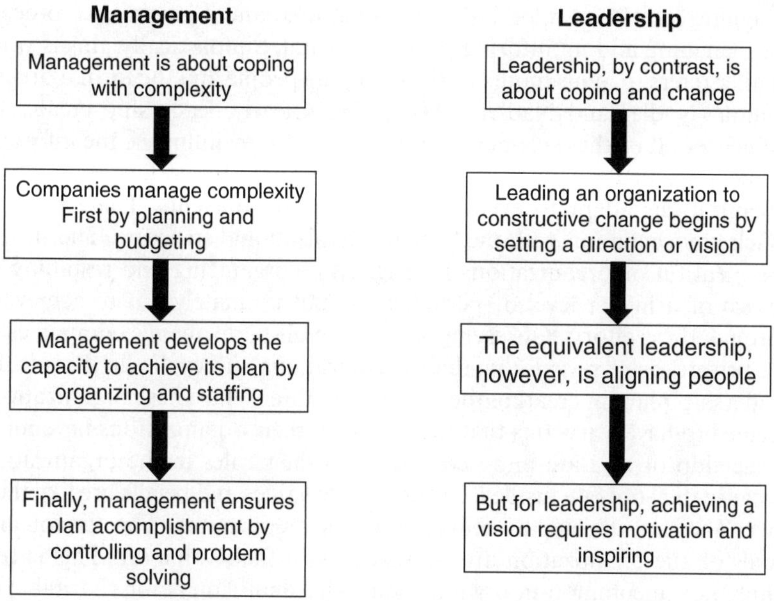

Figure 9.1 The difference between management and leadership

SUSTAINING ORGANIZATIONS

Anyone who has been part of an organization or group that they thought was well managed perceived things as occurring in an orderly fashion. The goals were clear and shared by all; the interaction with teammates or co-workers was mostly cordial and supportive and there was a feeling of clear direction. In other less positive experiences, confusion about the mission, friction, frustration and other such adjectives were probably more commonplace. What is the difference? The authors believe that the difference is clearly the quality of leadership. What and how leaders can make such a difference is better understood if a distinction is drawn between management and leadership as depicted in Figure 9.1. Management is simply all about organizing and directing the necessary tasks of the organization. Leadership is about creating for the organization a culture which enables it to fulfill its mission and, therefore, sustain and perpetuate itself. In this context, leadership is not a hierarchical concept, but can come from anywhere in the organization.

For any organization, a fundamental design or structure must be developed; criteria for selecting talent and continuing to support it is necessary;

assigning individual roles and tasks is required and a governance process for managing and monitoring progress is vital. Simplistically, this is what Nadler refers to in his model as the work, the people and the formal organization (Nadler and Nadler, 1997, p. 34). Clearly, leadership creates or influences all of these elements. However, how they influence the informal element, particularly the culture, is most important for understanding how organizations adapt to events, including extreme events. There is often much preoccupation with the formal organizational structure. The stories are plentiful of organizations redesigning, reorganizing and restaffing in search of a higher level of performance, but ultimately not to achieve it through these efforts. Observing so-called high-performance companies in the private sector reveals that these organizations' success lies in the role the leadership plays in creating the winning culture. What these organizations seemed to have learned is that the people in their organizations have more ownership of and are more connected to their roles in the organization because they operate inside a culture whose values and beliefs are like their own. They are affirmed by being part of the organization. The mission and goals of the organization are consistent with their value set. This is certainly not uncommon in organizations whose mission is for charitable or public service purposes. It is less common in public and private organizations where the intent appears to be of a less different nature.

In their book, *Execution: The Discipline of Getting Things Done*, Larry Bossidy, a senior executive of several global enterprises and Ram Charan, a consultant and Harvard Business School professor, define a company's culture as 'the sum of its shared values, beliefs and norms of behavior' (2002, p. 89). In their view, behaviors are beliefs turned into action. A commonly held set of values and beliefs produces or permits behaviors that are seen to be consistent with the purpose or mission of the organization and through which a context is created for understanding the events that occur within the organization. Bossidy and Charan describe what Nadler calls his informal organization as the 'social operating mechanisms', or social software of an organization 'these are formal or informal meetings, presentations, even memos or e-mail exchanges – anywhere that dialogue takes place' (ibid., p. 99). They describe them as the integrative process where the leader's beliefs and behaviors are spread through the organization.

Since any organization at its point of origin brings certain cultural aspects of the larger culture in which it exists, the relevant question is how leaders cause the organization to form a culture which is a best fit for their needs. So-called start-up companies are often formed by small groups of individuals who already share certain values, interests and capabilities. The early culture often, therefore, reflects these common traits and is formed by their 'doing what they do'. In more established institutions, surveys and studies are often conducted to inform leaders of 'how it really is around here'. Leaders of well-performing organizations pay careful attention to these assessments, realizing how alignment around the desired culture is critical to success. It is not uncommon for companies to structure a portion of the performance awards for leaders on how the employees evaluate the company. How an employee answers the question about how satisfied he/she is with their employment experience is a surrogate for how aligned they feel with the culture they are working within. When corporate mergers and acquisitions occur, the first headline in the news often is how well the two cultures will integrate. The notion of winners and losers within the employee groups is often a question of whose values prevailed. The failure to integrate these cultures into a common one embraced by all will most often cause the merger to fail its original purpose.

Since sustaining and perpetuating itself to fulfill its mission is the primary reason for an organization's existence, it is the role of the leadership to create a culture for the organization which optimizes its capability. Leaders must not only help create the culture, they must also insist on and enforce the implementation of it. How then is this best achieved? An old saying may be appropriate which states that the heart and mind are most captivated by that which the hand helped produce. Engaging the organization in creating, defining and assuring the culture is critical. The key to that is to build an underlying value set which is grounded in the concept of mutually held trust. If the leaders understand and believe that all people need to be valued and want to be valuable, the groundwork is in place for creating such a culture. Traditionally, organizations establish their goals and then begin to distribute them across the organization as team or individual objectives. The assumption is that individuals within the organization will understand and accept accountability for their assigned responsibilities because they trust and believe in their leadership and the organization. All too often, that is not the case. As depicted in Figure 9.2, the organization must first create a basis of trust. When that sense of trust is shared and felt across the organization, accountabilities are more readily accepted, action is undertaken in a more coherent manner and goals are more likely to be achieved. In addition, the organization is able to adapt continually in a positive manner to the ever-changing conditions that present themselves. If that value set is truly 'baked in', the organization's ability to withstand even an extreme event is greater.

The most fundamental asset, or liability, of leaders in their role of building the organizational culture is their personal behavior. Leaders are more watched than listened to, or as sometimes said, it is not what you do, so much as how you do what you do. Behavior trumps words. Observing, and subsequently interpreting, another's behavior is heavily, if not totally,

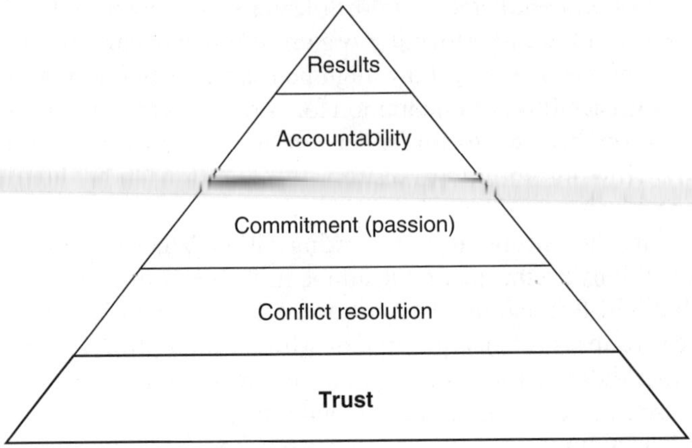

Figure 9.2 Elements in an organization necessary to achieve optimal results

influenced by the context in which the person is observing. In organizations, one of the leader's key roles is as the lead communicator of organizational issues and direction. The leader's personal communication style plays heavily into how the organization hears the news. In the last decade, many organizations experienced significant layoffs or forced reductions. No single issue had the organizations sitting on the edge of their seats as much as this one. With this pressure, many leaders chose to curtail broad employee communication sessions, since often the future plans were uncertain and they believed that communicating uncertainty was more harmful. This lack of communication almost always produced even more uncertainty. Other leaders chose to continually provide updates, admit where plans were uncertain and faithfully commit to provide updates as available. While there was still organizational anxiety, the context of trust was upheld.

The use of communication skills as the example is particularly appropriate since no other single skill or trait is more central to a leader's ability to influence the culture. Executive coaches work with their clients to encourage them to allow themselves to be more open and even vulnerable. The old image of the rigid, tough-it-out executive is well replaced by a style that is viewed as approachable and one that seeks answers rather than portraying that they always have them. There exists a whole arsenal of strategies that leaders can use to influence the organizational culture. Many lie in the domain of the human resources department, including selection, performance evaluation, training, compensation and reward, recognition and many others. Ultimately, however, it falls back to a central question: what collective set of values and beliefs will cause behaviors which, in total, are most likely to produce the outcomes to which the organization is committed, including outcomes for employees, clients and customers, and shareholders and stakeholders.

There are many resources available to those who may wish to learn more about risk and crisis communications. Two websites that may be especially helpful are Peter Sandman (http://www.psandman.com/) and the Center for Risk Communication (http://www.centerforriskcommunication.com/staff.htm). In addition, public officials may benefit from *Communicating in a Crisis: Risk Communication Guidelines for Public Officials* (USDHHS, 2002).

ORGANIZATIONAL BEHAVIOR IN EXTREME EVENTS

As discussed earlier, the key task for leadership in an organization is to help establish a culture which promotes behaviors which are aligned with the core mission of the organization. Maintaining that alignment is a continuous process requiring individuals in the organization to adapt to ever-changing conditions and challenges. That adaptation will be a positive force for the organization if it is determined according to the beliefs and values of the organization. Making these changes, in fact, strengthens the organization over time because it is seen that the values are ratified and are an enduring element of the organization. Conversely, if the value system is not well entrenched, adapting to the ongoing challenges creates a level of stress which can become destructive.

Organizations face stress all the time. The strategy of an organization with a positive organizational culture will incorporate the new stresses into its routine problem-solving processes and not view this as an extraordinary task. Service provider organizations, such as utilities, emergency services and others, have longstanding cultures that incorporate response in times of crisis. Following the 2005 devastating storms in the Gulf of Mexico, the media reported on many instances where employees in those organizations continued to work even while their own families were struggling to cope with severe devastation at home. While not a universal experience, the willingness of many workers to choose occupational role over personal and family roles was documented early in the field of disaster behavioral health (USDHHS, 1985). The occupational and organizational culture is, without question, a significant factor in those behavioral choices.

In addition to the 'service culture' response, there are examples of how companies or institutions reflected their cultural values in responding to a severe threat. Today, cases of product recall in the food, medical and toy industries are well recognized. However, such was not always the case. In 1982, Johnson & Johnson, a well-known medical and healthcare products company probably established the benchmark for product recall when a severe problem arose. Several bottles of their Extra Strength Tylenol had been laced with cyanide, undetectable to the purchaser. Their leading pain care product, on shelves in grocery and drug stores worldwide, had been tampered with in selected locations. Their corporate response is well documented (Jones et al., 2002; Dean, 2004). In the first few hours of discovery, without even knowing the full extent of the problem or who was behind it, or why it was happening, the senior leaders of the company ordered an immediate 100 percent recall, which involved 31 million bottles and a cost of over $100 million. In the immediate aftermath, Johnson & Johnson stock fell seven points and their share of the non-prescription pain reliever market fell from 35 to 8 percent (Lewin, 1986). While hindsight clearly validates the decision, the extent of the recall was questioned at the time. The potential for damage extended well beyond the immediate financial impact. The decision could have resulted in permanent damage to the entire reputation of an established and respected company. The CEO, however, had a different context for viewing the problem. His company's values were reflected in their marketing approach which was that supporting the health and safety of our customers and consumers *is* our business. To tolerate any ambiguity in this situation contradicts who we are and, therefore, any other course of action is out of the question. Fortunately, good often does win out and the public, the customers and the investors ultimately celebrated his decision and the company continues to thrive. What both of these examples attempt to show is that organizations with a well-established value system as the basis for their response to extraordinary events and challenges, see this response as a natural, albeit not normal, extension of how they operate. The Johnson & Johnson *Credo* (Johnson & Johnson, 2007) reflects several of the principles discussed in this chapter and can be found in Box 9.1.

As discussed earlier, these values are often developed as an organization is forming, or when new leadership is put in place. Very often, they are conceived as a response to known problems or issues in an earlier version of the organization, or from earlier experiences of the leadership. In one example, a global cellular company was consolidating several independent companies, each located in a different Latin American country, into a single organization. The presidents of each of these companies would form the new leadership team. The obvious issues of different countries, different cultures, lack of personal knowledge and experience with one another and other issues were present. The first task the new team took on was to 'invent' their value structure. In early discussions, it was recognized that one

BOX 9.1 JOHNSON & JOHNSON CREDO

We believe our first responsibility is to the doctors, nurses and patients, to mothers and fathers and all others who use our products and services. In meeting their needs everything we do must be of high quality. We must constantly strive to reduce our costs in order to maintain reasonable prices. Customers' orders must be serviced promptly and accurately. Our suppliers and distributors must have an opportunity to make a fair profit.

We are responsible to our employees, the men and women who work with us throughout the world. Everyone must be considered as an individual. We must respect their dignity and recognize their merit. They must have a sense of security in their jobs. Compensation must be fair and adequate, and working conditions clean, orderly and safe. We must be mindful of ways to help our employees fulfill their family responsibilities.

Employees must feel free to make suggestions and complaints. There must be equal opportunity for employment, development and advancement for those qualified. We must provide competent management, and their actions must be just and ethical.

We are responsible to the communities in which we live and work and to the world community as well. We must be good citizens – support good works and charities and bear our fair share of taxes. We must encourage civic improvements and better health and education. We must maintain in good order the property we are privileged to use, protecting the environment and natural resources.

Our final responsibility is to our stockholders. Business must make a sound profit. We must experiment with new ideas. Research must be carried on, innovative programs developed and mistakes paid for. New equipment must be purchased, new facilities provided and new products launched. Reserves must be created to provide for adverse times. When we operate according to these principles, the stockholders should realize a fair return.

of the biggest challenges they faced was inadequate and inaccurate financial data. On the surface the issue appeared to be the result of trying to integrate financial systems from several different countries. In reality, the core issue was a cultural history among all the countries of not reporting 'bad news'. This crippled their ability to make timely and informed financial decisions. After much discussion, they set up their values, the first being a commitment to 'Confront the Brutal Facts'. The new CEO and his team made it clear to the entire organization that those who brought forward facts, no matter how negative, and could offer thoughtful alternative plans would be acknowledged and validated. Those continuing to cover up bad news would face consequences. While changing behavior of this type is not an overnight task, having the leadership consistently stand for this credo ultimately engages the entire organization.

BUILDING SHARED VALUES AND TRUST

Figure 9.3 portrays several of the critical points in previous sections and lays the groundwork for the remaining sections of this chapter. Key beliefs that are central to the thesis of this chapter are shown in this figure and in the following sections. These include:

- Organizations and all those who work in them function in the context of the communities in which they are located and live.
- Organizational culture is the context in which all work-related activities of the organization and its members exists.
- All members of the organization and their families are affected by what is happening in both the organization and the community context in which they live. Each impacts on the other.
- Leaders (formal and informal) are a subset of the broader organizational membership.
- Leadership is central to all elements of this model.
- Leaders have families too and must manage role conflict.
- Human and organizational continuity planning has significant overlap but also independent elements.
- All members of the community (including/especially leaders) and their families have an interest in successful and integrated organizational and human continuity planning.
- Shareholders/stakeholders have an interest in all aspects of this model. They also have interest beyond the organization, its members and the communities in which both exist.

Several of these important elements were identified and reinforced in a study regarding workplace preparedness for terrorism (Ursano et al., 2006) where interviews with corporate leaders were conducted. Among their findings are:

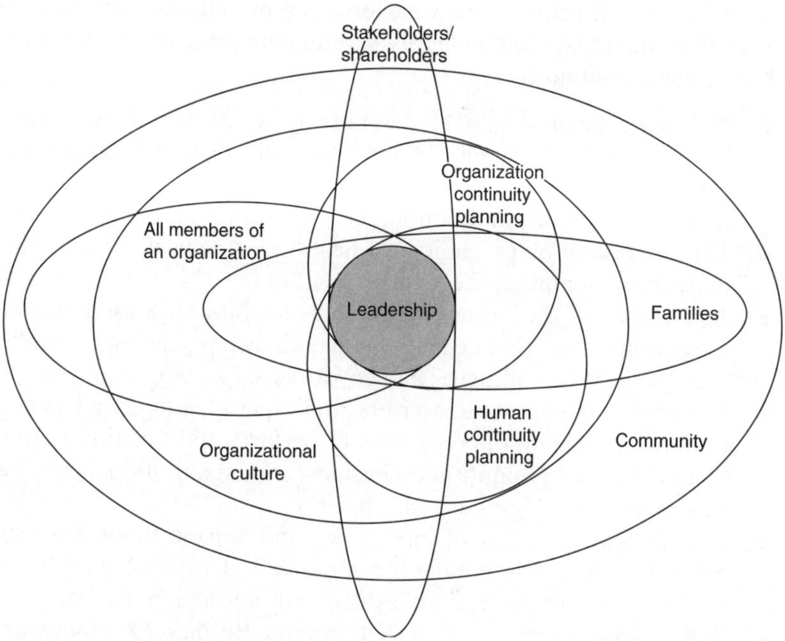

Figure 9.3 Overlapping roles and responsibilities of stakeholders in extreme events

- The importance of recognizing that corporations exist within larger communities.
- The importance of corporate culture in determining the content and effectiveness of disaster plans.
- A corporate culture of preparedness is established by many parts of the organization including senior corporate leadership, corporate security, personnel/human resources, occupational medicine, and divisions of employee assistance. The degree to which these components were internal or outsourced, substantially impacts levels of attention given and participation in the preparedness process.

FUNDAMENTAL ELEMENTS FOR SUSTAINING AND ENHANCING HUMAN CONTINUITY IN EXTREME EVENTS

If one accepts the premise that organizational survival is the chief goal of any organization facing adversity, then fostering, maintaining, and enhancing the well-being of the workforce is a primary task in achieving that goal. Flynn et al. (2005) identify several core principles necessary to achieve human continuity:

- People are the most valuable asset in any business or organization.
- The stability and viability of business and organizations and their people before, during, and following adverse events must be a core of Business Continuity and Continuity of Operations Programs. (p. 190)
- This continuity of psychological health has medical, ethical, business, and economic consequences. (p. 190)
- All members of the business or organization have responsibilities and opportunities to promote and assure their own psychological health, and that of their colleagues and co-workers. (p. 190)
- Executives, managers, supervisors, and other identified leaders have responsibilities and opportunities to perform and function in ways that protect and promote good business and organizational practices as well as psychological health. (p. 190)
- Policies and practices of businesses and organizations play significant roles in determining the continuity of psychological health as well as worker perception, loyalty, and satisfaction. (p. 190)
- Businesses and organizations that promote Business Continuity and Continuity of Operations with Human Continuity are more likely to be functioning effectively after adverse events.

Another fundamental belief is shared responsibility. Individuals and organizations share responsibilities in the protection and promotion of psychological well-being before and following extreme events. Neither individuals nor organizations will succeed without the other meeting their responsibilities. Safeguarding the health (in every definition of the word) should be a primary, non-negotiable priority of every organization. At the same time, individuals have a primary responsibility for their own health and well-being that should never be delegated. This shared responsibility and opportunity is described by Shultz et al. (2007). They describe and propose a 'tool-kit' for both individuals and organizations to promote worker resiliency before, during, and following extreme events. Specific suggestions include the following.

Individual Responsibilities/Opportunities

Pre-event

- Establishing and maintaining individual and family emergency plans. Having such a viable plan will not only safeguard workers and their

families but also mitigate against worker distractibility, preoccupation with family safety, and absenteeism. Weisaeth (1989) found that, in workplace crises, an individual's level of preparedness was the strongest predictor for optimal behavior during the event.
- Maintaining a health and balanced lifestyle as a buffer against multiple stressors. Include strategies in many life domains such as physical, emotional, cognitive, behavioral, social, and spiritual domains.
- Learn and practice stress management techniques.
- Learn to break cycles of inaccurate thinking. Examples include resisting catastrophizing events and situations and self-correcting inaccurate absolutes (such as 'never' and 'always').
- Nurture family and social supports. Strong social support is a powerful mitigating factor in adverse psychological impact of extreme events (Norris et al., 2002).

During an event

- Learn self-monitoring of stress. In the midst of emergency events, other systems to monitor stress (for example, colleagues and supervisors) may not function optimally.
- Take necessary steps to optimize personal safety (for example, use personal protective equipment (PPE).
- Practice self-care and self-management. Eat properly and get as much sleep/rest as necessary. Keep cognitively oriented toward work tasks.
- Establish and maintain family and support system contact to the extent possible.

Following an event

- Recognize that returning to normal work and family life is seldom an event but a process. This process can be both rewarding and stressful. The process can be lengthy.
- Participate in after-action reviews to learn from the adversity.
- Review with family how family disaster plans can be improved.
- Continue to monitor stress recovery in all domains noted earlier.
- Use the crisis to gain self-knowledge and adjust priorities.

Organizations

Pre-event

- Plan for safety and security for all members of the organization.

- Develop business continuity and COOP plans with human continuity in mind. Remember issues of leadership discussed earlier.
- Conduct drills and simulations including realistic human stresses (include all levels of the organization).
- Make arrangements for assessment and referral of any member of the organization who may need help with stress-related issues or may benefit from a formal psychological evaluation.
- Develop policies and procedures that reflect an organizational culture of health and concerns. Address issues that both reflect the culture as well as anticipate worker stressors (for example: if I am ordered to shelter in place (at home) will I get paid? If the organization shuts down for a period of time, will I be paid? Can I bring my family to the worksite with me? How will families be notified if there is an injury or death at work?).
- Provide and/or encourage training in a variety of topics such as psychological first aid, risk and crisis communication, leading in a crisis, incident command System and National Incident Management System (FEMA, 2007b).
- Promote workplace well-being as an organization value.
- Establish stress management as a job skill.
- Focus on strengthening teams. Workplace teams become part of the social support and stress monitoring function.
- Plan with an eye toward recognition of the importance of safety and security of family members.
- Have information available regarding the effects of various types of disasters so that it does not have to be developed in the midst of an event.
- Integrate planning efforts with key community partners (for example, schools, police, public health and so on).

During an event

- Place high priority on the safely and security of members of the organization and their families.
- Monitor stress of members of workforce.
- Communicate clearly, honestly, and frequently. Many good resources exist regarding risk communication and communicating effectively during a crisis. Several are listed at the end of the previous section, 'Organizations as Systems'.
- Remember the importance of team briefings, end of shift briefings, and so on as ways not only to optimize continuity of response but also to elicit and monitor psycho-social impact and status.

- Use a buddy system to optimize social support and as another set of eyes to monitor stress levels.
- Establish family support centers when appropriate. These are locations where families can be assured of safety and security, obtain basic needs (for example, meals), obtain timely and accurate information, and gain support.

Following an event

- Identify ongoing needs of members of the organization and their families.
- Provide support for members of the organization (including those who are ill or injured, or families of the deceased).
- Provide helpful and accurate information about the status of the organization and its future.
- Modify business continuity and COOP plans to reflect lessons learned.
- Assist in family and organizational reintegration of deployed workers.
- Enhance linkages with community partners not only to meet the needs of workers but also to enhance integration of services in future events.
- Evaluate organizational performance.
- Continue to monitor stressors in the workplace and larger community.

EARLY INTERVENTION FOLLOWING AN EXTREME EVENT

There are important shifts occurring in the practice of early interventions to reduce or prevent adverse psychological consequences following extreme events. These shifts are well known to those in the behavioral sciences who regularly deal with trauma and crisis. They may be less well known to those in the behavioral sciences who do not routinely deal with trauma. They may understandably be unknown to organizational leaders who have a different background and responsibilities. These changes are worth noting in this chapter as they point to some real opportunities to positively impact on psychological comfort and outcome while implementing sound and efficacious organizational principles and practices.

The first significant shift is away from prescriptive early psychological interventions. The field of disaster and emergency behavioral health has long sought interventions that could be provided shortly after a traumatic event that could be widely applied, reduce adverse and uncomfortable symptoms, and prevent the development of clinical conditions (such as post traumatic stress disorder; PTSD). In the early 1980s, critical incident stress debriefing (Mitchell, 1982, 1983) gained broad acceptance as such an intervention. Originally conceived for use with first responders (police, fire, rescue) it soon was being utilized for a wide spectrum of people who had experienced all manner of potentially traumatizing events. As more studies of its efficacy were conducted and more evidence gathered, many became convinced that it did not live up to its promise and, in fact, in some cases, might impede the recovery process and cause harm (Litz and Gray, 2002; Bisson, 2003).

Passions still run high on this topic but there is an undeniable trend away from these single-session intense group experiences toward a broader range of interventions that focus on provision of basic needs, calming, and support. The current trend is clearly to embrace the wide variety of psychological first aid models that are emerging (Everly and Flynn, 2005; NCCTS and NCPTSD, 2006). While each is somewhat different, most share common elements and approaches. Perhaps the most comprehensively documented model is that jointly developed by the National Center for Child Traumatic Stress (NCCTS) and the National Center for PTSD (NCCTS and NCPTSD, 2006), which focuses on the following priorities: contact and engagement; safety and comfort; stabilization; information gathering: current needs and concerns; practical assistance; and connection with social supports.

In a related article (Hobfoll et al., 2007) reviewed the empirical literature on early intervention in mass trauma situations. They derived from this analysis five essential elements for early intervention: provide a sense of safety; calming; sense of self- and community-efficacy; connectedness; and hope.

So, what does all this behavioral health discussion have to do with the function of organizations during and following extreme events? It means that the interventions currently believed to be the most important (even essential) in reducing psychological distress and potentially preventing the development of psychiatric disorders are entirely compatible with the priorities and practices of healthy organizations. It means that, in doing its work well, an organization simultaneously can be of major benefit to its members. It means that providing state-of-the-art behavioral health intervention is entirely consistent with, and can be integrated in, sound business and organizational survival strategies.

The next section will begin to explore ways in which human and organizational continuity can come together such that each is enhanced.

HOW DOES HUMAN CONTINUITY AND ORGANIZATIONAL CONTINUITY COME TOGETHER IN AN EXTREME EVENT?

Promoting Adaptive Individual and Organizational Coping

All individuals and all organizations (as noted earlier) cope with adversity and stress all the time. The key question therefore is not *if* they are coping but *how* they are coping. Depending on the circumstances, coping can be adaptive (adjusting in positive ways to promote well-being) or maladaptive (behaving in ways that diminish well-being). Promoting adaptive coping and reducing or eliminating maladaptive coping requires four elements as follows.

Determining what behavior is adaptive and what is maladaptive

Contrary to common perception, it is not always simple or easy to distinguish adaptive from maladaptive behaviors. In some cases, the adaptive behavior is obvious. For example, in the case of an intensifying building fire, rapid and orderly evacuation is the obvious adaptive (desired) behavior. If there also exists a cloud of toxic fumes outside in the area where evacuees will be sent, determining the desired behavior is not as simple. When there are issues of unfamiliar threat, complex alliances/allegiances, lack of experience, poor communication, unclear direction, and so forth it is important to understand that neither individuals nor organizations may intuitively behave in the ways that are most likely to promote well-being and reduce adverse consequences. In another health-related example, an employee who has a contagious illness may feel conflict regarding coming to work anyway to accomplish work and earn money, and staying home to speed recuperation and reduce exposure of co-workers.

Suggestions for enhancing desired decision making include:

- Pre-event anticipation. Think and talk about the possibilities and options. Utilize a positive organizational culture as a backdrop for these considerations
- Align desired outcome with policies and procedures. For example, if the organization wants sick employees to stay home when they have a communicable illness, ensure that leave and leave notification policies support that decision/policy.

- Reward/praise positive individual and collective adaptation.

Communicate effectively what should (adaptive) and should not (maladaptive) be done

- Maintain open, clear, and understandable communication that provides unambiguous guidance about the desired behavior during and following a crisis.
- Relate the communication efforts to the organization's value set.
- Get immediate expert advice if direction is unclear.
- Communicate using established risk and crisis communication principles.

Ursano et al. (2006) note that corporate leaders interviewed in their study cited the importance of 'communicating to employees a recognition of their value to the corporation – of instilling in employees the idea that they "mattered" – was important to preservation of function in times of crisis' (p. 3).

Adaptation monitoring

- Monitor how individuals and groups are adapting to the crisis.
- Provide feedback and corrective or reinforcing guidance.
- Monitor even when the crisis has passed.

Reinforcing or modifying behavior

To aid individuals and organizations in thinking about promoting adaptive coping, Table 9.1 identifies elements that might characterize coping at various event stages.

Understanding Key Principles of Changing/Reinforcing Behavior in an Organization

Confirmatory bias

How do our beliefs (accurate or inaccurate) become solidified and strengthened? One of the primary factors is our tendency to see and believe those things that confirm or ratify our beliefs and overlook and deny those things that do not support our beliefs. This phenomenon is called 'confirmatory bias' and is portrayed in Figure 9.4.

Why is this important when considering individual and organization response to a crisis? It is central because, especially under times of stress, people are less able to process new information and may rapidly resort to

Table 9.1 Characteristics of coping: pre-event, during and post-event

	Individual	Organizational	Leadership
Pre-event			
Adaptive	Family/personal preparedness	Meaningful organizational and human continuity planning	Communicates a culture of shared values Highlights human continuity in continuity of operations planning
Maladaptive	'It can't happen here'	Preparedness and continuity is mainly an IT function	Focus only on protecting the business without regard for needs/ concerns of workers
During			
Adaptive	Practice self-care Provide psychological first aid for co-workers	Effectively implement sound organizational and human continuity plans	Be present, visible, honest Practice sound risk and crisis communication skills
Maladaptive	Abandon occupational and/ or family roles	Mismatch workers and mission Overwork members Put members at extreme/ unacceptable risk	Be absent Fail to communicate to members
Post-event/recovery			
Adaptive	Understand psycho-social recovery patterns Seek help when needed	Implement flexible work schedules Assure availability of needed services Monitor health status	Focus on positive recovery Communicate recovery successes and failures Reinforce positive values/culture
Maladaptive	Cope with alcohol/drugs Reject needed services	Make unreasonable demands on members Discourage referral to needed services	Remain silent Ignore human concerns Isolate from other community stakeholders

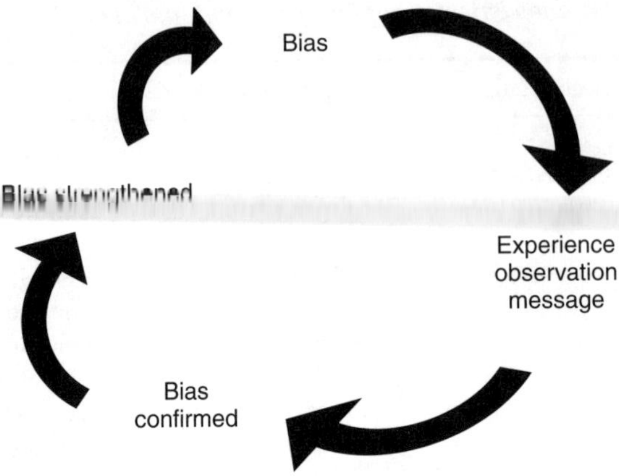

Figure 9.4 Understanding the power of confirmatory bias

basic, often unconscious biases. Those biases may be either adaptive or maladaptive. For example, if an organization asks its workers to continue to come to work in anticipation of a hurricane, people will process that request based on a wide variety of factors (including concerns for personal and family safety, financial factors, and personal experience). If the worker has a bias toward viewing the organization as having an organization that values safety and equitably shared sacrifice, he/she may be more likely to report to work. On the other hand, if the worker has a bias toward viewing the organization as having an organizational culture that is unconcerned about worker well-being and focused only on maintaining operations for profit not shared by the worker, he/she may be more likely to not report. What happens after making that choice, the experiences, observations, and messages (written, verbal, behavioral) received from leadership and co-workers will either reinforce and strengthen the bias or weaken the bias (as illustrated in Figure 9.5). This phenomenon is what was referred to in a previous section as the 'social software'.

Times of crisis and high stress are key points when biases are either strengthened or modified. Strategies to use understanding about confirmatory bias positively include:

- Assess the basis, nature, and strength of biases prior to an extreme event.
- Anticipate how continuity of business and continuity of operations plans either strengthen or weaken bias.
- Be aware of confirmatory bias in all communication pathways including correspondence, verbal communications, and behavior.

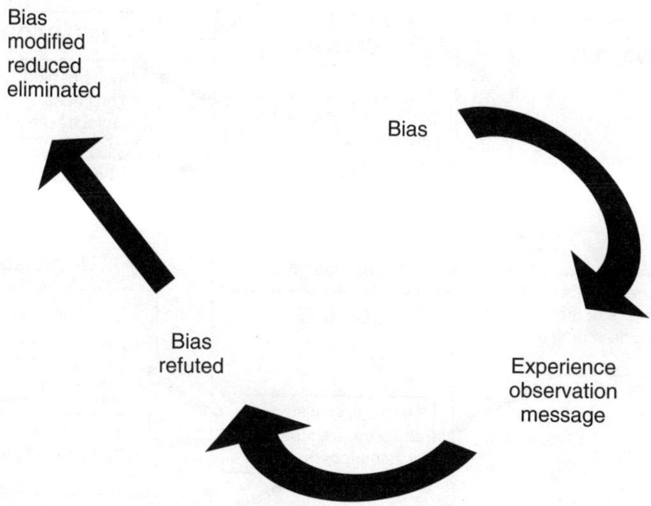

Figure 9.5 Modifying bias

- Use times of crisis as opportunities to establish and strengthen positive values and culture.

Behavior changes cognition

Organizational consultants often remind their clients that, 'we don't think ourselves into a new way of acting, we act ourselves into a new way of thinking' (Bossidy and Charan, 2002, p. 99). In other words, acting differently causes us to think differently more than thinking differently causes us to act differently. Perhaps unknowingly, this advice is based on basic principles of behavioral psychology's approaches to achieving and maintaining behavioral change.

Changing beliefs (or thoughts) as a way to change behavior is a lengthy process often requiring assistance and intervention. Smoking cessation efforts over the years that focus on changing people's beliefs about smoking provides discouraging case studies. In addition, beliefs are often not easily visible so we are often not aware of the underlying beliefs held by members of an organization. Behavior on the other hand is visible and arguably more easily and appropriately dealt with in an organization. An organization is hard-pressed to tell its members what to think or believe. It is, however, far more entitled to prescribe what behaviors it wants its members to engage, or not engage in.

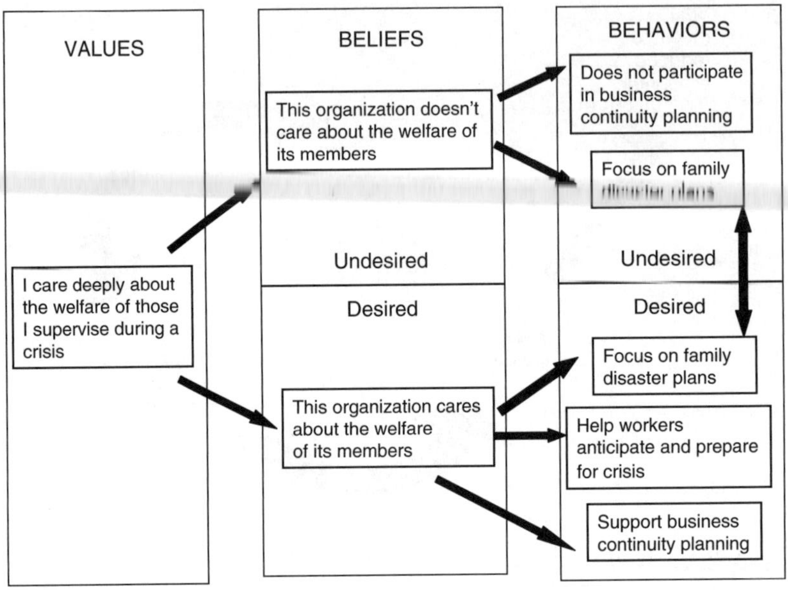

Figure 9.6 *Values, beliefs and behaviors*

As illustrated in Figure 9.6, values play a large role in determining beliefs and beliefs cause us to behave in various ways. Values are not easily changed. Our beliefs are based upon the values we hold. Those beliefs then lead to behaviors that are the vehicle for expressing values and beliefs. Changing behavior often leads to changed beliefs. As illustrated in Figure 9.6, if the undesired behavior can be stopped, and the worker can be compelled to engage in the desired behaviors, then there is an increased likelihood that the undesired beliefs may change to the desired beliefs.

There are two additional reminders in the figure:

- Values are very resistant to change and are by and large positive. Undesirable beliefs frequently cloud positive values.
- Sometimes the desired behavior (for example, focus on family disaster planning) can result from both desired and undesired beliefs. For this reason, it is important to place primacy on behavior and recognize that there are many roads to the desired result.

Uncertainty impacts on both individuals and organizations

For most individuals and organizations, uncertainty produces stress. At the same time, there are enormous variations and exceptions to this. Some

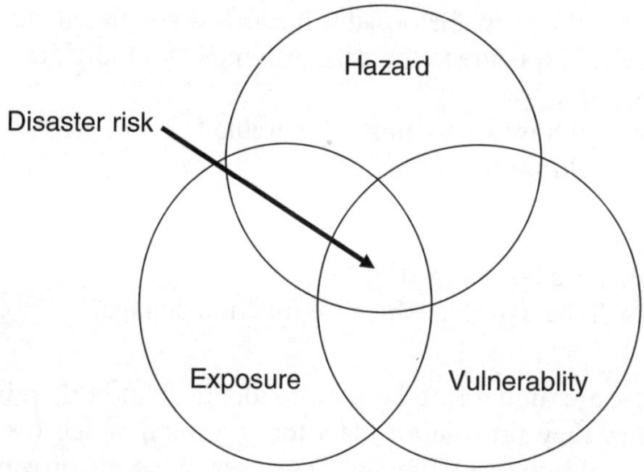

Source: Adapted from Shultz et al. (2007).

Figure 9.7 Disaster risk

individuals and organizations thrive on dealing with the unexpected and meeting new challenges. However, for the most part this approach is based on high reward and manageable risk. The potential of disaster typically unbalances this equation, threatening catastrophic risk and no reward.

The impact of uncertainty and change is described by Shultz et al. (2007) in their discussion of disaster ecology. Their conceptualization has application for both individuals and organizations. They describe the forces of harm that attend disasters and extreme events as composed of:

- exposure to a hazard: wind, water, disease, fire, explosion and so on;
- loss: loss of life, home, workplace, income, social support and so on; and
- change: change in work/living location, social structures.

They then discuss the affected populations in harm's way and their vulnerabilities. They complete the model by discussing how the forces of harm and the affected populations in harm's way play out in the context of individuals and families, communities, and society. They describe disaster risk as the intersection of the hazard itself, exposure, and vulnerability, as displayed in Figure 9.7.

This formulation provides a wonderful opportunity for individuals and organizations to review and assess these elements in the service of sound preparedness and response that integrated human and business continuity. The key questions for both individuals and organizations are virtually identical:

- What are the probable/possible hazards we might encounter?
- What does exposure to these hazards mean for individual, family, and organization?
- Where and how are we most vulnerable?
- How should we best prepare?
- What do I/we do when it happens?
- What will recovery look like?
- How will we be changed?
- How will the world in which we function change?

Because these are fundamental questions identical for both individuals and organizations, they provide an ideal focus around which discussion and planning can take place. Optimally, these discussions will provide an opportunity to reinforce and align positive personal and organization values, and in so doing enhance integrated preparedness, response and recovery.

ENHANCING ORGANIZATIONAL RESILIENCE

Prepare relentlessly
(Giuliani, 2002, p. 51)

While sound preparedness requires commitments of time, energy, and funding, there is no substitute for a credible and ongoing process of hazard assessment and examination of resulting human and organizational sequelae. These efforts, as described below, can yield benefit even if an extreme event never occurs.

Plan Development and Implementation

Effective planning does not just happen. It is a process. That quality of that process will likely be as important in determining a quality outcome as the technical components of the resulting plan itself.

Much information is available regarding the process and content of effective disaster preparedness (USDHHS, 2003; CDC, 2007; FEMA, 2007a). Little guidance, however, is available to help understand the organization *context* necessary to implement a viable process and outcome. Figure 9.8 reflects the commitment ladder that organizations must scale to optimize the potential for a successful disaster preparedness process and product.

Producing a plan is the practice of ratifying the culture. All organizations have cultures. Ideally, the planning process reinforces and reflects a healthy culture:

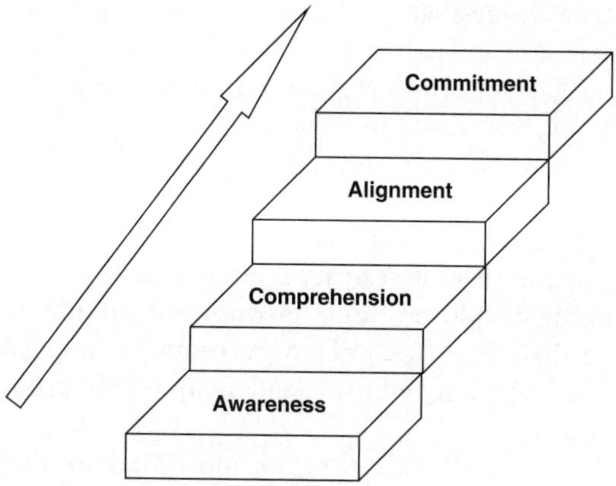

Figure 9.8 The commitment ladder

- Planning is a win–win situation, but at a cost. There can be no denying that viable, integrated, and sustainable planning requires resources: time, people, and money. All organizations must understand this reality from the start. For some organizations, preparing for extreme events is optional. For others, it is required by law and regulation. While acknowledging the costs, those organizations choosing to embark on this process can achieve a classic win–win scenario. If planning results in an improved response and recovery, there are clear benefits to both individuals and the organization. Examples include minimizing/reducing damage, injury, disability, more rapid and functional return of the workforce, and accelerated business recovery. At the same time, the other win occurs even if there is no extreme event. The planning process itself will increase understanding by all members of the roles, reasonabilities, assets, and challenges of others. A meaningful planning process becomes the vehicle for reinforcing a positive organization culture. It provides an opportunity for leadership (formal and informal) to emerge and be displayed.
- Use outside experts and consultants for their specific expertise but never outsource the development of either human or business continuity planning. If the organization does not retain full ownership of both the planning process and product, its relevance and utility will likely be significantly reduced. Therefore, own both the process and the plan.

Specific Guidance/Suggestions

Ursano et al. (2006) report several interesting corporate findings in their interviews:

- 'The corporate security offices emerge as the most critical corporate voice in identifying preparedness vulnerabilities to leadership, transmitting the message of security awareness and practices (behaviors) to corporate employees, and providing a culture of anticipation of future disasters' (p. 2). They recommend that corporate security should be positioned for a leadership role in promoting human continuity.
- All levels of the organization are motivated by threat awareness, assessment, and perception. Organizational history with crisis plays a significant role.

In an article discussed earlier, Flynn et al. (2005) suggest specific steps that organizations can take to enhance the integration of human and business continuity. These include establishing a human continuity team with the following roles (p. 191):

- Review and analyze business continuity plans to ensure that human continuity needs are addressed and/or gaps are identified.
- Ensure documentation and assessment of these plans, as well as regular testing, modification, and revision.
- Ensure commitment to preparation for and provision of services in support of human continuity.
- Ensure that steps are taken to prepare for provision of services in support of human continuity.
- Provide consultation and leadership regarding human continuity in the event of an incident.
- Monitor and evaluate efforts addressing human continuity both during and following events.
- Provide feedback following events on ways to enhance the workplace response to human continuity needs.

Have Appropriate Resources Ready

An effective and integrated human and business continuity plan in the context of a wide range of potential extreme events requires many types of resources. A few organizations will have the full range of needed services and experience in-house or easily available. Many, perhaps most, will not. Since few organizations will have everything they need to prepare, respond and recover, in the business and human continuity context, this represents an incentive and opportunity to link with other community stakeholders who have the needed expertise, services, or complementary needs. Knowing what is needed and knowing how to get it is a critical preparedness task. Some specific suggestions follow:

1. Pre-event

   - Identify and assess business and human continuity expertise.
   - Assess local and regional emergency management plans, expertise, resources.
   - Acquire knowledge of governmental and business group advice, checklists, templates that can be adapted.
   - Anticipate a full range of physical, behavioral, and public health consequences and know where and how to access expertise. Have your experts ready. Self-proclaimed experts emerge very quickly in extreme events – just when an organization is least able to assess the legitimacy of the expertise. Having the required expertise identified beforehand will help reduce bad choices during an event.
   - If a human continuity team has been established as described earlier, ensure that they are part of identifying needed expertise.

2. During event

   - Continuously monitor the environment for the nature and intensity risks to people.
   - Implement communications plans. Ensure access to specialized expertise on risk and crisis communications.
   - Ensure rapid access to a broad range of physical and behavioral health resources.
   - Link with community partners to ensure optimal access to and utilization of services and expertise.

3. Post-event

   - Continue to monitor the environment for risks.
   - Identify organizational, individual and family recovery needs.
   - Be aware of the scope and limits of governmental, voluntary, and philanthropic resources that have the potential of ameliorating human and organizational problems.

- Identify needed expertise in restoring and preserving business and human continuity.
- Memorialize losses and the lost. Symbols and rituals can be very powerful in expressing very complex and mixed emotions of loss, hope, and anger. Often these rituals and symbols are closely tied to culture and sometimes to occupation (for example, police, firefighters).
- Focus on strengths and resilience. Modify plans based on experience.

CONCLUSION

If this chapter has accomplished its goal, readers will have gained an enhanced appreciation for the many and different ways in which organizations, individuals, and their environments interact. We hope that the relationships among these elements, while supported by scholarly work, also make intuitive and practical sense.

This chapter has attempted to integrate knowledge and information about organizational and behavioral health factors and to convince the reader of the merits of proceeding to plan and respond based on that integrated approach. There are enormous opportunities when the professional and occupational boundaries of behavioral health and organizational design and function are crossed. The potential rewards are great, as is the price of failure.

The Price of Failure

While this chapter is part of a scholarly text, this is not an academic exercise. There is the potential for serious, even catastrophic, loss if organizational and human continuity are not both fully developed and integrated in the context of a healthy organizational culture.

The organizational cost of failure, in the extreme, can be the failure of the organization and the resulting loss for members of the organization, stakeholders and shareholders, and those served by the organization. In a less extreme scenario, there is less catastrophic loss but loss nonetheless. These losses could be industry position, competitive advantage, reduced market share, lower profits/greater losses, and loss of customer confidence and loyalty.

It is also important to remind ourselves that organizations do not exist in isolation. Organizations are all part of a mosaic of industries, service sectors, communities, and nations. The success or failure of one impacts on all. In the case of failure of multiple organizations in a sector or community, there will likely be a cascading effect that adversely impacts on all.

At the start of this chapter we were reminded that the primary goal of any organization is to preserve itself. We hope that the information included in this chapter will assist organizations in achieving that goal and enable them to reduce adverse human and business consequences and promote human and organization health.

The Promise of Success

The main thesis of this chapter is that if organizations, leaders in organizations, and all members of an organization are able to unite around a shared value of promoting health and reducing adverse consequences, all parties will not only survive but also have the potential of thriving. The chapter is based on not just the adverse consequences of failure but also the promise (both individual and collective) of successful integrated preparedness and response.

The era of top-down, autocratic, paternalistic organizational models is vanishing. Organizations and their structures (the *engineering* discussed at the beginning of this chapter) and processes are becoming more transparent and inclusive. The shared and inclusive process of preparing for and responding to extreme events becomes a platform from which to exercise modern organizational principles and to communicate and strengthen positive organizational culture.

Knowledge, and unfortunately experience, regarding the full extent of adverse behavioral health consequences of extreme events is rapidly expanding. Expanded understanding supports the notion that roles and relationships are numerous and varied, and, in the face of crisis, each influences others. We are also learning that, through fostering resilience, adverse behavioral outcome is not a foregone conclusion.

Understanding and adapting the suggestions provided here will help both individuals and organizations relate and interact in healthy ways even in the absence of extreme events. Shared responsibility and healthy interdependence is the path to promoting resilience in both individuals and organizations of all types.

REFERENCES

Bisson, J.I. (2003), 'Single-session early psychological intervention following traumatic events', *Clinical Psychology Review*, **23**(3), 481–99.

Bossidy, L. and Charan, R. (2002), *Execution: The Discipline of Getting Things Done*, New York: Crown Business (Division of Random House).
Centers for Disease Control (CDC) (2002), 'Impact of September 11 attacks on workers in the vicinity of the World Trade Center – New York City', *Morbidity and Mortality Weekly Report*, **51**, 8–10.
Centers for Disease Control (CDC) (2007), *Get Informed. Get Prepared*, http://www.pandemicflu.gov/, June 17, 2007.
Danieli, Y., Brum, D. and Sills, J. (eds) (2005), *The Trauma of Terrorism: Sharing Knowledge and Shared Care, An International Handbook*, New York: Hayward Press.
Dean, D.H. (2004), 'Consumer reaction to negative publicity: effects of corporate reputation, response, and responsibility for a crisis event', *Journal of Business Communication*, **41**, 192–211.
Everly, G.S. and Flynn, B.W. (2005), 'Principles and practice of acute psychological first aid after disasters', in G.S. Everly and C.S. Parker (eds), *Mental Health Aspects of Disaster: Public Health Preparedness and Response*, Vol 1, 2nd edn, Baltimore, MD: Johns Hopkins Center for Public Health Preparedness, pp. 68–76.
Federal Emergency Management Agency (FEMA) (2007a), *FEMA: Business* http://www.fema.gov/business/index.shtm, June 16, 2007.
Federal Emergency Management Agency (FEMA) (2007b), *Welcome to the National Incident Management System*, http://www.fema.gov/emergency/nims/index.shtm, June 16, 2007.
Flynn, B.W., Flanigan, S. and Everly, G.S. Jr (2005), 'Fostering human continuity: an essential element in workplace crisis intervention', *International Journal of Emergency Mental Health*, **7**(3), 187–93.
Giuliani, R.W. (2002), *Leadership*, New York: Hyperion.
Halpern, J. and Tramontin, M. (2007), *Disaster Mental Health Theory and Practice*, Belmont, CA: Thompson.
Hobfoll, S.E., Watson, P., Bell, C.C., Brant, R.A., Brymer, M.J. and Friedman, M.J. (2007), 'Five essential elements of immediate and mid-term mass trauma intervention: empirical evidence', *Psychiatry*, **70**(4), 283–315.
Institute of Medicine (IOM) (2003), *Preparing for the Psychological Consequences of Terrorism: A Public Health Strategy*, Board on Neuroscience and Behavior (NBH), Washington, DC: Institute of Medicine.
Johnson & Johnson (2007), *Credo*, http://www.jnj.com/our_company/our_credo/index.htm;jsessionid=RQUXI1QGKCCKQCQPCCGSU0A, June 16, 2007.
Jones, G.H., Jones, B.H. and Little, P. (2002), 'Reputation as reservoir: buffering against loss in times of economic crisis', *Corporate Reputation Review*, **3**(1), 21–9.
Lewin, T. (1986), 'Tylenol maker finding new crisis less severe', *New York Times*, February 12.
Litz, B.T. and Gray, M.J. (2002), 'Early interventions for mass violence: what is the evidence? What should be done?', *Cognitive and Behavioral Practice*, **9**(4), 266–72.
Mankin, L.D. and Perry, R.W. (2004), 'Terrorism challenges for human resource management', *Review of Public Personnel Administration*, **24**(1), 3–17.
Mitchell, J.T. (1982), 'Recovery from rescue', *Response Magazine*, Fall, 7–10.
Mitchell, J.T. (1983), 'When disaster strikes . . . The critical incident debriefing process', *Journal of Emergency Medical Services*, **8**(1), 36–9.

Nadler, D.A., Gerstein, M.S. and Shaw, R.B. (1992), *Organizational Architecture*, San Francisco, CA: Jossey-Bass.

Nadler D.A. and Nadler, M.B. (1997), *Champions of Change*, San Francisco, CA: Jossey-Bass.

National Center for Child Traumatic Stress (NCCTS) and National Center for PTSD (NCPTSD) (2006), *Psychological First Aid: Field Operations Guide*, 2nd edn, http://www.ncptsd.va.gov/ncmain/ncdocs/manuals/nc_manual_psyfirstaid.html, June 16, 2007.

Norris, F. (2002), 'Psychological consequences of disasters', *PTSD Research Quarterly*, National Center for Post Traumatic Stress Disorder, **13**(2), 1–3.

Norris, F., Friedman, M., Watson, P., Byrne, C., Diaz, E. and Kaniasty, K. (2002), '60,000 disaster victims speak, Part 1: An empirical review of the empirical literature, 1981–2001', *Psychiatry*, **65**, 207–39.

Owens, D. (2003), 'Managing a crisis', paper presented to the International Association for Counterterrorism and Security Professionals, New York City, November.

Shultz, J.M., Espinel, Z., Flynn, B.W., Hoffman, Y., and Cohen, R. (2007), *DEEP PREP-All-Hazards Disaster Behavioral Health Training*, Miami, FL: Disaster and Extreme Event Preparedness (DEEP) Center, Miller School of Medicine, University of Miami.

Shultz, J.M., Espinel, Z., Galea, S. and Reissman, D.B. (2007), 'Disaster ecology: implications for disaster psychiatry', in Ursano et al. (eds) (2007), pp. 69–98.

Stein, B.D., Elliot, M.N., Jaycox, L.H., Berry, S., Collins, R., Klien, D.J. et al. (2003), 'A National Longitudinal Study of the Psychological Consequences of the September 11, 2001 Terrorist Attacks: reactions, impairment, and help-seeking', RAND Corporation.

Ursano, R.J., Fullerton, C.S. and Norwood, A.E. (eds) (2003), *Terrorism and Disaster: Individual and Community Mental Health Interventions*, Cambridge: Cambridge University Press.

Ursano, R.J., McCaughey, B.G. and Fullerton, C.S. (eds) (1994), *Trauma and Disaster: The Structure of Human Chaos*, Cambridge: Cambridge University Press.

Ursano, R.J., Vineburgh, N.T., Gifford, R.K., Benedek, D.M. and Fullerton, C.S. (2006), 'Workplace Preparedness for Terrorism (report of findings to Sloan Foundation)', Center for the Study of Traumatic Stress, Department of Psychiatry, Uniformed Services University School of Medicine, Bethesda, MD, http://www.centerforthestudyoftraumaticstress.org/downloads/CSTS%20Sloan%20Workplace.pdf, June 17, 2007.

Ursano, R.J., Fullerton, C.S., Weinsaeth, L. and Raphael, B. (eds) (2007), *Textbook of Disaster Psychiatry*, Cambridge: Cambridge University Press.

US Department of Health and Human Services (USDHHS) (1985), *Role Stressors and Supports for Emergency Workers*, Washington, DC, DHHS Publication No. (ADM) 85-1408.

US Department of Health and Human Services (USDHHS) (2002), *Communicating in a Crisis: Risk Communication Guidelines for Public Officials*, Washington, DC: Department of Health and Human Services.

US Department of Health and Human Services (USDHHS); Flynn, B.W. (2003), *Mental Health All-Hazard Disaster Planning Guidance*, DHHS Publication No. SMA 3829, Rockville, MD: Department of Health and Human Services.

Virgo, J.M. (2001), 'Economic impact of the terrorist attacks of September 11', *Atlantic Economic Journal*, **29**(4), 353–7.

Weisaeth, L. (1989), 'Stressors and the post-traumatic stress syndrome after an industrial disaster', *Acta Psychiatrica Scandinavia*, **80** (Supplement 355), 25–37.

10. Maintaining work motivation during trying times

Parbudyal Singh[1]

INTRODUCTION

Motivating employees is one of the most important and challenging activities that managers perform, even in normal times. Essentially, employee motivation involves getting people in the workplace to exert sufficient levels of effort to attain organizational goals. Over time, several theories of motivation have been proposed, including needs theory (McClelland, 1961), goal-setting theory (Naylor and Illgen, 1984), the job characteristics model (Hackman and Oldham, 1975), equity theory (Adams, 1965), and expectancy theory (Vroom, 1964).

However, an examination of these theories reveals that they are only partially useful in explaining human behavior and employee motivation under 'abnormal' conditions, such as circumstances during and after a terrorist attack. For instance, the job characteristics model (JCM), arguably the most relevant of the management theories in these circumstances, identifies five key job characteristics, their interrelationships, and their impact on employee productivity and motivation. In summary, the JCM posits that job characteristics (skill variety, task identity, task significance, autonomy and feedback) are related to critical psychological states (such as experienced meaningfulness of the work and experienced responsibility for outcomes of the work) and that these lead to personal and work outcomes, such as work motivation and satisfaction. Even though JCM provides some guidance for managers for job design in terroristic times, including the development of feedback mechanisms, it is limited in its usefulness in understanding human behavior and motivation in traumatic events. The same criticism may be applied to other theories of motivation.

Thus, in this chapter, a systematic study of this issue using a grounded theory approach is attempted. In essence, this methodological approach relies on examining the data first, and then allowing for propositions and theories to emerge. A traditional approach was first contemplated, whereby an established theory would be tested for conformity under traumatic conditions. However, the grounded theory approach is better suited, given the exploratory nature of the research and the unprecedented nature of the situation. Furthermore, an examination of extant theories revealed that none was particularly applicable and that it may be better to inductively study the situation in a process that allows for theories to emerge.

The New York Presbyterian Hospital (NYPH) was selected for two main reasons. First, a major center of the hospital is located in close proximity to 'Ground Zero', and the entire hospital system had to respond to the emergency situation. Second, hospitals generally have plans to deal with disasters and thus, arguably provide the best units to use as 'benchmark cases' for other organizations. The NYPH also lost some of its employees in the September 11 attack, and had to respond to its own workforce like other organizations in lower Manhattan. In addition, the hospital had to respond to the escalating demand for its services from those in need, putting its workers under further duress. Overall, it was very challenging to motivate its staff; nevertheless, drawing from the inner strength of the workers and using a range of tactics, management was reasonably successful in effectively dealing with the tragedy.

METHODOLOGY/PROCEDURE

The Study Design

This study utilizes a grounded theory design, or the 'discovery of theory from data' through the general method of comparative analysis (Glaser and Strauss, 1967). This approach, which has its roots in sociology, is based on two important principles (Corbin and Strauss, 1990). First, since a phenomenon is not considered as static but continually changing in response to fluid conditions, there is an emphasis on design strategies that acknowledge and deal with such change. Second, strict determinism is rejected, as is non-determinism. Actors are seen as having the means of controlling their destinies by their conscious responses to conditions (ibid.).

There are three basic elements of grounded theory: concepts, categories and propositions (Glaser and Strauss, 1967; Corbin and Strauss, 1990; Strauss and Corbin, 1990). Concepts are the basic units of analysis. As Corbin and Strauss (1990, p. 7) state:

> [A] theorist works with conceptualizations of data, not the actual data per se. Theories can't be built with actual incidents or activities as reported; that is, from 'raw data.' The incidents, events, and happenings are taken as, or analyzed as,

> potential indicators of phenomena, which are thereby given conceptual labels . . . [o]nly by comparing incidents and naming like phenomena with the same term can a theorist accumulate the basic units for theory. In the grounded theory approach such concepts become more numerous and more abstract as the analysis continues.

Categories, the second element of grounded theory, are formed through the grouping of similar concepts; not all concepts become categories. As Corbin and Strauss (p. 7) further point out:

> Categories are higher in level and more abstract than the concepts they represent. They are generated through the same analytic process of making comparisons to highlight similarities and differences that is used to produce lower level concepts. Categories are the 'cornerstones' of a developing theory. They provide the means by which a theory can be integrated.

Propositions, the third element of grounded theory, indicate generalized relationships between a category and its concepts and between discrete concepts. Originally termed 'hypotheses' by Glaser and Strauss (1967), it is felt that 'propositions' is more appropriate since this term implies conceptual relationships, whereas hypotheses require measured relationships (Whetten, 1989).

The emphasis in grounded theory is on inductive reasoning; however, it does allow for deductive derivations, especially in developing research questions and in determining the sample (Strauss and Corbin, 1990). That is, researchers can use existing research to frame the broad parameters of their studies. The emphasis on inductiveness contributes to a significant difference between the grounded theory and the traditional quantitative approach: the placement of the 'literature review'. In the positivist tradition, previous research and established theories are reviewed at the beginning of the study, before the data and results are introduced, and then the results are examined for conformity or deviations. However, in grounded theory, the data are first examined and the emerging theory is then compared and contrasted with the extant literature and theory (Glaser and Strauss, 1967; Strauss and Corbin, 1990; Turner, 1990).

Data Sources and Collection

Several sources of data were used for this study. First, press reports on the September 11 tragedy were examined. Following the disaster, there was extensive coverage of the effects of terrorism, including the impact on workers. *Fortune* magazine even organized a forum of some of the country's top leaders to discuss the issue. Second, 10 interviews were conducted with five top executives of NYPH, and five with employees. The interviews lasted approximately 45 minutes to 1 hour each. The executives were directly responsible for the organization's response to the terrorist attack on New York. The interviews were held in April 2002, six months after the attacks. Third, archival data and internal communications of the organization were examined, including a booklet titled 'Days of Unity' which contained stories from approximately 25 executives and employees about their work experiences following the September 11 tragedy. Finally, the extant literature was analyzed, including the relevance of current theories to the situation.

THE NEW YORK PRESBYTERIAN HOSPITAL CASE

Background Information and Event Description

The NYPH, one of the world's leading medical centers, was formed in 1997 through the merger of the Society of the New York Hospital (affiliated with Cornell University Medical College) and the Presbyterian Hospital in the City of New York (affiliated with the Columbia University College of Physicians and Surgeons). The merged hospital (the NYPH) has five sites in the City of New York and one in Weschester County. In 2001, the NYPH had combined revenues of $1.8 billion, 14,000 employees and over 6000 physicians. The New York–Cornell Center is located about five miles from the World Trade Center and the Columbia Hospital is about 11 miles away.

Like most people in the rest of the country, employees at NYPH went to work on September 11, 2002, with little thought about terrorism or Osama bin Laden on their minds. However, around mid-morning their lives would be changed forever as the planes crashed into the Twin Towers.

Even though the hospital had responded to prior disasters and traumatic events, nothing was on this scale. Given the unprecedented nature of the unfolding events, there was inconsistent information from the authorities at first. However, within seven minutes of the first airplane striking the World Trade Center, the Hospital's Disaster Plan was made operational. The Plan 'delineates the authorities, responsibilities and procedures to be activated by hospital staff as well as volunteers and community organizations in response to a natural or manmade disaster, involving multiple casualties, occurring within either this geographical area or the hospital itself' (NYPH, 2002). Implementation of the Plan involves, among other things, the expeditious notification of all relevant internal and external agencies and key employees, assignment of personnel to cover all necessary staff positions, provision of space, supplies, and security, establishing back-up communication systems, activating alternative sources of utilities, suspending elective surgery, cancellation of clinic sessions, and augmenting supplies, pharmaceuticals, and biomedical equipment in the Emergency Department in contemplation of an influx of victims.

As people in the city, country and world watched in horror as the Twin Towers collapsed, the hospital staff prepared for the inevitable process of treating victims and employees as well. At the same time, at the request of the Fire Department of the City of New York (FDNY), the Hospital's Emergency Medical Services department sent a dozen ambulances and support vehicles, along with 27 Emergency Medical Technicians and Paramedics to 'Ground Zero'. They were assembled as one of the medical triage units when the buildings collapsed. Three of the workers lost their lives and 14 were injured. Over the next few days, the NYPH received and treated over 125 victims and 24 of the most severely burned were admitted to the Randolph Hearst Burn Center at the New York Cornell Center.

The psychological and emotional effects of the disaster were immediate. As were other people with loved ones working near the World Trade Center, employees were worried about their families and friends who may have been in the area. Many screamed and cried openly. There was a demand to use phones, and desks were in danger of going unmanned as both visitors and employees tried to get information from various sources. Management had to react fast, and with compassion, so as to acknowledge real concerns without spreading confusion and chaos.

The hospital's social workers took the lead in setting up counseling centers in several areas in the hospital system. They offered advice and support for employees and visitors who needed help. They also established special telephone lines as many families were calling the hospital to locate loved ones. This was a sensitive task that required expert handling. Management also allowed employees to congregate and talk about their fears and concerns provided that this did not affect the overall operation of the hospital.

There are many effects felt to this day. Apart from the three Emergency Medical Service (EMS) workers who died, several other employees lost their spouses and loved ones, and there are many cases of post-traumatic stress disorder. One year later, two NYP*EMS workers were still out of work due to injuries sustained on 9/11. As one manager sums it up: 'A day does not go by without hearing some comment about "nine eleven" or reference to it. I believe that the lives of New Yorkers are changed forever. We once felt invincible; we now realize how vulnerable we are'.

Concepts and Categories

During and after the attacks, employees displayed an extraordinary ability to continue their work. Many employees walked to work – sometimes a distance of several miles – since most of the city's transportation system was immobilized. Employees who lived nearby provided overnight accommodation for their colleagues, many slept at the hospital, and most of the executive and senior management staff remained on continuous duty for the first 36 hours of the response. How were employees motivated to continue to work, and work even harder than usual?

There are several answers, and associated categories, that emerge from the data collected. First, they were given astute leadership under stressful conditions. Leaders in the organization were trained to handle traumatic events and their tasks were made relatively easy because they had a Disaster Plan, a document that helped in operationalizing emergency procedures. As a result, employees received reasonably clear instructions. Leaders were also integrally involved with working with staff to deal with the crisis through leadership by example. As one executive noted, 'our patient care directors and managers were visible, involved, and participated with their staff. Our executive group was vigilant with communication on status updates'. It appears as if emotions had to be kept private so as to perpetuate a sense of control. As a senior executive stated, 'as the events unfolded I was rocked with raw emotions but I could not show this in front of employees, as it would send the wrong message. But as soon I closed the doors of my office, I cried and cried'.

Second, motivating employees was made easier because of the nature of the organization and profession. In general, hospitals provide care, and healthcare workers are devoted to their profession. Many employees saw their responses as natural, given their training. Helping in a time of need came naturally to them. As one technician remarked, 'I am trained to help. My duty is to treat patients and they come first. I am committed to the profession and this means that I put my work first. This also helped to take my mind off the tragedy'. One manager noted that 'our job in healthcare is to help and doing our job was our way of helping in the bigger scenario'.

Third, many, if not all, employees were driven by a sense of patriotism, as well as a love for New York. Many saw the attack as one on fellow Americans and the country's system of governance. Like other Americans, many took flags to work. The institution itself arranged to have flags fly in conspicuous positions. Many were inspired by speeches coming from political leaders, especially Mayor Rudy Giuliani. As one worker remarked, 'we wanted to help fellow Americans and work hard for our country and our city'. An executive remarked, 'our response was in a large part patriotic since our country was at war. We were under attack'.

Fourth, in what may be termed a 'spirit of communality', many wanted to simply help, in any way possible. Many felt helpless in the face of the enormity of the situation and wanted to assist, to be part of the response, even if, in some situations, it meant risking their lives. This desire to help went beyond patriotism and encompassed a genuine desire to help their fellow humankind. Many donated blood. Others worked long hours, even sleeping at the worksite when they became tired. Others walked to work, sometimes over the bridges, hoping that their unselfish services would help in some way to restore some semblance of order. This sense of communality was pervasive and extended to the union as well (only some units at NYPH are represented by organized labor). As one manager remarked, 'the union was very helpful in our response. There was a sense of oneness. At the chapel service, the head of the union squeezed my shoulder and said I could get anything I wanted from them at this time'.

Fifth, and somewhat related to the above two points, there was a noticeable change in the attitude of many workers. People were more polite to each other and showed renewed respect for the work of their colleagues. This changed attitude has resulted in a new respect for many professionals, including security officers and social workers. As one security officer remarked, 'people even say good morning and smile now when they enter the buildings. They do not object to us searching their bags or requesting their ID cards. In fact, you sense that they want us to screen them'. Furthermore, many were willing to work in groups. As one worker recalled,

> The most wonderful thing to come out of this for me, was the devotion, caring and dedication I witnessed in my fellow social workers. We took care of each other, whether we were answering the phone for an exhausted colleague, offering words of support, or bringing in food for the crew. Without their support, I would never have been able to do the work I did.

Management explicitly took advantage of this phenomenon by organizing work in teams. This allowed workers to talk to each other and share mutual concerns. It also proved to be a more efficient management system, as employees complemented each other's strengths.

For some, the events of September 11 may have impacted on their spirituality. This is, however, not a pervasive sentiment. Many executives reported no increase in spirituality in the workplace. Nevertheless, according to Reverend Raymond Lawrence Jr, the Director of Pastoral Care at NYPH, there was an increase in spirituality immediately following the tragedy. He observed that an unprecedented 1000 people attended a hospital chapel service immediately following the attack, noting 'after the service was over, nobody wanted to leave. We couldn't have gotten 100 people into chapel service prior to September 11th' (Media Advisory, NYPH, November 14, 2001). Such renewed spiritualism may have motivated many employees in the workplace.

Finally, organizational leaders must deal with the psychological effects emanating from traumatic events. To effectively manage in the aftermath of a disaster, management must understand the range, magnitude, and duration of the effects. A review of the literature (Norris, 2001) suggests that the range of outcomes includes specific psychological problems (such as post-traumatic stress), health problems, and behavioral problems (such as anxiety). The magnitude of the effects can range from minimal impairment (approximately 9 percent of employees) to very severe impairment (16 percent), and the duration of the effects varies. In the event of extreme and widespread damage to property and loss of human life, as occurred in New York and Washington in September, the evidence suggests that there can be severe, lasting and pervasive psychological effects (ibid.). A review of the management and sociological and psychological literature suggests a series of critical initiatives that management should implement following disasters. For instance, the United States Office of Personnel Management (1996) suggests that following a disaster (terrorism was not considered but the suggested principles are applicable), management, among other initiatives, should modify work rules, allow employees to congregate and talk about their concerns, and take steps to prevent accidents and illnesses. Many scholars and practitioners have made similar recommendations. For instance, Sullivan (2001) suggests that managers should, among other things, be more flexible in requests for using sick leave and vacation; involve the workers or union in the process of alleviating anxiety in order to lessen their fears; allow stressed workers to work from home or to use their sick days until their anxieties subside; and provide counseling services.

Propositions, Emerging Theories and the Extant Literature

An emerging theory from the above is related to what may be termed 'political work commitment'. There are several forms of work commitment discussed in the extant literature: career commitment, job involvement, affective organizational commitment, continuance organizational commitment, and work ethic endorsement (for a review, see Morrow, 1993). Political work commitment is different. Essentially, it is largely an extra-work variable that impacts on work attitudes by an appeal to an employee's patriotism and his/her sense of righteousness about the state's system of governance. This form of work commitment may be intrinsically motivated and will vary with an individual's affinity with patriotic values. It may also be extrinsically triggered through persuasion by political and organizational leaders. Charismatic leaders would be especially effective in this respect. Political work commitment is not restricted to the response to traumatic events in the workplace but may also be evident in the face of external threats or competition.

Another relevant theory relates to professional commitment. This is not a new concept as it has been around for some time (Blau, 1985; Morrow, 1993). Essentially, professional commitment refers to one's identification with, and the level of involvement in, his/her profession. One's profession is related to an occupational community – that is, a group of people who are employed in the same line of work (Van Maanen and Barley, 1984) – and it transcends employing organizations (Blau, 1999). There are a few existing scales that measure professional commitment (Aranya et al., 1981; Blau, 1999), and they include such items as: 'If I could do it all over again, I would choose to work in this profession', and, 'I would recommend a career in this profession to others' (ibid.). In general, the concept of professional commitment, and its related scales, measures the intensity with which someone identifies with a profession through such indicators as membership in professional organizations, acquaintance with relevant literature, acceptance of the goals of the profession, and a willingness to exert significant effort to uphold the tenets of the profession. Many professions, such as those in the medical field, as in this case dealing with terrorism and the NYPH, display high levels of professional commitment.

Third, the concept of 'involved management' was very evident as a factor in the hospital's success in dealing with the tragedy. Involved management is used here to represent the concepts that straddle aspects of various types of managerial styles (such as 'participatory management', 'management by walking around', and 'leading by example') but do not fit neatly into any one of these categories. Essentially, involved management is used here to reflect a more hands-on approach to management, without getting bogged down in the intricacies of the details. The presence of the managers is 'felt' through their physical appearances or open communication channels. Managers are also actively involved in operational issues. As in the NYPH case, or Mayor Giuliani's example, for management to succeed under traumatic conditions, organizational leaders must be in the forefront of the response and display calm and intellect in decision making.

Finally, as above, it was difficult to classify the effective leadership style, identified in the literature and emanating from the case, using extant categories. This style may be referred to as 'compassionate autocracy' and it straddles aspects of several leadership styles, such as 'contingency leadership', 'transformational leadership', 'autocracy', and 'servant leadership'. Essentially, a compassionate autocrat follows strict rules and procedures without being a dictator, and at the same time, shows an understanding of, and deals sensitively with, employee concerns. This concept is eloquently captured by Lichteim (1973), as cited by Bass (1995, p. 50) in an essay on the qualities of good leaders through history: 'In 2300 B.C. in the Instruction of Ptahhotep, three qualities were attributed to the Pharoah – authoritative utterness is in thy mouth, perception is in thy heart, and thy tongue is the shrine of justice'.

The role of the compassionate autocrat is temporally transcending and lasts for a relatively short period. It is contingent on the situation. That is, a crisis situation such as a terrorist attack may require leaders to take an affirmative lead for a while, as effective military leaders do in the middle of combat. Furthermore, these leaders are able to motivate subordinates, or transform them, to work at a higher level by making them more aware of the importance of their tasks, and inducing them to transcend their own self-interest for the sake of their team, organization or society (Burns, 1978; Bass, 1985). Subordinates admire, respect and trust these decisive leaders, even though they display temporarily autocratic qualities. They may operate somewhat like temporary benevolent dictators. These leaders, however, continue to respect employee concerns and ideas, appear as 'leaders of the people', and serve the masses (Greenleaf, 1995).

IMPLICATIONS FOR MANAGEMENT

As evident from this study, employees at the NYPH were motivated beyond normal levels during the crisis. This may have not been uncommon in many New York organizations in the face of the terrorist attack (Kondrasuk, 2004). Lewis et al. (1999) document similar 'heroism' following the Oklahoma City attack. A logical and relevant issue that may be raised relates to the sustainability of increased motivation levels for an extended period. This is an issue that has faced society before, as evident in the levels of motivation and mobilization during wars and natural disasters. As Etzioni (1968) notes, in times of crisis, individual people and units in a system are willing to give their assets and energy to a larger collective cause. However, as the immediate threats of the crisis recede, so does motivation and commitment. Burke (2005a) notes that with the passage of time, adverse reactions are also significantly reduced for most, and that people generally tend to return to their typical pattern of behaviors. While it is of significant theoretical and practical importance to address the issue of how to sustain increased motivation levels over time, this paper focused on the immediate response to traumatic events, such as a terrorist attack, and the following implications must be viewed in this context.

A review of the literature and the NYPH case suggests that there are three distinct responses following a disaster that have implications for management (Lincoln, 2002; Desplaces and Comerford, 2003; Mankin and Perry, 2004; Burke, 2005b; Larson et al., 2006). Despite the fact that some time has elapsed since the tragedy, these implications (and the recommendations that follow) continue to be relevant for all organizations. Furthermore, the relevance transcends spatial considerations. That is, even though the events occurred in New York, the implications and recommendations are equally relevant for similar situations in Belfast, New Delhi, Tel Aviv, London, Paris, or elsewhere. First, organizational leaders must respond to the logistical demands of the situation. That is, as in NYPH case, there must be a Disaster Plan that outlines a step-by-step process of the organization's response. Human resource (HR) managers should be integrally involved in the development and implementation of this plan.

Second, organizational leaders must demonstrate intellect and decisiveness in their response. Managers must ensure effective communication of the necessary details of the situation. They must also assist in implementing systems that promote employee participation. This may include group work, cooperation with the union and open discussion forums.

Third, organizational leaders must effectively deal with the psychological effects of traumatic events such as terrorist attacks (Burke, 2005a). Even though HR managers are generally not trained in this area, they must be trained to be aware of the issue, and what to do in crisis situations. It is advisable that effective relationships be established with the social work department in the organization, if it has one, or with an external agency, if it does not. Many HR departments train their personnel in counseling and handling stress, and this may be useful in dealing with an immediate situation; nevertheless, professionals must be contracted at the earliest opportunity.

Based on the results of the case study and a review of the literature on the above three impacts on management, the following specific recommendations are suggested for practitioners:

1. Ensure that you have an effective Disaster Plan. Test the plan to ensure that it is easily implementable. Stick to the plan during the disaster. Have back-up offices for staff and store data at a back-up site.
2. Ensure that you have as many facts as possible on the situation and act decisively. Keep internal and external communication lines open. Inform employees of the situation as it evolves.
3. Involve employees in the response effort and encourage participation.
4. Work cooperatively with employee unions. Labor–management cooperation, regardless of prior labor relations, has the potential of bringing the two sides together for the good of the organization.
5. If relevant, use patriotism wisely. Do not stifle employees' sense of loyalty to the state and government.

6. Take care of your own people first. That is, organizations should locate their own employees first and communicate with their families. Relief centers may be considered since traditional disaster relief agencies may be limited in their resources.
7. Modify office rules and procedures that are counterproductive. For instance, organizations may have to grant administrative leave to employees who need time off to normalize their family situation.
8. Take steps to prevent accidents and illness.
9. Provide opportunities for employees to talk about their experiences and/or the situation. This may be encouraged through formal or informal avenues.
10. Organize for professional counseling services to be made available to employees, and the public, if relevant.

It is hoped that we shall never again experience a terrorist attack like that of September 11, 2001. Nevertheless, we must be prepared to deal with any eventuality. This chapter is a modest attempt in this direction.

NOTE

1. The author is grateful for the cooperation of executives and employees at the New York Presbyterian Hospital. Thanks to Dr Dale Finn and Dr Allen Sack for reviewing a draft of the chapter. All errors are the responsibility of the author.

REFERENCES

Adams, J. (1965), 'Inequity in social exchanges', in L. Berkowitz (ed.), *Advances in Experimental Social Psychology*, Vol. 2, New York: Academic Press, pp. 267–99.

Aranya, N., Pollock, J. and Amernic, J. (1981), 'An examination of professional commitment in public accounting', *Accounting, Organizations and Society*, **6**, 271–80.

Bass, B. (1985), *Leadership and Performance beyond Expectations*, New York: Free Press.

Bass, B. (1995), 'Concepts of leadership: the beginnings', in T. Wren (ed.), *The Leader's Companion*, New York: Free Press, pp. 49–52.

Blau, G. (1985), 'The measurement and prediction of career commitment', *Journal of Occupational Psychology*, **58**, 277–88.

Blau, G. (1999), 'Early-career job factors influencing the professional commitment of medical technologists', *Academy of Management Journal*, **42**(6), 687–95.

Burke, R. (2005a), 'Effects of 9/11 on individuals and organizations: down but not out!', *Disaster Prevention and Management*, **14**(5), 629–38.

Burke, R. (2005b), 'International terrorism and threats to security: implications for organizations and management', *Disaster Prevention and Management*, **14**(5), 639–43.

Burns, J. (1978), *Leadership*, New York: Harper & Row.
Corbin, J. and Strauss, A. (1990), 'Grounded theory research: procedures, canons, and evaluative criteria', *Qualitative Sociology*, **13**(1), 3–21.
Desplaces, D. and Comerford, R. (2003), 'The impact of September 11 on the IT sector in Rhode Island: a partial assessment of human resource management practice', *Employee Responsibilities and Rights Journal*, **15**(2), 45–55.
Etzioni, A. (1968), *The Active Society*, New York: Free Press.
Glaser, B. and Strauss, A. (1967), *The Discovery of Grounded Theory: Strategies for Qualitative Research*, Chicago, IL: Aldine.
Greenleaf, R. (1995), 'Servant leadership', in T. Wren (ed.), *The Leader's Companion*, New York: Free Press, pp. 18–23.
Hackman, J. and Oldham, G. (1975), 'Development of the job diagnostic survey', *Journal of Applied Psychology*, April, 159–70.
Kondrasuk, J. (2004), 'Effects of 9/11 and terrorism on human resource management: recovery, reconsideration and renewal', *Employee Responsibilities and Rights Journal*, **16**(1), 25–33.
Larson, R., Metzger, M. and Cahn, M. (2006), 'Responding to emergencies: lessons learned and the need for analysis', *Interfaces*, **36**(6), 486–504.
Lewis, C., Tenzer, M. and Harrison, T. (1999), 'The heroic response to terror: the case of Oklahoma City', *Public Personnel Management*, **28**(4), 617–36.
Lincoln, R. (2002), 'Dealing with HR issues following the 9/11 terrorist attacks', *Employee Relations Today*, **28**(4), 1–7.
Mankin, L. and Perry, R. (2004), 'Terrorism challenges for human resource management', *Review of Public Personnel Administration*, **24**(1), 3–10.
McClelland, D. (1961), *The Achieving Society*, New York: Van Nostrand Reinhold.
Morrow, P. (1993), *The Theory and Measurement of Work Commitment*, Greenwich, CT: JAI Press.
Naylor, J. and Illgen, D. (1984), 'Goal setting: a theoretical analysis of a motivational technique', in B.M. Staw and L.L. Cummings (eds), *Research in Organizational Behavior*, Vol. 6, Greenwich, CT: JAI Press, pp. 95–140.
New York Presbyterian Hospital (NYPH) (2002), *NYPH Emergency Preparedness Plan*, New York: NYPH.
Norris, F. (2001), 'The range, magnitude, and duration of effects of natural and human-caused disasters: A review of the empirical literature', http://www.ncptsd.org/facts/disasters/fs_range.html, accessed June 16, 2005.
Strauss, A. and Corbin, J. (1990), *Basics of Qualitative Research: Grounded Theory Procedures and Techniques*, Newbury Park, CA: Sage.
Sullivan, J. (2001), 'How should HR and managers react in the aftermath of terrorism events?', *Erdaily*, December 3, http://www.erexchange.com, accessed March 20, 2005.
Turner, B. (1990), 'The use of grounded theory for the qualitative analysis of organizational behavior', *Journal of Management Studies*, **20**(3), 333–48.
United States Office of Personnel Management (1996), *Handling Traumatic Events: A Manager's Handbook*, Washington, DC: Office of Workforce Relations.
Van Maanen, J. and Barley, S. (1984), 'Occupational communities: culture and control', in B.M. Staw and L.L. Cummings (eds), *Research in Organizational Behavior*, Vol. 6, Greenwich, CT: JAI Press, pp. 287–365.
Vroom, V. (1964), *Work and Motivation*, New York: John Wiley.
Whetten, D. (1989), 'What constitutes a theoretical contribution?', *Academy of Management Review*, **14**, 490–95.

11. Worksite emergency preparedness: lessons from the World Trade Center Evacuation Study

Robyn R.M. Gershon, Kristine A. Qureshi, Briana Barocas, Julie Pearson and Stephanie A. Dopson

INTRODUCTION

The events of September 11th, 2001, and other recent workplace fires, explosions, natural disasters, violent assaults with deadly weapons and other emergencies, have underscored not only the potential vulnerability of worksites, but the importance of preplanning for disaster response. Since the World Trade Center Disaster (WTCD), the role of emergency preparedness, in terms of mitigating the adverse impacts of disaster events in the workplace, has gained in importance and recognition, both nationally and internationally. In the United States (US), the leading workplace safety agency, the National Institute of Occupational Safety and Health (NIOSH), has identified emergency preparedness and response as an important cross-sector research program in recognition of this significance (NIOSH, 2007).

Worksite readiness for a wide range of emergencies is important, not only because it can help reduce disaster-related morbidity and mortality among workers and visitors, but it can also help to preserve organizational assets and maintain corporate functions and processes. Preplanning and response, therefore, is essential to business continuity and viability. While numerous recent events heightened both interest and efforts in preplanning, a number of challenges to preparedness have hampered these efforts. Perhaps the most important of these is the general lack of consensus of what constitutes 'preparedness'. There is also lack of agreement on how best to measure and achieve this, and this is true at a number of levels, including the community, business, public health, and public safety sectors. From a management perspective, a serious challenge to preparedness is the competition for limited resources; low probability/high consequence events can, and often do, play a lesser role when there are other, more pressing, issues at hand. The question facing work organizations of all sizes and types after addressing: 'Are we prepared for emergencies?' and 'How do we know that we are prepared?' is 'How can I achieve this in the most efficient and cost-effective way possible?' This is the question that is also of concern to emergency planners, disaster researchers, worker representative groups, and workers.

This chapter provides an overview of the role of workplace emergency preparedness in protecting the health and safety of workers and work organizations. It also presents evidence from the World Trade Center Evacuation (WTCE) Study on the factors that influence employees' response to a workplace disaster, along with simple and relatively low-cost recommendations that address these factors.

Disasters

Life-threatening workplace disasters may be naturally occurring, human-made, or technologically related. Often, the term, 'emergency', is used interchangeably with the term, 'disaster', which has been defined as 'any destructive event that disrupts the normal functioning of a community' (Veenema, 2006). Some examples of naturally occurring disasters are wild fires, floods, hurricanes, tornadoes, earthquakes, and extreme weather or temperature conditions, and so on. Examples of human-made disasters (either intentional or unintentional) include fire, explosions, major hazardous materials release (for example, chemical, biological, and radiological), exposure to physical hazards (for example, electrical and so on), and acts of violence (including, in its most extreme form, terrorist attacks or threats of such attacks). Technological disasters include nuclear power plant releases, chemical plant explosions, oil spills, and equipment failures. Disasters are generally declared when the event reaches a certain magnitude that is beyond the capability and capacity of the local emergency response agencies. The statutory framework for the declaration of a major disaster is provided by the Robert T. Stafford Disaster Relief and Emergency Assistance Act (2005).

Human behaviors during disasters

From a worksite preparedness perspective, a particular aspect regarding disasters is of special importance, namely human behavioral responses during these types of events, as it can provide guidance to the planning process. Scientists have long been fascinated by human response to disaster or emergency, and some of their findings are informative here. Careful study of how individuals and groups respond under threatening conditions has led to some surprising and counterintuitive conclusions. For example, although panic is believed to be common during disasters, this is, in fact, rather uncommon. The conditions for panic have been identified by Quarantelli (1997) and others and include the following:

- Ineffective crisis management (leaves people feeling completely on their own).
- Pre-existing beliefs in a group that certain kinds of situations will lead to panic (that is, during a fire people will stampede exits).
- Immediate threat of entrapment (panic does not develop when people know they are trapped, but rather when they sense that their chances for escaping danger are dwindling).
- There is a sense of complete social isolation – that is, that there is no one else in the setting upon whom they can depend.
- There is no possibility of survival, except through flight.

These findings support the important role that emergency preparedness can play in helping to manage the emergency, thereby lessening the chance of panic. In the WTCD, panic behaviors were rare. This is most likely due to the fact that the conditions for panic were absent; people were familiar with one another and felt supported, there was some leadership provided, and escape was possible for those below the impact zones. Importantly, most people did not know how serious the impacts were to the structural integrity of the buildings and thus did not know that the buildings would collapse (Gershon et al., 2007). Other research has shown that people usually behave in active and adaptive ways during a disaster, with altruistic behavior not uncommon.

Another area of study that also has implications for worksites is the resistance some people have to evacuation during disasters or the threat of disasters. Social scientists (Lindell and Perry, 1992) have identified four key questions that people ask themselves before they decide to evacuate:

- How believable is the threat?
- What is the feasibility of my escape (for example, stairway access)?
- Do I personally have the ability to escape?
- How effective will my attempts be to actually escape?

The responses to these questions have been found to be affected by various situational and social factors such as age, ethnicity, socioeconomic status, family status, personality characteristics, experience and so on. This model has been tested and found to explain significant variance in protective and proactive responses to disasters. The important message to management from these findings is that accurate communication during an emergency, along with training and drilling, can result in appropriate action. These findings also highlight the role of senior leadership in setting an example by following evacuation orders.

Health and safety impacts

The adverse impact of disasters on the health and well-being of those affected, even indirectly, can be great, and these can have important implications for the affected worksite. Mental health problems, such as post-traumatic stress disorder (PTSD), depression, and anxiety, are especially common in the aftermath of disasters. Importantly, beyond the immediate negative physical and mental health effects related to disasters, long-term consequences can also result. In a review of more than 160 disaster reports, with roughly 10 percent classified as violence related, a number of co-morbidities of mental health disorders, including elevated substance use, non-specific distress, chronic difficulty coping with the activities of every-day life, and psycho-social resource losses, were identified (Norris et al., 2002). Consequences appear to be most severe when the disaster has the following characteristics: (i) widespread property damage, (ii) ensuing and ongoing financial crises, (iii) high rates of trauma or fatalities, and (iv) intentional human violence.

Terrorism in particular, including the threat of terrorism, has been shown to have the greatest adverse impact on survivors' well-being, in comparison to other types of disasters (Ursano, 2002; Stith Butler et al., 2003). These findings have been confirmed by Galea et al. (2005) in a review that focused on the determinants of PTSD after disasters. The prevalence of PTSD in the first year, after human-made/technological disasters, has been documented as high as 75 percent (Galea et al., 2005). In comparison, studies of adult survivors of disaster have estimated the prevalence of PTSD as high as 60 percent (Brooks and McKinlay, 1992; Green et al., 1992; Dooley and Gunn, 1995; Selley et al., 1997; North et al., 1999; Arata et al., 2000; Yule et al., 2000; Galea et al., 2005). In further comparison, the lifetime risk of PTSD within the general public is 10 percent (Kessler, 2000). Although there has been less study on other psychopathologies following disasters, data on post-disaster depression (for example, Kuo et al., 2003; Morgan et al., 2003; Salcioglu et al., 2003), anxiety (Ben-Ezra, 2004; Centers for Disease Control and Prevention, 2004; Verger et al., 2004), and panic disorders (Deering et al., 1996; Bystritsky et al., 2000) indicate that survivors of disasters are at risk. However, despite high levels of psychological distress, many disaster victims fail to seek mental healthcare (Tierney et al., 2001). Investigators have suggested that much of the work needed to discern the appropriate post-disaster mental health response should involve survivors on a variety of levels, including those 'directly' and those 'indirectly' affected (Everly, 1999). Thus, disaster management plans should include services to primary (direct survivors), secondary (families of survivors), and even tertiary victims (the neighboring population) (Green and Lindy, 1994), since stress and maladaptive coping mechanisms are frequently underestimated in all groups (Burkle et al., 1994). The delivery of mental health services to appropriately screened populations is one of the most pressing, and promising, areas of post-disaster research. Since the workplace may be directly or indirectly affected by terrorism and other disasters, the workplace can make an effective target for reaching at-risk individuals for screening and treatment.

Terrorism

Terrorism is a type of disaster, even though, in some cases, it may affect only a small number of individuals. Terrorism has been defined as an act of violence or threat of violence against civilian populations to achieve political, religious, or ideological objectives (FBI, 2001). In the workplace, terrorist attacks can take many forms (Cameron et al., 2000); as any of the more typical life safety threats (for example, fires, workplace violence, hostage situation, bombs and bomb threats, hazardous material release and so on) may involve acts of terrorism. For example, terrorism can manifest overtly, such as a bomb attack (for example, such as the bombings of abortion clinics), or covertly (for example, the mailing of anthrax letters to news centers). Sometimes there is a fine line between terrorism and felonious assault; mass shooting by a disgruntled worker can be construed as a terrorist event or as an act of violence. From a management perspective, preparedness enables the worksite to effectively prepare and respond to a range of terrorist events, regardless of the form it takes.

Terrorism in the workplace can take on special meaning as it not only disrupts the work environment, but can also lead to a destabilization of trust in the work organization. Because work is a central organizing factor and key source of identity in adults (Schouten et al., 2004), workplace terrorism can disturb deeply held beliefs. Further, by targeting government and iconic structures, including high-rise buildings (for example, the Murrah Building in Oklahoma City, the World Trade Center (WTC), and embassies in Africa), terrorists can even affect the trust that the general public has in their government, with respect to their security and safety. The attack on the WTC was an example of the far-reaching effect of targeting an iconic (that is, symbolic) structure. Because the towers were a symbol of capitalism, the attack served to assault the US financial system, the culture, and the country itself (ibid.). Further, extensive damage can result when the worksite targeted is in a large urban area, resulting in the greatest possible damage and disruption. Similarly, attacks on train stations and other mass-transit systems disrupt commuters (to and from work) and the general public, as well as the area in the immediate vicinity of the attack. Both mass-transit and large worksites are likely targets because they provide a convenient source of large numbers of people at predictable times and places.

Emergency Preparedness

Emergency preparedness, sometimes referred to as emergency management or disaster management, is the process of preparing for, and recovering from, disasters. Emergency preparedness is categorized into four components: mitigation, preparedness, response, and recovery (Veenema, 2006), and is considered the proactive planning for each of these categories. With respect to the workplace, the National Fire Protection Agency (NFPA) published a standard that establishes a common set of criteria for emergency preparedness, and management and business continuity that applies to both the private and public sectors (NFPA, 2007). While earlier editions of the standard focused on the continuity of operations and the property, facilities and infrastructure of businesses, and delivery of services, the newer editions stress the importance of emergency preparedness that protects the health and safety of persons in the affected area, as well as personnel responding to the incident. The 2007 version emphasizes preparedness in terms of worksites preplanning collaboratively at the local (for example, emergency services), state (for example, state agencies), and federal (for example, Department of Homeland Security, Federal Emergency Management Agency (FEMA), FBI and so on) levels.

Increasingly, the emergency preparedness and response community is moving toward a concept referred to as an 'all hazards' approach, which posits that preparedness in general will improve responsiveness to a wide range of emergencies. The elements of a basic 'all hazards' emergency response plan for worksites include the following: (i) the development of an incident management system, (ii) a written emergency plan, and (iii) a business continuity plan.

The incident management system is simply a way to identify who will be in charge (that is, the leader or 'incident commander') during an emergency. It also identifies who will assist this person and how all of these individuals can be contacted in case of emergency. The leader is the key decision maker during an emergency and also serves as the link to professional responders when they arrive at the scene.

A written emergency plan can range from a simple document to a highly complex and detailed one, depending on the size and type of the work organization. In general, for most small to mid-size businesses, this can be relatively easy to develop, usually by a small committee. The committee can help identify the potential risks that could threaten the worksite (for example, fires, earthquakes, hurricanes, bombings and so on). Once these are identified, a plan is prepared for responding generally to disaster events and, more specifically, to certain types of events (for example, floods versus fires). The plan, in essence, serves as the blueprint for how the worksite will manage a disaster. Key elements in the plan generally include the following:

- How to report emergencies, both externally and internally to the leadership.
- How to evacuate the workplace.
- How to help individuals with special needs evacuate or move to a safe location.
- How to shut down critical workplace equipment or procedures – and who is authorized to do this.
- How to account for all employees after an evacuation.

Some workplaces have found it helpful to identify and train a cadre of workers to act as a response team (that is, 'the emergency response team'). The teams can be effective in terms of assisting the incident commander or leader, or by helping co-workers evacuate, or by providing immediate first aid prior to the arrival of professional responders.

The third key aspect of worksite emergency preparedness is the continuity of operations plan. This is a written document that can help the affected business survive the disaster. The quicker the business can become functional after a disaster, the better the outcome will be for the employees. The major elements of a continuity plan (FEMA, 1993; Chao and Henshaw, 2001) are listed below:

- Identify staff, materials, processes, and equipment that are essential to the business.
- Prepare a back-up plan for these essential elements.
- Prepare contact lists for all key support systems and keep back-up copies at several locations.
- Plan for continuation of operations in alternative locations.
- Determine how employee contact will be maintained if the work location is inaccessible.
- Determine how payroll will be maintained if normal services are no longer functioning.

RECENT ACTS OF TERRORISM IN THE US WORKPLACE

These case studies provide insights, with respect to the impact these events had on the involved worksites, as well as the surrounding communities. Most of these worksite terrorist events resulted in changes in practice at multiple levels, especially including worksites, seen as vulnerable or likely targets for terrorism. In addition, both the public safety and public health sector response plans were greatly affected by these events.

World Trade Center Bombing, 1993

On February 26, at 12:18 p.m., a terrorist bomb containing approximately 1200 pounds of urea nitrate exploded in an underground parking garage at the WTC in New York City (Fahy and Proulx, 1995). The explosion resulted in a five-storey crater that was 24–36 meters wide (Hinman and Hammond, 1996). The bomb incapacitated the fire alarm and power systems, and a back-up generator quickly overheated and shut down. The resulting fire, smoke, and structural damage caused more than 1000 injuries to building inhabitants (mostly employees) and the death of six workers. A mass evacuation of the seven fully occupied (approximately 100,000 occupants) high-rise building complex (including the two 110 storey towers) was immediately attempted, but took over six hours to fully complete. Hours after the bombing, occupants were found by rescue workers still sitting at their desks, unaware of the evacuation orders (Fahy, 1995).

Several studies reported on various aspects of this event, including the fire safety aspects (ibid.), morbidity risk factors (Quenemoen et al., 1996), the effect of group size and social cohesiveness on the evacuation behavior of WTC occupants (Wenger, 1987; Aguirre et al., 1998), and the human behavior factors associated with the evacuation (Proulx, 2001). From these studies, valuable information on high-rise fire safety and optimal group size and group characteristics for adaptive and pro-social behavior in workplace disasters were obtained. For example, social cohesion was shown to affect pro-social behaviors, thus reducing risk of group panic, and thereby supporting the evacuation process. These studies also provided important insights as to the role of worksite preparedness on evacuation behaviors of occupants, which, in turn, affected survivability.

These studies also noted some serious response flaws. For example, the most serious response problems involved the following: delays in notification that a mass evacuation was in effect, locked egress routes, inoperable elevators (some workers were physically incapable of exiting 100 or more flights of stairs), poor lighting in the stairwells, and 'turning-back' behaviors of workers. Here, employees who were afraid of exiting through smoky stairwells turned back or sought refuge on the nearest floors. Incomplete or delayed evacuation of the buildings occurred, as many workers stayed where they were because they were waiting for instructions. Even though they were capable, some of these workers were not evacuated until hours later when fire personnel reached their locations. Many workers also reported that they were unaware of the nearest fire notification (that is, pull boxes) stations and many (90 percent) reported that they had *never* participated in an evacuation drill and, therefore, were unfamiliar with the location of the stairs. This was especially important since many of the people who reportedly did not know where the stairs were located also stated that they were assigned as fire wardens for their floor. While 80 percent of those surveyed stated that they had received prior safety information, most felt that it had simply prepared them to wait for instructions, which in some cases either never came or came many hours after the bombing had occurred. Importantly, many building occupants were unaware of the poor helicopter access to the rooftop, because of the communication towers that were placed there, and ascended for a rooftop rescue.

The Port Authority of New York and New Jersey, which managed the WTC at the time of the 1993 bombing, made numerous improvements to fire safety at the complex after this event. For instance, reflective egress signage was added and new overhead intercom systems were installed throughout the buildings. Many businesses that leased space in the towers similarly increased and improved their emergency readiness plans. Unfortunately, some important recommended changes were not implemented, such as widening stairwells and adding stairways.

Oklahoma City Bombing, 1995

On April 17 at 9:02 a.m., the detonation of 4000 pounds of explosives in front of the nine-storey Murrah federal building in Oklahoma City resulted in massive structural damage. The blast resulted in a 30-foot wide and 8-foot deep crater that blew away nearly half of the building (Hinman and Hammond, 1996). The bombing caused 168 deaths and more than 800 injuries among building occupants, those in adjacent locations, and on the street (North et al., 1999). The fatality rate (46 percent or 168/361) of Murrah building occupants, mainly government employees, was extremely high (ibid.). The occupant-injury rate (47 percent) was also high and the combined fatal and non-fatal injury rate of Murrah occupants alone was 92 percent (331/361) (ibid.). The force of the explosion was sufficient to damage or destroy approximately 800 buildings in the area, causing property damage in excess of $625 million.

In the aftermath of the bombing, several assessments were conducted, including a study of injuries to rescue workers (Dellinger et al., 1997). A study of psychiatric disorders among survivors approximately six months after the disaster by North et al. found that 45 percent of 182 adult survivors had a post-disaster psychiatric disorder and 34.3 percent had PTSD. Predictors of mental health sequelae included exposure factors, female gender, and history of pre-disaster psychiatric disorders. However, details on the preparedness planning for the occupants of the building were not provided. Other studies related to this disaster, including disaster-related substance use, focused on first responders and community members.

The Murrah building was targeted because it was a symbol representing the federal government. It was a relatively easy target since access was readily available (that is, the bomb-laden truck was able to park directly in front of the main entrance, even though this was a no-standing zone). Following the attack, the General Services Administration ordered the more than 1300 US federal buildings to increase security and surveillance. Inspection of packages and vehicles was instituted and, in some buildings, parking restrictions and the installation of barriers occurred. Barriers were also placed at critical points at many other non-federal buildings that were seen as potential targets.

Another aspect of the Murrah building bombing was that it affected a first-floor daycare center for young children of employees. When the bomb exploded, many children were injured and 19 were killed (Pfefferbaum et al., 2003). After this event, many at-risk buildings housing similar daycare centers moved children to safer areas of their buildings. Some businesses now refuse to house daycare centers on their premises.

Pentagon Attack, 2001

At 9:38 a.m. on September 11, 2001, terrorists crashed a Boeing 757 carrying 20,000 gallons of fuel into the Pentagon. The Department of Defense reports that 125 people on the ground and 64 in the hijacked airliner died, while 76 people in the Pentagon were injured (Grieger et al., 2003). Studies on the mental health of survivors (ibid.) examined PTSD, alcohol use, and perceived safety in 77 survivors several months after the attack. They noted a probable PTSD rate of 14 percent, with 13 percent of respondents reporting increased use of alcohol. Women were more than five times more likely than men to have PTSD and almost seven times more likely to report increased use of alcohol. Women were also more likely to have a lower perceived sense of safety compared to men. Furthermore, those with lower perceived safety several months after the attack were more likely to have PTSD and increased alcohol use. In another study of Pentagon employees ($n = 120$), more than one year after the attack, the prevalence of PTSD was even greater at 23 percent (Grieger et al., 2004). Anecdotal reports suggest that Pentagon survivors felt isolated from their community and the larger grieving US community, which was highly focused on the WTC attack. Since the attack, there are reports that emergency drills have increased, but no other information regarding preparedness efforts has been forthcoming.

Anthrax Attacks, 2001–2002

Letters contaminated with a microbial pathogen (anthrax) were mailed from Trenton, New Jersey one week after the September 11th terrorist attacks (CDC, 2001). Five separate letters were mailed on September 18, 2001 to different media organizations, although only two of these letters were obtained by the authorities. A short time later, on October 9, two additional contaminated letters were mailed to members of the US Senate. These later two letters differed from the first wave of mailings in that they contained highly refined 'weaponized' grade anthrax spores. Altogether, 22 people contracted anthrax through contact with the letters and, in at least two cases, the point of contact was never identified. Five out of the infected individuals subsequently died, all from inhalation anthrax, a deadly form of the disease.

This terrorist event had an enormous economic impact. It resulted in millions of dollars of clean-up costs and it effectively shut down the Brentwood postal facility in the District of Columbia for over two years. Total damages are believed to have exceeded one billion dollars. The impact on the workers who may have been exposed was great; more than 10,000 people were placed on prophylactic medication and, years after the attack, reports of chronic anxiety and other physical and mental health problems were reported (Quinn et al., 2005). This event also spurred interest in bioterrorism in general. Significant change occurred at the health sector level as healthcare employees needed training and education on bioterrorism signs, symptoms, and treatment, since patients with known or suspected bioterrorism-related infection rarely, if ever, had been previously encountered. Some workplaces developed specific training and education programs on bioterrorism for their staff, although, for the most part, this was not done.

At the government level, this event led to very significant changes. For example, the Strategic National Stockpile, a large supply of medicine and supplies the US government stores in case of a public health emergency, was greatly expanded. The US also increased the number of both trained personnel who can respond to a bioterrorism event, and the entire bioterrorism response infrastructure. However, there is now a growing realization that, in addition to the need for a public health and Department of Homeland Security response, businesses will need to be involved in preparedness planning and response as well. Specifically, the government is considering approaching large companies to develop a response plan that would include being able to receive assets, whether in the form of prophylaxis or vaccine, develop the capability to set up a 'clinic' to vaccinate, and provide prophylaxis to workers. This would remove some of the burden on the local health departments during a large-scale emergency since there is still limited workforce at the federal, state and local levels. The role of management, in addition to receiving, staging, storing, and distributing assets to their workers, would include developing an emergency management plan with a policy addressing continuity of operations. Companies would also need to provide their employees with the technology to work at home, thereby limiting contact in public gathering places, such as schools and workplaces, during a contagious outbreak that might result from a bioterrorist event.

WTC Disaster, 2001

At 8:46 a.m. on September 11, 2001, WTC Tower 1 was attacked by terrorists who used an American Airlines Boeing 767 passenger plane as a weapon (Centers for Disease Control and Prevention, 2003). The impact occurred on the north elevation of the building between the 94th and 98th floors. At the time of impact, the airplane was traveling at 470 mph and carried an estimated 10,000 tons of fuel. Seventeen minutes later, at 9:03 a.m., terrorists crashed a United Airlines Boeing 767 into the south elevation of Tower 2 between the 78th and 84th floors. This plane was traveling at 590 mph and also carried an estimated 10,000 gallons of fuel. The collapse of Tower 2, 56 minutes after impact, was followed less than an hour later by the collapse of Tower 1 at 10:28 a.m., 102 minutes after impact.

An estimated 100,000 people visited or worked in the seven buildings that formed the WTC complex, on the morning of September 11, 2001, however, roughly half that number were present. For a variety of reasons (for example, it was the first day of public school, a primary mayoral election was being held), approximately 58,000 people were in the WTC buildings 1–7 and concourse (including the subway and train stations), with an estimated 17,400 people in WTC Towers 1 and 2. The Observation Deck, which was typically crowded with tourists, was empty at the time of impact since it normally opened at 9:30 a.m. With nearly 2800 deaths related to the WTC attacks (including 157 individuals on the two airplanes) (ibid.), this was the worst terrorist attack in US history (the Pearl Harbor death toll stands at 2403) (Aguirre et al., 1998). Data collection on injuries and fatalities began within hours of the WTC attack and, in the CDC's review of victims' injuries (ibid.) it was found that of 810 individuals treated at five local hospitals after the attack, 16 percent required hospitalization and 0.4 percent died while receiving emergency care. Only 14 percent of the patients suffered non-injury conditions (for example, cardiac, respiratory, or psychiatric), with most victims suffering from exposure to smoke, dust, debris, and fumes. Other common injuries included burns, fractures, head injuries, and crush injuries. The percentage of WTC inhabitants treated was much lower than that in the Oklahoma City bombing because most WTC victims either escaped or died (no one in Tower 1 above the 92nd floor survived, and only 18 people above the 78th floor in Tower 2 survived). The evacuation effort in the WTC attacks was remarkable (especially in comparison to the 1993 bombing noted above) in that most of the people who were at their work stations that morning escaped in a mere 102 minutes between the first impact and the collapse of the second tower. The rapidity with which the buildings were evacuated has been credited to two factors: first, the low occupancy rate and, second, because of the numerous fire safety enhancements that were made following the 1993 bombings. A number of problems were identified in the studies that examined the evacuation (Proulx and Fahy, 2002; NIST, 2005; Gershon et al., 2007) and these are discussed in the final section of this chapter.

EMERGENCY STANDARDS AND GUIDELINES RELEVANT TO WORKPLACE EMERGENCY PREPAREDNESS

Codes and Standards

Management should be aware that there are numerous standards, codes, and guidelines that address workplace safety, some of which are particularly relevant to the discussion of emergency preparedness for the worksite. Standards differ from guidelines in that standards are enacted under the law, whereas guidelines are advisory recommendations usually applied through consensus or convention. Standards have the weight of the law and infractions can result in penalties.

With respect to worksite preparedness, codes and standards for fire safety and building design, construction, are most applicable here. In the US about half of the states have adopted the National Fire Protection Association life safety code (NFPA, 2000) which establishes the minimum requirements for providing a reasonable degree of protection from fire.

With respect to building codes, the National Institute of Standards and Technology (NIST) produces standards and technical reports that refer to buildings. In addition, in 1994 the US International Code Council (ICC) developed a comprehensive set of model standards and codes to standardize the myriad building design and construction codes that exist. These include the Building Officials and Code Administrators (BOCA) International, which publishes the National Building Code (ICC, 2003), the International Conference of Building Officials (ICBO), which publishes the Uniform Fire Code (ICBO, 1994), and the Southern Building Code Congress International (SBCCI), which publishes the Standard Building Code (SBCCI, 1993; ICC, 2003). There are also industry standards and codes, such as the American National Standards Institute (ANSI) codes, that serve an important role in the safety and integrity of building structures (Society of Fire Protection Engineers, 2003). Other countries have also developed similar standardized codes.

In terms of worker protection related to worksite emergencies, regulations that are applicable here have been promulgated by the federal Occupational Safety Health Administration (OSHA). The general industry standards for fire safety and emergency evacuation are found in two subparts of the OSHA regulations at 29CFR 1910, Subparts E and L. The OSHA regulations, 29CFR 1910.36, 37, and 38, detail the requirements for emergency egress, basic fire safety, emergency response, and fire prevention (FEMA, 2002). All employers covered by the general industry standard are *required* to comply with *all* aspects of 1910.36 and 37 (fire exits, alarm systems, visibility and signage for egress routes and so on). Section 38, which describes OSHA's requirement for emergency response and fire prevention, applies only if the workplace is required by another OSHA regulation to have an emergency action or fire prevention plan. For example, workplaces that comply with the OSHA Hazard Communication Standard would also be required to comply with section 38. In addition, most employers in office, retail, and manufacturing establishments are subject to section 38 in order to be in compliance with OSHA standards related to fire detection systems and equipment (Subpart L, 29 CFR 1910.156–65) (Eachempati et al., 2004). Under section 38, employers have an obligation to develop an 'emergency action plan', which is a written policy for evacuation of the worksite in case of an emergency. In addition, employers need a written fire prevention plan, which describes safe handling and procedures for controlling workplace fire hazards. To help employers meet the requirements of these standards, OSHA developed a number of fire protection training courses, including fire safety videos, and has extensive information available to both employers and employees on their website (FEMA, 2002). OSHA also conducts inspections to determine whether the standards are being met; between October 1999 and September 2001, the federal OSHA issued in excess of 4000 citations for violations with elements of these two standards, with roughly $1.4 million in penalties. Violations typically were related to lack of fire extinguishers, blocked egress, including blocked stairwells, and lack of fire prevention and protection programs (Department of Labor, 1987).

The key elements of OSHA Emergency Preparedness Standards are listed below.

1. A written emergency action plan.
2. A written fire prevention plan.
3. Employee training on the plan, including when employees are first hired.
4. Establishment of a chain of command and the designation of an emergency response team coordinator.
5. Emergency communications systems, including alarm systems.
6. Emergency escape procedures and emergency escape routes.
7. Procedures to account for all employees after emergency evacuation has been completed.
8. Assignment of rescue and medical duties and training for those employees who are to perform them (emergency response teams).
9. The means of reporting fires.
10. The capability for rapid provision of emergency medical assistance on- or off-site.
11. Security of the affected worksite, including maintenance of important records and documents (for example, listing employees' relatives to be notified in case of emergencies).
12. Unobstructed egress from all parts of the workplace at all times it is occupied.
13. Clearly marked exits and exit routes.
14. Where applicable (by size), fire alarms.
15. Exits which discharge directly to a street or other open space.

In addition to federal and state OSHA standards that are enforced for businesses with 11 or more employees, there are a number of state and local fire safety code regulations which apply to businesses, including egress regulations and fire supervision regulations. These state and local regulations generally apply to all businesses, regardless of their size. There are also guidelines that are published to help improve disaster readiness, such as the FEMA guidelines entitled 'Emergency management guide for business and industry', for emergency planning, response, and recovery for companies of all sizes (FEMA, 1993). As guidelines, they serve as recommendations, not regulations, and therefore there are no reporting requirements. Similarly, the CDC Fire Safety Guidelines, published in the *CDC Health and Safety Manual, 1997*, are also recommendations (CDC, 2007).

In response to the WTC disaster, 2001, the first major New York City (NYC) fire safety code regulation in 30 years was legislated to improve emergency preparedness of high-rise structures. This new law (Local Law 26, 2004), entitled 'Building Safety', requires that the NYC Fire Commissioner promulgate rules establishing standards, procedures, and requirements for the orderly evacuation of occupants from any office building or building classified in occupancy group E (high-rise office space buildings) including, but not limited to, the evacuation of persons necessitated by fire, explosion, biological, chemical, or hazardous material incident, or release arising from natural disasters, or other emergencies, or threat of such events (Local Law 26, 2004).

As a result of Local Law 26, and the new requirements that resulted, the NYC Fire Department (FDNY) passed the following rule in 2006: 3 RCNY § 6-02, entitled 'Office Building Emergency Action Plans' (FDNY, 2005). This new rule requires that all high-rise buildings in NYC designated as office space must have an approved emergency action plan (EAP) that must be implemented in the building (ibid.). The EAP for each building must include provisions for the following: sheltering in place, in-building relocation, and partial and full evacuation of the building as a result of different emergency situations occurring either within the building or the building's vicinity (ibid.). This rule also stipulates that in addition to having an EAP for all class E high-rises in NYC, each of these buildings must also have a Fire Safety/Emergency Action Plan Director (FS/EAPD) who is responsible for implementing the EAP. Certification as a FS/EAPD requires training that is in addition to the training required for certification as a fire safety director. The training requirements for certification as a FS/EAPD are documented in the proposed rule 3 RCNY §9-08 entitled 'Training Courses for Fire Safety/Emergency Action Plan Directors' (2006).

Other cities have adopted, or are planning to adopt, similar high-rise emergency codes. Details of these may be obtained by contacting local fire departments.

Compliance Issues

Code violations at both the employee and employer level have been well documented. Researchers have examined the reasons for this and the research sheds important light that may be informative, with respect to compliance with emergency preparation codes and standards. For example, several studies have shown generally poor compliance with the OSHA

Blood and Body Fluid Standard (Department of Labor, 1998), which was developed to help minimize the risk of worker (primarily healthcare workers) exposure to infectious diseases transmitted through blood and other body fluids. One of the most, if not the most, important correlates of compliance at the employee level was found to be 'safety climate', which is defined as employees' perception of their organization's commitment to safety (Gershon et al., 1995, 2000).

In addition to safety climate, a number of psycho-social variables have also been significantly correlated to employee compliance, such as inaccurate perception of risk, maladaptive fear response, and negative influence of subjective norms (co-workers). In addition, certain individual factors, such as risk-taking personality profile, as well as socio-demographic factors (male gender), have also been associated with lower rates of compliance with workplace regulations (Becker and Maiman, 1983; Gershon et al., 1995, 1999; Michalsen et al., 1997). Organizational variables that affect employees' perceptions of safety climate (for example, weak safety culture, such as lack of safety supplies and resources, inadequate training and educational programs, lack of safety leadership and so on) have also been directly (as well as indirectly through safety climate), correlated with compliance (Gershon et al., 1995).

At the management level, generally, the employer's knowledge of the regulations and an understanding of the cost-effectiveness of risk reduction measures (that is, the ability of preventive measures to offset the cost of exposure) have been shown to correlate with regulatory compliance (Gershon et al., 2007).

HIGH-RISE EVACUATIONS

The WTCD resulted in the full-scale evacuation of all the buildings in the WTC complex, including the towers. High-rise buildings, including the older ones, have robust construction designs and redundant fire safety systems, thus making mass evacuations generally unnecessary. When a high-rise emergency does occur (usually a fire event), the building occupants are simply relocated to safe areas only one or two floors away from the affected area. Yet certain events can result in danger to all occupants and, in these rare cases, full building evacuation may be necessary. This occurred, as noted earlier, in the 1993 WTC and Murrah building bombings. Other recent serious fires include the Chicago high-rises fires in 2003 and 2004 ('Fire in Chicago high-rise office building kills six', 2003; Ciokajlo et al., 2004). In each of these situations, the complexity and lengthy time required to fully evacuate increased the risk to the occupants.

To better understand the process of evacuating large numbers of occupants from high-rises in extreme conditions, a study was conducted in the aftermath of the WTCD. The pertinent findings are described in the next subsection.

World Trade Center Evacuation Study Background

The World Trade Evacuation Study was designed as a three-year and five-phase examination of the factors that helped or hindered evacuation of the towers on September 11, 2001. The study examined a number of potential factors, using both qualitative and quantitative epidemiologic methodologies. There were three major constructs of interest that were targeted: individual, organizational (that is, workplace), and structural (building environment) factors.

Data were collected from over 1400 individuals, or roughly 10 percent of the individuals who evacuated from Towers 1 and 2. Individuals participating in the study completed an extensive questionnaire, available as a paper or web-based version. Almost all of the findings from this study have emergency preparedness implications and are not necessarily limited to high-rise structures. The findings underscore the important role that management plays with respect to preparing employees to respond quickly and effectively to emergencies, including those that result in mass evacuation. Factors from each of the three major categories are discussed in detail below, along with the management implications.

Individual Factors that Served as Barriers to Initiation of Evacuation

While most participants were motivated to initiate their evacuation process, delays in getting started were not uncommon. The primary reason for delaying the initiation process was the lack of information; many occupants did not know what had happened that caused the building to sway (which was the first indication for most occupants that an event had taken place). Consequently, occupants spent time in the immediate few minutes following the impact trying to obtain information, generally, by seeking out others, or by trying to make phone calls. Some individuals who were in Tower 2 when Tower 1 was attacked saw the plane as it hit Tower 1 and, in some cases, occupants felt the heat of the fireball that resulted. In these cases, the occupants understood immediately that something very serious had occurred and, even though they were in Tower 2, which had not yet been attacked, they immediately began their descent. This quick action saved the lives of those situated on high floors, as only 17 minutes later Tower 2 was attacked by the second plane and these individuals would have been above the point of impact. In Tower 1, the first building to be attacked, occupants facing away from the point of impact did not know what had happened and many people did not find out until they reached the ground level and could see for themselves how heavily damaged the building was. Because of the loss of most infrastructure in Tower 1 (for example, power, elevators, PA system and so on), communication was completely disrupted and, therefore, the normal channels of notification of occupants could not be used. Individuals learned about what had happened, either from people who had seen the attack (that is, those on lower floors that were facing the impact), or from others who had learned about what had happened via battery operated radios or a landline phone call. Hand-held devices were not widely used at that time, but the few who had them also learned about what had happened in this manner. Without information, people spent time seeking out others; a well-known behavior pattern in other types of emergency situations, as people tend to seek out groups to share information and feel safety or support. From a management perspective, a back-up system that is not dependent on the building's electrical supply in order to operate, is important to consider so that information can be readily transmitted during emergencies. Many companies now routinely provide their employees with hand-held devices for just this purpose.

Another factor that led to delays in initiation was the performance of tasks. This included making outside phone calls, trying to find telephones that worked, trying to contact the security desks of the towers, collecting personal items, trying to save computer files and collecting hard files, completing last-minute work activities (for example, shutting down computers), seeking permission to leave, and trying to decide whether to take elevators or stairs. Delays were also reported by individuals who were concerned about their physical ability to walk down multiple flights of stairs. Others delayed because of their general lack of familiarity with the building layout, including uncertainty regarding where the stairwells were located and where they terminated (that is, whether or not they would end at street-level exits). From a management perspective, most of these delays could have been prevented through appropriate education of employees regarding the building design. This should include practice on the use of stairs (which would also help the occupants know where stairwells are located), provision of written policies that explicitly state that permission is unnecessary for evacuation, final appraisals of individuals' capability of evacuating, and planning for those requiring special assistance. From an employee perspective, compliance with recommended or required training regarding emergency preparedness will clearly be beneficial, regardless of the type of building structure of the worksite.

Individual Factors that Served as Barriers to Progression

Footwear played an important role in progression of evacuation, with many reports of uncomfortable shoes (for example, high heels, slip-ons) slowing people down. In some cases, the footwear had to be removed, which was a problem for some evacuees when they encountered broken glass and debris in the lobby area. Some individuals were injured in this way and had to be carried out of the building. This problem could be prevented by advising building occupants, and most especially high-rise occupants, to have adequate footwear readily available in case rapid egress is needed.

Occupants were also slowed down by physical limitations, health conditions, and disabilities. Again, physical capabilities should be appraised before full-scale evacuation (without use of elevators) is ever required of occupants. Preplanning, including practice drills using a buddy system or fire safety team, should take place. This is an ongoing challenge for many high-rise buildings. In the WTCD, evacuation of an occupant, who used a wheelchair with an additional breathing support mechanism, took place. A rotating team of eight people took turns carrying the occupant and the extremely heavy chair down several flights of stairs. All of these individuals evacuated safely, although there are anecdotal reports of long-term musculoskeletal problems in the persons who assisted. Specialized equipment is available to assist the evacuation of persons with limited mobility and, in some cases, this equipment was used in the WTCD evacuation. In other cases, individuals would have used the equipment, but they did not know that it was available. In addition, some individuals knew that equipment was available, but did not know its location. Many of these special evacuation devices had been purchased after 1993, but, over time, they had been placed in inaccessible storage areas. Management should have full knowledge of the persons they employ who may require assistance during an emergency, in order for every option to assist them to be explored and planned as needed. Frequent training is required so that all occupants are familiar with any special equipment that can be used to assist persons with disabilities.

Organizational Factors that Influenced Evacuation

By far the most important factor, at the organizational level, that affected both the initiation and progression of evacuation, was leadership. In an emergency, people turn toward leaders for guidance since most people are unfamiliar with emergency operations. During the WTCD, lack of knowledgeable and trained leadership led to delays. In some cases, new leaders emerged; individuals with special knowledge or experience from their prior military service or experience as an emergency responder stepped in to fill this role. By providing training to *all* employees on emergency operations, management can ensure a ready supply of trained and knowledgeable individuals. This is important during a large-scale event, as fire safety personnel may be occupied elsewhere and additional trained staff will be able to extend their usefulness.

Another important organizational aspect, which will help occupants of high-rise buildings or large complex structures, is providing them with training on the layout of the building, the location of the exits, including the terminal exits, and the policies regarding elevator use and roof-top egress. Management should institute practice drills to reinforce classroom, web-based, or written training materials. In the WTC, fire drills often consisted of gathering people in hallways and simply pointing out the fire exits. Many of the evacuees reported that they had never even entered a stairwell before the disaster occurred. The inconvenience and disruption created by full-scale drills is well recognized, but this can be ameliorated by having frequent, smaller, drills so that disruptions are kept at a minimum. Drills should include going down one or two flights of stairs and then descending again only one or two flights leading to the lobby level. This will shorten the drill and prevent untoward consequences for occupants located on high floors. Walk-through orientation of the entire building structure should take place frequently and be mandatory for all new employees. Senior management should support these educational efforts by participating and publicizing them.

Environmental Factors that Affected Evacuation

Environmental sensory cues, such as loud noises or fuel odors, served as strong motivators to initiate evacuation. Many individuals reported that the buildings' physical safety features, such as adequate lighting on the staircases and handrails, and steps marked with reflective tape, also facilitated the rapid and orderly evacuation. Staircases maintained their integrity, for the most part, and this also supported rapid egress.

Building conditions that impeded progress of the evacuation included: structural damage such as debris on the stairways and collapsed interior walls that blocked egress routes; heavy congestion on certain stairways due to the sheer number of people descending and counterflow of first responders moving upwards with equipment; slowly moving individuals (which caused a back-up of those behind them); debris and glass in the lobby; smoke and water conditions on the stairs, which intensified towards the end of the evacuation; and, in a few cases, locked staircase doors (that is, doors leading back onto floors). Slow-moving individuals, including people with various types of conditions, general poor health or fitness, obesity, persons who use wheelchairs and guide dogs, and the elderly, slowed down the progress on the stairwells. From a management perspective, some of these structural factors may not be within their purview, especially with respect to businesses located in high-rise structures. In these types of buildings, there are generally many leaseholders and the fire safety program is managed by the building's owner and building's management firm. However, worker safety is the responsibility of employers and knowledge, awareness, and assessment of the building's capabilities should be well within the scope of management. Many high-rise buildings provide extensive training and information to major leaseholders. In some cases, major leaseholders have their own in-house fire safety director and fire safety team. Regardless of whether management relies on the building manager or their own fire safety staff, they should be aware of any structural impediments to the safe movement of occupants during a disaster response, while management should encourage the implementation of effective interventions that will address any gaps.

Recommendations

As part of the WTCE Study, a group of evacuees from the towers met extensively with the research team to develop a set of recommendations for improvement in terms of high-rise evacuations. This process, referred to as Participatory Action Research (PAR), was very productive, as it led to the development of an extensive list of action items. Many of these (provided in summary form in Appendix Tables 11A.1–3) may be helpful to managers as they develop their own emergency preparedness plans.

In addition to recommendations based on the WTCD Study, other recommendations have been developed to address the important area of disaster mental health. These are discussed in the next subsection.

Disaster Mental Health Strategies

There is growing recognition that the workplace provides a vital link to post-disaster mental healthcare. In the event of any type of disaster, there is an increased reliance on familiar social supports which embody safety and security. In the trauma literature, the availability of social support has been associated with more favorable outcomes (Kaniasty and Norris, 1993; Irving et al., 1997; Andrews et al., 2003). Isolation and the absence of social support, conversely, can serve to exacerbate unfavorable responses to trauma and disaster. Thus, in preparing for and following disasters, workplaces can play a critical role in providing workers with general social support, as well as specific services (Schouten et al., 2004). Assistance with basic needs and efforts to provide social support are beneficial and mechanisms that provide social support to employees need to be built into the core of an organization's emergency plans (Tyler and Rogers, 2005). Such mechanisms can enhance the resilience of employees in the face of a terrorist attack at the worksite and reduce the ensuing psychological sequelae. As Schouten et al. (2004) note, it is the offer of organizational support and services in the workplace that is particularly important to workers, regardless of actual utilization. Employees view the offer of support services as a compassionate, timely, and much appreciated gesture by the employer. Informal support, in the way of peer support provided by co-workers, can also be useful in facilitating recovery and return to normal functioning.

While larger organizations often have detailed plans for crisis response and business continuity, smaller organizations are often not in a position to have well-developed preparedness plans or to provide resources for response to mitigate the impact of an event at the worksite. Yet, according to Pohlman (1999), in these smaller organizations, employee commitment is more likely to be influenced by intangible perceptions of preparedness and response efforts that can impact on an organizational performance. Thus, smaller organizations may face a greater risk in terms of impact if they are unprepared to react quickly and appropriately.

In many workplaces today there are efforts in place, in the form of employee assistance programs, to address a variety of work, family, and personal problems. The goal is to enhance employee well-being, to promote job performance, increase organizational efficiency, and help facilitate work–family balance (Masi et al., 2002; Akabas and Kurzman, 2005). In the past few decades, more and more workers have begun to seek help with work-related stress and other emotional problems and there is a growing need for a broad array of services to assist them. The knowledge and utilization of these services may be seen as part of a continuum of preparedness for a disaster response to mitigate the impact of psychological sequelae that is likely to follow.

Employee assistance program designs and services vary. Programs can be sponsored by employers and/or unions (Masi et al., 2002). Employee assistance services can be provided internally by staff members of the organization or externally where services are outsourced to an outside provider (Straussner, 1988). Services can also be provided through a mixed model of internal and external providers, as well as both on- and off-site service provision. EAPs in the workplace have evolved and expanded over the past 30 years to include both professional and paraprofessional service providers and a myriad of support services (Masi et al., 2002). It is interesting to note that employee assistance services are required to be provided by federal agencies for their employees. However, considerable latitude in structuring federal agency employee assistance programs is left to the discretion of each agency to meet its own mission and organizational culture (US Office of Personnel Management, 2004).

Disaster mental health interventions in the workplace can be provided by employee assistance programs or other individuals or groups that offer social support and support services. Worksite interventions should foster community resilience; the ability of the workplace to return to normal and adaptive functioning after a disaster (Benedek et al., 2006). While the need to provide mental health interventions in the workplace has been established, the efficacy and appropriateness of the interventions has been the subject of debate. The empirical research on the critical incident and stress management model (CISM), an intervention designed for first responders, but also adapted in other workplace settings, is inconclusive. Benedek et al. suggest an approach of multidisciplinary teams using a program of health education and informational briefings which occur at the worksite and outside of traditional healthcare settings. With an emphasis on early intervention, the goal of disaster mental health services is to reduce the severity and/or duration of event-related distress and to promote the restoration of normal functioning. Early interventions include meeting basic needs, psychological first aid, assessing needs, monitoring the rescue and recovery environment, outreach and information dissemination, technical assistance, consultation and training, fostering resilience/recovery, triage, and treatment (National Institute of Mental Health, 2002). Interventions that are tailored to address individual, community (for example, the workplace), and organizational cultural needs and characteristics are most likely to be helpful. In addition, interventions should be matched to the phase of the event, to the setting, and to the survivor. Clearly, one size does not fit all. There is no ubiquitous mental health intervention that can address the needs of all parties.

Mental health personnel, including EAP staff, have key roles to play in workplace disaster preparedness (ibid.). They can assist in the coordination of service provision in order to ensure that mental health is an integrated element of a comprehensive workplace disaster preparedness and management plan. Furthermore, mental health expertise can guide and facilitate the implementation of interventions to achieve a positive mental health outcome for individuals in the workplace affected by a disaster (ibid.).

Workplace disaster preparedness requires the coordination and integration of various efforts, including mental health. More specifically, it requires the coordination and integration of efforts between the areas of security, human resources, employee assistance programs, and medical care (Benedek et al., 2006). While employers may be interested in introducing disaster preparedness, there may also be reluctance for fear of raising undue anxiety among employees (Vineburgh, 2004). Vineburgh suggests that 'a workplace model that reframes disaster preparedness as resiliency training can educate employees about disaster behaviors in the pre-event phase, as well as promote awareness of the psychological consequences of terrorism, help-seeking, and employee assistance response in the post-event phase' (p. 143). This approach reframes negative attributes, and adverse consequences, associated with trauma and disaster and provides an opportunity for promoting empowerment.

In terms of disaster mental health preparedness in the workplace, it is essential that there is consistency within all levels of an organization (for example, individual employees, managers/supervisors, and organizational systems) in adhering to the commitment of preserving psychological health and well-being of all workers (Flynn et al., 2005). This idea needs to be integrated into overall organizational continuity planning.

CONCLUSION

Emergency preparedness in the workplace can help minimize the negative impact of disasters. Terrorism, or the threat of terrorism, can result in severe disruptions, in addition to loss of lives. Many workplaces have taken steps since the WTCD, and subsequent anthrax attacks, to protect their employees and material assets, but the challenge in developing and implementing emergency preparedness plans is believed to have kept many other workplaces, especially smaller ones, with minimal plans. Few studies have actually assessed workplace preparedness planning, but anecdotal reports indicate that lack of preparedness is widespread in the business world. A number of agencies and organizations have prepared simple templates and guidelines that can be helpful in this regard. Special attention to these, as well as the guidelines on mental health services for those affected by work place emergencies, can help businesses meet their responsibilities.

REFERENCES

Aguirre, B.E., Wenger, D. and Vigo, G. (1998), 'A test of the emergent norm theory of collective behavior', *Sociological Forum*, **13**, 301–20.

Akabas, S.H. and Kurzman, P.A. (2005), *Work and the Workplace: A Resource for Innovative Policy and Practice*, New York: Columbia University Press.

Andrews, B., Brewin, C.R. and Rose, S. (2003), 'Gender, social support, and PTSD in victims of violent crime', *Journal of Traumatic Stress*, **16**(4), 421–7.

Arata, C.M., Picou, J.S., Johnson, G.D. et al. (2000), 'Coping with technological disaster: an application of the conservation of resources model to the *Exxon Valdez* oil spill', *Journal of Traumatic Stress*, **13**, 23–39.

Becker, M.H. and Maiman, L.A. (1983), 'Models of health-related behavior', in D. Mechanic (ed.), *Handbook of Health, Health Care, and the Health Professions*, New York: Free Press, pp. 539–68.

Benedek, D.M., Ursano, R.J., Fullerton, C.S., Vineburgh, N.T. and Gifford, R.K. (2006), 'Responding to workplace terrorism: applying military models of behavioral health and public health response', *Journal of Workplace Behavioral Health*, **21**(3/4), 21–33.

Ben-Ezra, M. (2004), 'Traumatic events take their toll on mental health', *Nature*, **430**, 611.

Brooks, N. and McKinlay, W. (1992), 'Mental health consequences of the Lockerbie disaster', *Journal of Traumatic Stress*, **5**, 527–43.

Burkle, F.M.J., Newland, C., Orebaugh, S. and Blood, C.G. (1994), 'Emergency medicine in the Persian Gulf War: part 2. Triage methodology and lessons learned', *Annual Emergency Medicine*, **23**, 748–54.

Bystritsky, A., Vapnik, T., Maidment, K., Pynoos, R.S. and Steinberg, A.M. (2000), 'Acute responses of anxiety disorder patients after a natural disaster', *Depression and Anxiety*, **11**(1), 43–4.

Cameron, G., Pate, J., McCauley, D. and DeFazio, L. (2000), '1999 WMD terrorism chronology: incidents involving sub-national actors and chemical, biological, radiological, and nuclear materials', *Nonproliferation Review*, **7**(2), 157–74.

Centers for Disease Control and Prevention (2001), 'Update: Investigation of bioterrorism-related anthrax and interim guidelines for clinical evaluation of persons with possible anthrax', *Morbidity and Mortality Weekly Report*, **50**(43), 941–8.

Centers for Disease Control and Prevention (2003), *Public Health and Aging: Health Related Quality of Life among Low-income Persons aged 45–64 Years, 1995–2001*, New York.

Centers for Disease Control and Prevention (2004), *Mental Health Status of World Trade Center Rescue and Recovery Workers and Volunteers, July 2002–August 2004*, New York.

Centers for Disease Control and Prevention (2007), *Fire Safety Directory*, http://www.firesafety.gov/, October 5, 2007.

Chao, E.L. and Henshaw, J.L. (2001), *How to Plan for Workplace Emergencies and Evacuations*, http://www.osha.gov/publications/osha3088.html, September 1, 2007.

Ciokajlo, M., Heinzmann, D., Pang, K., Cholo, A.B., Vittachi, I. and Kellogg, A.P. (2004), 'Loop high-rise fire injures', *Chicago Tribune*, December 7, 25.

Deering, C.G., Glover, S.G., Ready, D., Eddleman, H.C. and Alarcon, R.D. (1996), 'Unique patterns of comorbidity in posttraumatic stress disorder from different sources of trauma', *Journal of Comparative Psychiatry*, **37**, 336–46.

Dellinger, A.M., Waxweiler, R.J. and Mallonee, S. (1997), 'Injuries to rescue workers following the Oklahoma City bombing', *American Journal of Industrial Medicine*, **31**(6), 727–32.

Department of Labor, Occupational Safety and Health Administration (OSHA) (1987), 'Code of federal regulations', Washington, DC: US Government Printing Office, Chapter XVII.

Department of Labor (1998), *Using Universal Precautions with Bloodborne Pathogens* (*1910.1030*), Washington, DC: US Government Printing Office.

Dooley, E.G. and Gunn, J. (1995), 'The psychological effects of disaster at sea', *British Journal of Psychiatry*, **167**, 233–7.

Eachempati, S.R., Mick, S. and Barie, P.S. (2004), 'The impact of the 2003 blackout on a level 1 trauma center: lessons learned and implications for injury prevention', *Journal of Trauma-Injury Infection and Critical Care*, **57**(5), 1127–31.

Everly, G.S., Jr (1999), 'Emergency mental health: an overview', *International Journal of Emergency Mental Health*, **1**(1), 3–7.

Fahy, R.F. (1995), 'The study of occupant behavior during the World Trade Center evacuation – preliminary report of results', paper presented at the International Conference on Fire Research and Engineering, Orlando, FL, September 10–15.

Fahy, R.F. and Proulx, G. (1995), 'Collective common sense: a study of human behavior during the World Trade Center evacuation', *National Fire Protection Association Journal*, **9**(2), 59–67.

Federal Bureau of Investigation (FBI) (2001), *FBI Policy and Guidelines: Counterterrorism*, http://jackson.fbi.gov/cntrterr.htm, October 5, 2007.

Federal Emergency Management Agency (FEMA) (1993), *Emergency Management Guide for Business and Industry*, Washington, DC: FEMA.

Federal Emergency Management Agency (FEMA) (2002), *World Trade Center Building Performance Study: Data Collection, Preliminary Observations, and Recommendations*, New York: FEMA Region II.

Fire Department New York (FDNY) (2005), 'Office Building Emergency Action Plans', 3 RCNY §6-02.

'Fire in Chicago high-rise office building kills six' (2003), http://www.cnn.com/2003/US/Midwest/10/18/chicago.fire.ap/, March 13, 2006.

Flynn, B.W., Flanigan, S. and Everly, G.S. Jr. (2005), 'Fostering human continuity: an essential element in workplace crisis intervention', *International Journal of Emergency Mental Health*, **7**(3), 187–93.

Galea, S., Nandi, A. and Vlahov, D. (2005), 'The epidemiology of post-traumatic stress disorder after disasters', *Epidemiological Reviews*, **27**, 78–91.

Gershon, R.R., Karkashian, C.D., Grosch, J.W., Murphy, L.R., Escamilla-Cejudo, A., Flanagan, P.A. et al. (2000), 'Hospital safety climate and its relationship with safe work practices and workplace exposure incidents', *American Journal of Infection Control*, **28**(3), 211–21.

Gershon, R.R., Karkashian, C.D., Vlahov, D., Kummer, L., Kasting, C., Green-McKenzie, J. et al. (1999), 'Compliance with universal precautions in correctional health care facilities', *Journal of Occupational and Environmental Medicine*, **41**(3), 181–9.

Gershon, R.R., Qureshi, K.A., Rubin, M.S. and Raveis, V.H. (2007), 'Factors associated with high-rise evacuation: qualitative results from the World Trade Center Evacuation Study', *Prehospital and Disaster Medicine*, **22**(3), 165–73.

Gershon, R.R., Vlahov, D., Felknor, S.A., Vesley, D., Johnson, P.C., Delclos, G.L. et al. (1995), 'Compliance with universal precautions among health care workers at three regional hospitals', *American Journal of Infection Control*, **23**(4), 225–36.

Green, B.L. and Lindy, J.D. (1994), 'Post-traumatic stress disorder in victims of disasters', *Psychiatric Clinics of North America*, **17**(2), 301–9.

Green, B.L., Lindy J.D., Grace, M.C. et al. (1992), 'Chronic posttraumatic stress disorder and diagnostic comorbidity in a disaster sample', *Journal of Nervous and Mental Disease*, **180**, 760–66.

Grieger, T.A., Fullerton, C.S. and Ursano, R.J. (2003), 'Posttraumatic stress disorder, alcohol use, and perceived safety after the terrorist attack on the Pentagon', *Psychiatric Services*, **54**(10), 1380–82.

Grieger, T.A., Fullerton, C.S. and Ursano, R.J. (2004), 'Posttraumatic stress disorder, depression, and perceived safety 13 months after September 11', *Psychiatric Services*, **55**(8), 1061–3.

Hinman, E.E. and Hammond, D.J. (1996), *Lessons from the Oklahoma City Bombing: Defensive Design Techniques*, New York: ASCE Press.

International Code Council (ICC) (2003), *International Code Council Government Relations Code Adoption Toolkit*, Washington, DC: International Code Council.

International Conference of Building Officials (ICBO) (1994), *Publications List*, Whittier, CA.

Irving, L.M., Telfer, L. and Blake, D.D. (1997), 'Hope, coping, and social support in combat-related posttraumatic stress disorder', *Journal of Traumatic Stress*, **10**(3), 465–79.

Kaniasty, K. and Norris, F.H. (1993), 'A test of the social support deterioration model in the context of natural disasters', *Journal of Personality and Social Psychology*, **64**(3), 395–408.

Kessler, R.C. (2000), 'Posttraumatic stress disorder: the burden to the individual and to society', *Journal of Clinical Psychiatry*, **61**, 4–12.

Kuo, C.J., Tang, H.S., Tsay, C.J., Lin, S.K., Hu, W.H. and Chen, C.C. (2003), 'Prevalence of psychiatric disorders among bereaved survivors of a disastrous earthquake in Taiwan', *Psychiatric Services*, **54**(2), 249–51.

Lindell, M.K. and Perry, R.W. (1992), *Behavioral Foundations of Community Emergency Management*, Washington, DC: Hemisphere Publishing.

Local Law 26 (2004), 'Building Safety', New York City Building Code, § 27-383.

Masi, D., Altman, L., Benayon, C., Healy, H., Jorgensen, D.G. and Kennish, R., et al. (2002), 'Employee assistance programs in the year 2002', in R.W. Manderscheld and M.J. Henderson (eds), *Mental Health, United States, 2002*, Rockville, MD: US Department of Health and Human Services.

Michalsen, A., Delclos, G.L., Felknor, S.A., Davidson, A.L., Johnson, P.C., Vesley, D. et al. (1997), 'Compliance with universal precautions among physicians', *Journal of Occupational and Environmental Medicine*, **39**(2), 130–37.

Morgan, L., Scourfield, J., Williams, D., Jasper, A. and Lewis, G. (2003), 'The Aberfan disaster: 33-year follow-up of survivors', *British Journal of Psychiatry*, **182**, 532–6.

National Fire Protection Association (NFPA) (2000), *Life Safety Code, Section 3-3.25.06*, Quincy, MA.

National Fire Protection Association (NFPA) (2007), *Standard on Disaster/Emergency Management and Business Continuity Programs*, Quincy, MA.

National Institute for Occupational Safety and Health (NIOSH) (2007), *NIOSH Safety and Health Topics: Emergency Preparedness for Business*, http://www.cdc.gov/niosh/topics/prepared/, October 4.

National Institute of Mental Health (2002), *Mental Health and Mass Violence: Evidence-based Early Psychological Intervention for Victims/Survivors of Mass Violence (NIMH Publication No. 02-5138)*, Rockville, MD: National Institute of Mental Health, National Institutes of Health.

National Institute of Standards and Technology (NIST) (2005), *Federal Building and Fire Safety Investigation of the World Trade Center Disaster: Final Report of the National Construction Safety Team on the Collapses of the World Trade Center Towers*, http://wtc.nist.gov/reports_october 05.htm, September 1, 2007.

Norris, F., Friedman, M., Watson, P., Byrne, C., Diaz, E. and Kaniasty, K. (2002), '60,000 disaster victims speak, Part I: An empirical review of the empirical literature, 1981–2001', *Psychiatry*, **65**, 207–39.

North, C.S., Nixon, S.J., Shariat, S., Mallonee, S., McMillen, J.C., Spitznagel, E.L. et al. (1999), 'Psychiatric disorders among survivors of the Oklahoma City bombing', *Journal of the American Medical Association*, **282**(8), 755–62.

Pfefferbaum, B., Sconzo, G.M., Flynn, B.W., Kearns, L.J., Doughty, D.E., Gurwitch, R.H. et al. (2003), 'Case findings and mental health services for children in the aftermath of the Oklahoma City bombing', *Journal of Behavioral Health Services and Research*, **30**(2), 215–27.

Pohlman, C. (1999), *Study on Workplace Satisfaction in Public, Private Sectors*, Toronto: Canadian Federation of Independent Business.

Proulx, G. (2001), 'Occupant behavior and evacuation', paper presented at the Proceedings of the 9th International Fire Protection Symposium, Munich, Germany, May 25.

Proulx, G. and Fahy, R. (2002), 'A comparison of the 1993 and 2001 evacuations of the World Trade Center', paper presented at the National Fire Protection Association World Safety Congress and Exhibition, Minneapolis, MN, May 21.

Quarantelli, E.L. (1997), 'Ten criteria for evaluating the management of community disasters', *Disasters*, **21**(1), 39–56.

Quenemoen, L.E., Davis, Y.M., Malilay, J., Sinks, T., Noji, E.K. and Klitzman, S. (1996), 'The World Trade Center bombing: injury prevention strategies for high-rise building fires', *Disasters*, **20**(2), 125–32.

Quinn, S.C., Thomas, T. and McAllister, C. (2005), 'Postal workers' perspectives on communication during the anthrax attack', *Biosecurity and Bioterrorism*, **3**(3), 207–15.

Robert T. Stafford Disaster Relief and Emergency Assistance Act (2005), P/L/93-288 as amended, and 42 U.S.C. § 5121-5206, 44 C.F.R. 206.31-206.48.

Salcioglu, E., Basoglu, M. and Livanou, M. (2003), 'Long-term psychological outcome for non-treatment-seeking earthquake survivors in Turkey', *Journal of Nervous and Mental Disease*, **191**(3), 154–60.

Schouten, R., Callahan, M.V. and Bryant, S. (2004), 'Community response to - disaster: the role of the workplace', *Harvard Review of Psychiatry*, **12**(4), 229–37.

Selley, C., King, E., Peveler, R. et al. (1997), 'Post-traumatic stress disorder symptoms and the Clapham rail accident', *British Journal of Psychiatry*, **171**, 478–82.

Society of Fire Protection Engineers (2003), *Engineering Guide to Human Behavior in Fire*, Bethesda, MD.

Southern Building Code Congress International (SBCCI) (1993), *Products Catalog*, Birmingham, AL.

Stith Butler, A., Panzer, A.M. and Goldfrank, L.R. (2003), *Preparing for the Psychological Consequences of Terrorism: A Public Health Strategy*, Washington, DC: National Academies Press.

Straussner, S.L.A. (1988), 'A comparative analysis of in-house and contractual employee assistance programs', *Employee Assistance Quarterly*, **3**, 43–55.

Tierney, K.J., Lindell, M.K. and Perry, R.W. (2001), *Facing the Unexpected: Disaster Preparedness and Response in the United States*, Washington, DC: Joseph Henry Press.

'Training Courses for Fire Safety/Emergency Action Plan Directors', 3 RCNY §9-08 (2006).

Tyler, M.P. and Rogers, J.R. (2005), 'A federal perspective on EAPs and emergency preparedness', *International Journal Emergency Mental Health*, **7**(3), 179–86.

Ursano, R.J. (2002), 'Terrorism and mental health: public health and primary care', paper presented at the 18th Annual Rosalynn Carter Symposium on Mental Health Policy, Atlanta, GA, November 6–7.

US Office of Personnel Management (2004), *Employee Health Services Handbook*, http://www.healthierfeds.opm.gov/healthierfedsmanual.asp, August 30, 2007.

Veenema, T.G. (2006), 'Expanding educational opportunities in disaster response and emergency preparedness for nurses', *Nursing Education Perspective*, **27**(2), 93–9.

Verger, P., Dab, W., Lamping, D.L., Loze, J.Y., Deschaseaux-Voinet, C., Abenhaim, L. et al. (2004), 'The psychological impact of terrorism: an epidemiologic study of posttraumatic stress disorder and associated factors in victims of the 1995–1996 bombings in France', *American Journal of Psychiatry*, **161**(8), 1384–9.

Vineburgh, N.T. (2004), 'The power of the pink ribbon: raising awareness of the mental health implications of terrorism', *Psychiatry*, **67**(2), 137–46.

Wenger, D. (1987), 'Collective behavior and disaster research', in R. Dynes, B. DeMarchi and C. Pelanda (eds), *Sociology of Disasters: Contributions of Sociology to Disaster Research*, Milan: Franco Angeli, pp. 213–37.

Yule, W., Bolton, D., Udwin, O., Boyle, S., O'Ryan, D. and Nurrish, J. (2000), 'The long-term psychological effects of a disaster experienced in adolescence: I: The incidence and course of PTSD', *Journal of Child Psychology and Psychiatry and Allied Disciplines*, **41**(4), 503–11.

Table 11A.1 Summary of PAR team recommendations: to address delays in initiation of evacuation

Risk factor	Recommendations for improvement by category (I = individual level, O = organizational level, S = structural level)
1. Seeking out others to form a group for mutual support and information sharing during exit	• Participate in group activities both within company and on the floor where office is located (I) • In the absence of complete information or leadership guidance, individuals may have to act based upon instinct and 'gut feelings' (I) • Have multiple sources of communication available (e.g., battery operated radio) (I) • Share information about unusual circumstances with co-workers (I) • Support pre-event group formation with programs such as workplace coffee hour, ongoing social activities across floors and among floors (O) • Building management/employer must provide accurate information early in the event (O) • Provide clear and consistent, accurate messages (O) • Provide wireless cell phones for the safety team (O) • Link company computers to television stations, emergency broadcast announcements (e.g., Office of Emergency Management) (O) • Limit non-emergency public address (PA) announcements to only those that are necessary (O) • Utilize multiple environmental cues, i.e., flickering lights to alert participants of an unusual occurrence (S) • Install back-up communication systems (S) • Equip buildings to allow for emergency radio broadcast announcements (S) • Equip elevators and staircases with systems for public address announcements (S) • All communication sources should link to one main broadcast site (S)
2. Personal concerns about own health and ability to evacuate	• Individual should conduct self-assessment of their capability and time needed to fully exit (I) • Individuals should inform the building's safety personnel regarding special evacuation needs they may have (I) • System to rapidly identify those who will require special assistance for evacuation (O) • Pre-planning for persons with disabilities (O)

	• Evacuation drills for persons with special needs for evacuation, including situations where elevator is not functioning (O)
3. Individual behaviors that delayed initiation of exit (e.g., gathering items, making phone calls, work-related tasks, waiting for directions or permission to leave, changing shoes)	• Maintain an emergency 'go bag' at desk (I) • Delay calls until fully exited (I) • Ownership of own safety actions (i.e., act independently) (I) • Wear sensible footwear that will facilitate rapid evacuation (I) • Maintain evacuation supplies by the exits that are not locked and are easily accessible (O) • Implement policies that support employee initiative for evacuation (O) • Install emergency supply storage receptacles near exits (S)
4. Uncertainty about route to take including: exit locations, staircase endpoints, roof access, locked re-entry points, when to use elevators	• Compliance with training and drills (I) • Facilitate a workplace climate for emergency preparedness (O) • Enforce training and education of all employees for evacuation (O) • Enforce mandatory drills that involve entry into the staircase and various routes for terminal egress (O) • Color code the exit doors (S) • Signage that indicates where staircases terminate (S)

Table 11A.2 Summary of PAR team recommendations: to address delays that served to increase length of time of evacuation

Risk factors	Recommendations for improvement by category (I = individual level, O = organizational level, S = structural level)
1. Lack of knowledge and poor emergency preparedness of workers	• Active participation in training and drills (I) • Conduct training and drills (O) • Incorporate entry into stairwells in fire drills (O) • Promote an organizational emergency preparedness safety culture/climate from top down (O) • Conduct specialized safety training for senior management (O) • Foster employer ownership of safety training and climate, accountability enforced by building owner (O) • Garner political and regulatory support for training and drills (O) • Clarify use of elevator vs. stairs during emergencies (O)
2. Footwear that did not facilitate rapid exit downstairs and through lobby debris field	• Encourage wearing sturdy, closed-toe, flat-soled, footwear in the workplace (I)
3. Staircase characteristics (e.g., width, design, access)	• Widen staircases where feasible (S) • Consider drainage for water on stairs (S) • Assign widest staircases for evacuation and accommodations of persons using wheelchairs (S) • Ensure that doors to floors off stairwells unlock during emergency (S)
4. Complex building design was confusing	• Design high-rise buildings that are intuitive and easy to navigate (S)
5. Poor or missing signage	• Improve signage (S)
6. Suboptimal workplace emergency safety climate	• Ensure an organizational safety culture including emergency preparedness from top down (O) • Conduct specialized emergency preparedness safety training for senior management (O)

	• Foster employer ownership of safety training and climate, with accountability to be enforced by the building owner (O)
7. Inadequate communication system infrastructure	• Ensure a working emergency generator and PA system (S) • Link buildings to TV station and PCs for better communication (S)
8. Transportation of responder supplies	• Place emergency supplies throughout the building (S)
9. Flow of traffic in certain areas led to congestion on stairs	• Assess building capabilities to support evacuation (S) • Stagger evacuation of occupants to reduce numbers on stairs (S) • Designate specific staircases for slow movers or those that need special assistance (S)
10. Procedures for evacuation of persons with health conditions or disabilities (PWHC/D)	• Co-workers or response team should practice use of evacuation chairs (O) • Preparedness planning for PWHC/D (O) • Code identification badges to permit rapid identification of PWHC/D (O) • Instruct PWHC/D to use elevators as directed by safety personnel (O) • Assign buddies for evacuation of PWHC/D (O) • Training for those assisting PWHC/D so that they can offer assistance in cases where carrying is necessary (O) • Pass new regulations to protect PWHC/D (O) • Ensure confidentiality of information about PWHC/D (O) • Special evacuation training /drills for PWHC/D (O) • Keep extra wheelchairs at lobby level for PWHC/D (O) • Designate widest stairwell for PWHC/D (S)

Table 11A.3 Summary of PAR team recommendations: risk factors associated with sustaining an injury

Risk factor	Recommendations for improvement by category (I = individual level, O = organizational level, S = structural level)
1. Weak emergency preparedness	• Support, encourage individuals to have a positive safety attitude (I) • Participate in training drills (I) • Promote a strong organizational safety culture/climate (O)
2. Stair/building environment	• Designate use of the widest stair for a person with a disability or injured (O) • Develop innovative methods for evacuation (D) • Issue flashlights to occupants (O) • Widen existing stairs (S) • Maintain evacuation supplies throughout the buildings (S) • Reduce possibility of broken glass (S)
3. Health condition or disability of evacuees	• Awareness of PWHC/D in the individual's work area (I) • Conduct self-assessment of capabilities (I) • Awareness of all occupants regarding location and use of safety equipment (I) • Promote organizational commitment for a disability emergency preparedness climate (O) • Seek out and provide specialized training for volunteer assistants (O)
4. Footwear	• Educate building occupants regarding use of sensible shoes that will support rapid evacuation (I)

12. Airline industry responses to September 11th*

Jody Hoffer Gittell, Kim Cameron, Sandy Lim and Victor Rivas

INTRODUCTION

The US airline industry faced devastating losses in the wake of the September 11th, 2001 terrorist attacks. According to Kevin Murphy (2001), airline industry analyst for Morgan Stanley, 'If there was ever a stress test for a good business, this is it'. The day after the attacks, the major airlines appeared in front of Congress seeking relief in the form of federal assistance. Fifteen billion dollars were allocated to the industry, some in the form of outright grants to cover the loss of operating revenue in the days after the attacks when the industry was shut down by federal order. The rest of the $15 billion allocation was made available in the form of loan guarantees to be allocated according to rules established by the Air Transport Stabilization Board. Even with this federal assistance, however, the industry continued to lose millions of dollars on a daily basis due to the slow rate of passenger return. In response to these losses, the major airlines cut their flights by an average of 20 percent and laid off an average of 16 percent of their workforces in the weeks following the attacks. Even though all of the major airlines were devastated about equally in terms of the initial decline in passenger traffic, however, they did not respond in the same way. Some airlines emerged from this crisis resilient and strong, whereas others languished and even confronted bankruptcy.

This chapter investigates the factors explaining the success of some airline companies after the 9/11 attacks and the struggles of others. It focuses on the roles played by relational reserves, financial reserves and the viability of the underlying business models in 10 major airline companies. Specifically, the study explains why managers must maintain and enhance

* This chapter draws substantially from Gittell, J.H., Cameron, K., Lim, S. and Rivas, V. (2006), 'Relationships, layoffs and organizational resilience', *Journal of Applied Behavioral Science*, **42**(3): 300–329.

strong employee relationships (relational reserves) during crisis in order to ensure commitment and productivity. But it also highlights the crucial role played by financial reserves (cash flow and low debt levels) and an organization's business model in the ability to respond effectively to crisis. Organizations are better able to cope with a crisis when they maintain strong relational and financial reserves and when they have business models that fit the needs of the existing competitive environment. But achieving a viable business model is itself a function of positive employee relationships. While layoffs enable organizations to respond to crisis in the short term, they also risk damaging the relationships that are needed for long-term recovery.

RELATIONSHIPS AS A SOURCE OF RESILIENCE

Most organizational theory has focused on the negative consequences of crisis such as threat rigidity, downward spirals, vicious cycles, and tipping points (Staw et al., 1981; Sitkin, 1992; Gladwell, 2002; Weick, 2003), yet some organizations demonstrate a remarkable tendency to flourish and thrive in the midst of crisis. These organizations demonstrate *resiliency*. Resiliency in everyday parlance refers to the capability to 'recover from or adjust easily to misfortune or change' (*Webster's Ninth New Collegiate Dictionary*). In organizational science it refers to (a) the maintenance of positive adjustment under challenging conditions (Weick et al., 1999; Worline et al., 2004), (b) the ability to bounce back from untoward events (Sutcliffe and Vogus, 2003), and (c) the capacity to maintain desirable functions and outcomes in the midst of strain (Edmondson, 1999; Bunderson and Sutcliffe, 2002). Resilience is a dynamic capacity of organizational adaptability that grows and develops over time (Wildavsky, 1988). It is not a static attribute that organizations do or do not possess. Rather, it results from processes that help organizations retain resources in a form sufficiently flexible, storable, convertible, and malleable to avert maladaptive tendencies and cope positively with the unexpected (Sutcliffe and Vogus, 2003; Worline et al., 2004).

These processes enable the maintenance of positive social relationships at work, which have been linked to resilience and recovery in individuals (Ryff and Singer, 2001; Seligman, 2002) as well as in organizations (Luthans, 2002; Cameron et al., 2004; Spreitzer et al., 2005). That is, a growing body of empirical evidence supports the notion that positive relationships at work – or relational reserves – are a prerequisite to organizational resilience.

These outcomes center on the nature of the relationships that develop among organization members. The levels of social, emotional, and moral support provided to one another, for example, are key components of organizational resilience (Carver et al., 1989). In a laboratory study where subjects were exposed to equivalent levels of stressors, Aiello and Kolb (1995) found that members of cohesive groups reported the least stress. Similar evidence found that social support among employees reduced the negative effects of work stressors on work outcomes (Schaubroeck and Fink, 1998; Moyle and Parkes, 1999). Positive relationships served as the key coping resources that enabled individuals and organizations to develop resilience in the face of work stress. Individually, children who had close relationships with caregivers, other competent adults, or peers, and those living in a closely knitted community, were more likely to cope effectively during times of adversity (Masten and Reed, 2002).

FINANCIAL RESERVES AND VIABLE BUSINESS MODELS AS SOURCES OF RESILIENCE

The predominant theme in the resilience literature is that resiliency results from the presence of both positive relationships and the access to adequate resources. Wildavsky (1988) argued that, in addition to social support, retaining financial reserves in a form that is sufficiently flexible to cope with unanticipated events was a key mechanism for developing resiliency. Similarly, in a study of hospital responses to an unexpected doctors' strike, Meyer (1982) found that slack resources worked as 'organizational shock absorbers' that buffered the impact of environmental jolts. The accumulation of substantial financial reserves during tranquil periods enabled one of the most resilient hospitals in the study to adapt to the crisis without the need to lay off employees. In another hospital, a strong organizational ideology emphasizing employee well-being greatly affected the organization's ability to respond to environmental jolts. This hospital had to suffer a short-term decline in profits by not laying off employees, but the strategy allowed it to preserve its commitment to its employees and contributed to its ability to readjust quickly after the crisis.

Building on the theme of adequate resources, we further hypothesize that having a business model that is viable in the existing competitive environment is another important source of resilience for surviving a crisis. Financial reserves help an organization to weather the storm by providing a buffer against losses, but a business model that meets the needs of the existing competitive environment is expected to enable an organization to minimize those losses and thus to recover more quickly. A viable business model, in addition to financial reserves, is expected to be particularly important in the face of a sustained crisis. Given the evidence cited above regarding the role of relationships in achieving productivity and quality outcomes, we hypothesize that relationships play another critical role in achieving resilience. Not only are relationships critical as a coping mechanism in the face of crisis according to this story, but they are also critical for creating viable business models that can survive a sustained crisis.

LAYOFFS AS A RESPONSE TO CRISIS

Paradoxically, a common organizational response to crisis – that is, layoffs – tends to undermine the very relationships that help organizations cope during periods of crisis. Many studies have reported the negative effects of downsizing on organizational relationships (Cameron et al., 1991; Cameron, 1994, 1998) including: (i) the destruction of interpersonal relationships, shared values, trust and loyalty, and commonality and strength of culture; (ii) reduced information sharing and increased secrecy, deception, and duplicity; (iii) increased formalization, rigidity, resistance to change, and conservatism; (iv) increased conflict, anger, vindictiveness, and feelings of victimization; and (v) increased selfishness and voluntary turnover, as well as deterioration in teamwork and cooperation (also see Cole, 1993; DeWitt, 1993; McKinley et al., 1995; Cascio et al., 1997). A large majority of firms that downsize experience these deleterious effects, although a few do not, primarily due to the *way* in which downsizing is implemented (Cameron 1994, 1998), paying special attention to the effects on relational reserves.

As a result of the negative impact on relationships, layoffs also have negative implications for organizational performance. Most organizations, for example, experience deteriorating profitability, product and service quality, innovation, and organizational climate after downsizing (Cameron, 1998). Three years after downsizing, the market share prices of downsized companies were an average of 26 percent below the share prices of their competitors at the beginning of the 21st century. Among companies with similar growth rates, those that did not downsize consistently outperformed those that did in the 2001 recession. Moreover, almost all organizations that downsized – in the public sector or in the private sector – experienced an emergence of the 'dirty dozen' as a result of downsizing (see Henkoff, 1990; Bennett, 1991; Cole, 1993; Mishra and Mishra, 1994; McKinley et al., 1995; Cascio et al., 1997; Cameron, 1998). The dirty dozen are 12 common deleterious outcomes associated with downsizing, such as declines in trust, networks, communication, commitment, and innovation along with increases in threat-rigidity, selfishness, conflict, and goal displacement (Cameron et al., 1987).

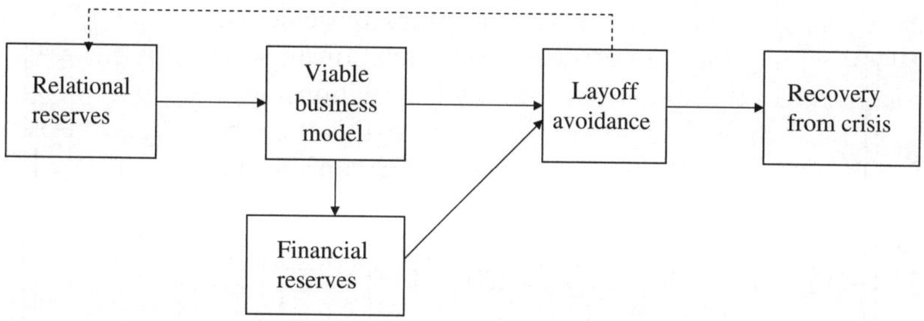

Figure 12.1 Conceptual model of organizational resilience

These negative performance effects of layoffs are consistent with theories of high performance work systems and mutual gains enterprise, which argue that employment security is essential for achieving sustained innovation and productivity that benefit both employees and stockholders (Kochan and Osterman, 1994). These negative performance effects of layoffs are also consistent with relational theory more broadly as summarized in the previous section. When layoffs are used as a primary coping response in a crisis, such as a sharp decline in the demand for the organization's products or services, the resulting weakened relationships mitigates the very resilience and recovery being sought.

CONCEPTUAL MODEL

This study investigates the extent to which the dramatic differences in resiliency and recovery in US airline companies after the tragedy of September 11th can be explained by these factors – the preservation of relational and financial reserves, and the existence of viable business models. The expected interrelationships of these factors are summarized in Figure 12.1.

Significant differences occurred in the strategies implemented by these 10 major airlines after the attacks of 9/11, as suggested by the percentage of employees laid off (see Figure 12.2). Stock price recovery after September 11th was slow in all the major airline companies, even after the immediate crisis passed. This was in part due to ongoing security concerns and the increased 'hassle factor', both of which contributed to the decline in demand for air travel (Sharkey, 2004). However, the recovery in stock prices varied substantially across the industry (see Figure 12.3). In particular, performance recovery since September 11th was fastest for Southwest Airlines, whose stock price was at 92 percent of its pre-9/11 level over the four years

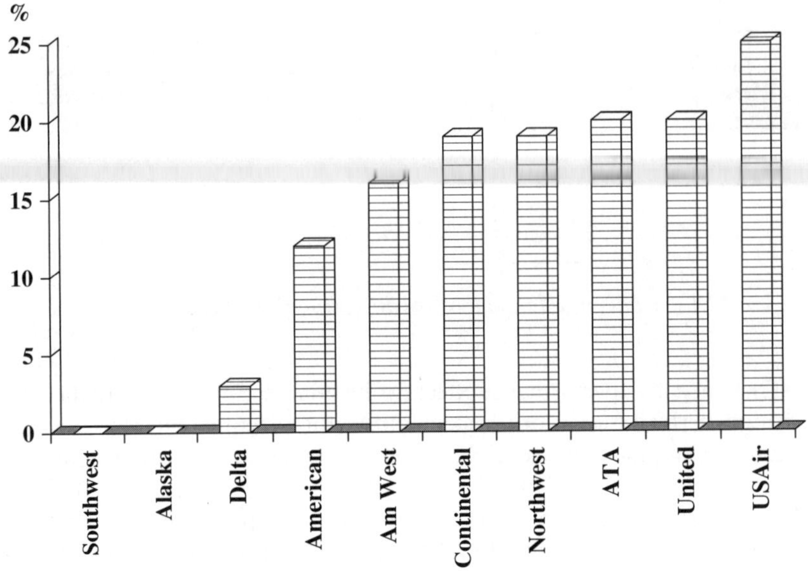

Source: Layoffs reported in press after September 11, divided by year-end employment for 2000 as reported by Bureau of Transportation Statistics.

Figure 12.2 Employee layoffs after September 11th

post-9/11, and slowest for United Airlines and US Airways, whose stock prices remained at 12 and 23 percent, respectively, of their pre-9/11 levels over the same period.

Figures 12.4 and 12.5 report cash on hand and debt–equity ratios of the major US airlines prior to September 11th. The figures show substantial variation in cash and debt levels among the major airlines prior to September 11th. As noted above, debt acts as a constraint in a downturn since interest payments are a fixed cost that must be paid regardless of revenues (Freear, 1980, p. 151). Low debt levels were thus expected to give companies flexibility in a downturn due to lower fixed costs. Furthermore, low debt levels provide companies with greater flexibility to take on new debt to survive the downturn. Cash on hand also provides flexibility in the face of a downturn in revenues, enabling organizations to pay expenses from past revenues to at least partially compensate for the shortfall in current revenues. However, cash on hand does not last long with extremely high debt levels.

In addition to the role of financial reserves in enabling airlines to forgo layoffs and recover more quickly, we need also to explore the viability of the business models themselves prior to the crisis of 9/11. Due to the

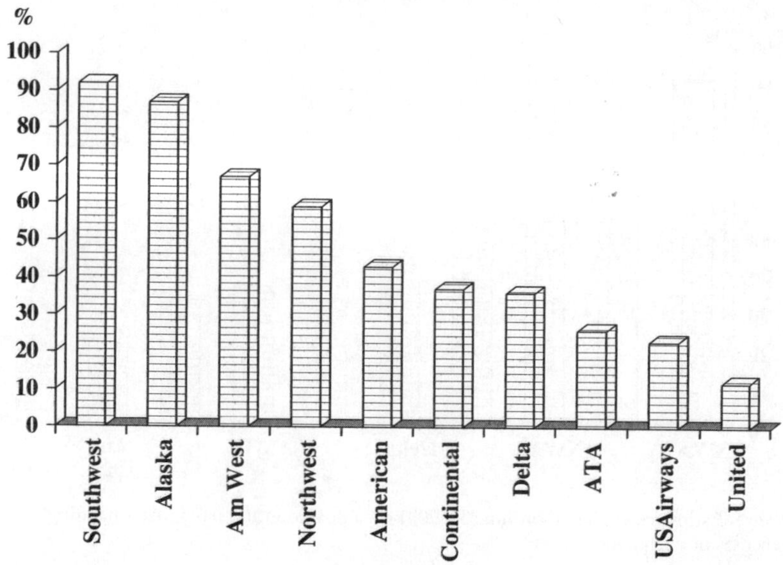

Note: Percentage change in stock price relative to September 10, 2001 levels, measured quarterly from December 10, 2001 to September 10, 2005.

Source: Yahoo Financials.

Figure 12.3 Average stock price recovery from December 10, 2001 to September 10, 2005, to September 10, 2001 levels

increasing threat and success of low-cost competition in the industry, there was a consensus just prior to 9/11 that one critical element of a viable business model in this industry was the achievement of low unit costs. As we see from Figure 12.6, there was significant variation in unit costs in this industry prior to 9/11, ranging from 7.5 cents per seat mile at Southwest Airlines to over 15 cents per seat mile at US Airways, with most carriers having unit costs around 9 to 10 cents per seat mile. But whereas a viable business model can be expected to play an important role in avoiding layoffs in the face of crisis, it is not the starting point for this story. Rather, we expect that a viable business model is enabled by the development and maintenance of relational reserves over time.

Results reported in Gittell et al. (2006) provide statistical support for this conceptual model of organizational resilience. We found that evidence suggesting that the development of relational reserves enables organizations to establish viable business models, therefore enabling them to build up their

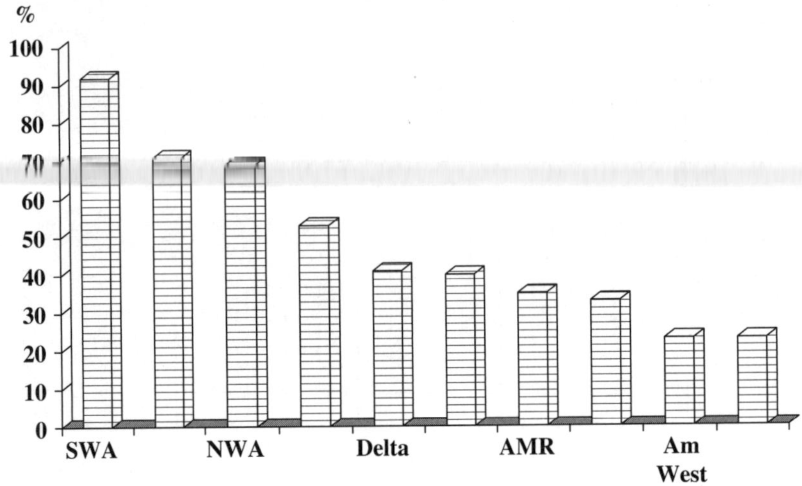

Note: Estimates are based on June 30, 2001 cash positions and daily cash operating expenses of each major airline.

Sources: CNNfn, Merrill Lynch & Co. and SEC filings.

Figure 12.4 Days of cash on hand preceding September 11th

financial reserves over time and minimize layoffs in times of crisis, relative to organizations with weaker relational and financial reserves.

US AIRWAYS VERSUS SOUTHWEST: A QUALITATIVE COMPARISON

In this section, we illustrate the above conceptual model by focusing on the contrast between Southwest and US Airways, given their similarity along several key dimensions, and their stark differences along other dimensions. Due to their focus on short-haul flights (Southwest's average flight length pre-9/11 was 481 miles and US Airways' was 576 miles, relative to an average of 781 miles for the major airlines), both airlines were more vulnerable than their competitors to the 'hassle factor' introduced by new airport security measures. The hassle factor is expected to affect disproportionately the recovery of short-haul travel because airport check-in procedures represent a higher percentage of total travel time for short-haul travel and because short-haul travel is more easily replaced by alternative forms of transportation such as trains, buses, and automobiles. In addition to anecdotal

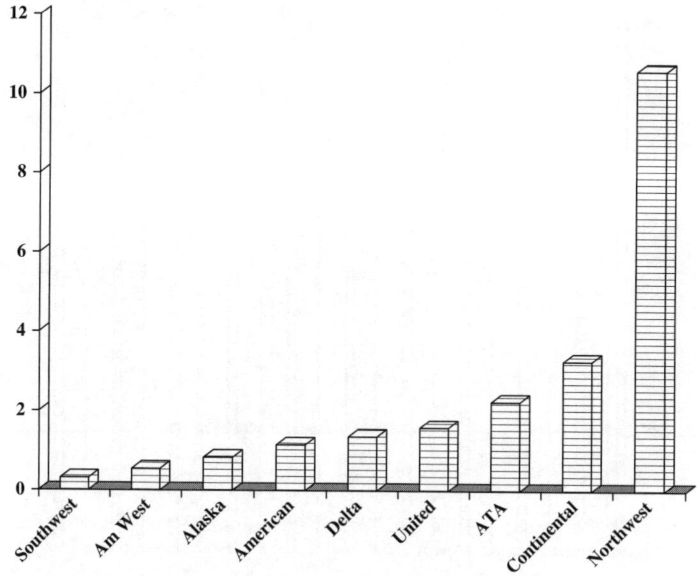

Note: US Airways was off the chart with a debt/equity ratio of 300.

Sources: Yahoo Finance and Thomson Financial.

Figure 12.5 Debt to equity ratios preceding September 11th

evidence (for example, Sharkey, 2004), data from the Federal Aviation Administration show that between December 2000 and December 2003, the number of short-haul flights dropped – that is, the number of flights shorter than 249 miles decreased by 20 percent, and flights between 250 and 499 miles dropped by 11 percent. Meanwhile, the number of long-haul flights increased – that is, the number of flights between 500 and 999 miles increased by 8 percent, and flights of 1000 miles or more increased by 1 percent. In addition, international routes proved to be more profitable than domestic routes during the post-September 11th period, due to less intense price competition. It might be expected, therefore, that airlines such as Southwest and US Airways that focused on short-haul, primarily domestic markets would experience the largest deterioration in demand and revenue in the period following the terrorist attacks, and would, therefore, have the hardest time recovering from the crisis of September 11th.

Moreover, Southwest and US Airways shared in common higher than average levels of unionization for the industry, with Southwest 89 percent unionized and US Airways 82 percent unionized pre-9/11, relative to an average of 69 percent for the major airlines as a whole. Unionization might

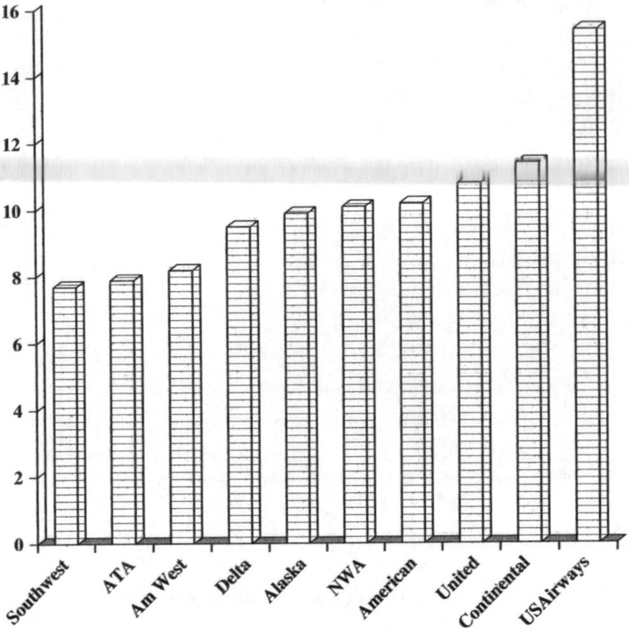

Note: Total operating costs (cents) per available seat mile.

Source: Form 41, US Department of Transportation.

Figure 12.6 Unit costs preceding September 11th

be expected to slow recovery, given the need to renegotiate contracts with one's unionized employees, or find legal grounds for abrogating those contracts.

The fact that Southwest recovered more quickly than any other major airline, while US Airways experienced one of the slowest recoveries, suggests a need to look at other factors driving recovery. We focus here on the factors that our previous analyses have shown to be important for driving recovery as they pertain to Southwest and US Airways – in particular, layoffs, the existence of financial and relational reserves, and the unit costs underlying their business models.

Layoffs as a Response to Crisis

US Airways' leaders conducted the highest level of layoffs in the industry, a 24 percent reduction compared to the industry average of 16 percent. This strategy was not surprising given the history of the leadership at US

Airways. When he became CEO in 1996, Stephen Wolf selected Rakesh Gangwal from United Airlines as president and began grooming him as his hand-picked successor (Meyer and Meyer, 2000). Wolf had a consistent and financially lucrative approach to managing airlines, though his approach did not appear to produce sustainable performance over time. As CEO of struggling Republic Airlines, Wolf threatened bankruptcy unless his pilots took a hefty pay cut. When they did, he restored it to profitability, then sold it to Northwest. He next took on Flying Tiger, a California-based cargo carrier on the verge of collapse, squeezing out labor concessions worth $50 million before selling the company to Federal Express. At United Airlines, he extracted $4.9 billion in wage and benefit concessions before the airline became employee owned, but labor demanded his departure as a condition of the buyout agreement.

In 2000 the *Denver Post* (May 28, p. 1.G) described the consequences of Wolf's strategy at United. At the same time United had achieved profitability under his leadership,

> There were no fond farewells from union leaders [at United] who had to negotiate contracts during the Wolf era. 'They hated him,' said Darryl Jenkins, an aviation professor at George Washington University. 'This is a person who everybody has an opinion on, and the opinions are always strong.' 'He managed through intimidation and fear,' said Ira Levy, a United employee and general chairman of the International Association of Machinists District Lodge 141 in Denver. 'I think there's still a lot of animosity toward him.'

Wolf's selection and tutelage of Gangwal at US Airways was, predictably, a perpetuation of the same strategic approach. Their expertise at tightening operations and extracting labor concessions was well known, so when they took the reins at US Airways, unions knew what to expect. 'Some people respected Wolf for having made United a champ. Most, however, were apprehensive. They knew Wolf's track record and they expected him to come after concessions with pliers in hand' (Meyer and Meyer, 2000, p. 247).

Gangwal became CEO in 1998, and when the tragedy of September 11th occurred, it is not surprising, in retrospect, that his leadership approach appeared to take advantage of the situation to accomplish goals he had not previously been able to accomplish (Barakat, 2001):

> Despite US Airways' huge losses, President Rakesh Gangwal said he is optimistic about the airline's future. Specifically, he said the September 11 attacks have allowed the airline to restructure and downsize in ways that would have been impossible otherwise. Specifically, the attacks allow the airline to invoke *force majeure* clauses in union contracts and eliminate unprofitable routes. *Force majeure* is the legal term for an uncontrollable event that releases a party from its contractual obligations.

> Gangwal said he expects the changes to be permanent. 'I don't want to take advantage of the situation, but we have to do what is right for the company,' Gangwal said in a conference call with analysts. 'And the events of September 11th have opened certain doors for the company that were pretty much closed before'.

Employees responded negatively to this apparent opportunism and disregard for human relationships on the part of US Airways' leadership, and their representatives filed a series of grievances against the airline related to its use of the *force majeure* clause. The head of the pilots' union noted, 'We've been saying all along that management has been using *force majeure* not as an opportunity to get through a crisis, but to take advantage of a crisis' (Barakat, 2001). Whether the actions taken by US Airways were legal or not, they are expected, based on our model, to do lasting damage to relational reserves as well as to undermine the credibility of its leadership. Indeed, US Airways' leadership was replaced in early 2002 due in part to its loss of credibility with employees as a result of its response to the crisis of September 11th.

The case of Southwest illustrates a different strategy for responding to the crisis. Southwest was determined to avoid layoffs altogether and couched its decision in terms of 'taking care of our people'. Traditional wisdom suggests that avoiding layoffs in the face of a dramatic decline in demand would jeopardize Southwest's short-term well-being. That is, investing in relationships by avoiding layoffs would put short-term survival at risk, as was articulated by the senior executives of US Airways. Indeed, Southwest was reportedly losing 'millions of dollars per day' (Trottman, 2001a) in the weeks following the terrorist attacks. Clearly Southwest could not afford to suffer losses of that magnitude indefinitely. Still, according to Southwest's CEO Jim Parker, 'we are willing to suffer some damage, even to our stock price, to protect the jobs of our people' (Conlin, 2001, p. 42). Southwest indicated a willingness to suffer these immediate losses in order to protect reserves and protect longer-term performance.

The result was that while other airlines shed both employees and unprofitable routes, Southwest maintained a steady presence in the wake of the attacks, refusing to lay off any of its employees. Instead, Southwest treated the crisis as an opportunity to increase its presence and expand the availability of its service to the flying public. According to an aviation consultant: 'They're doing what they do best, which is to shine in the hours of trouble' (Trottman, 2001b). Southwest used the crisis as an opportunity to strengthen rather than weaken employee relationships. Southwest had the most consistently positive employee relations of any airline in the industry, while at the same time being the most highly unionized of all the airlines (Gittell et al., 2004). As Southwest grew from an upstart to a major carrier, it continued to reach contractual agreements with its unions more quickly than any other airline (von Nordenflycht and Kochan, 2003). Just prior to 9/11, however, Southwest was embroiled in difficult labor negotiations. Southwest's no-layoff response to September 11th had the effect of reminding its employees of Southwest's tradition of caring. According to the president of the Transport Workers' Union local representing Southwest's ramp and operations employees (Trottman, 2001b, p. A1): 'What may have seemed like really big issues a month ago maybe aren't quite the big issues now . . . When it gets bad everywhere else, it's good here'.

Asked about Southwest's efforts to avoid layoffs in the wake of the September 11th attacks, a Southwest employee in the Office of Financial Analysis explained, 'It's part of our culture. We've always said we'll do whatever we can to take care of our people. So that's what we've tried to do' (Gittell, 2003, p. 242). Former Southwest's CEO Herb Kelleher explained his philosophy regarding layoffs in early 2001, before the crisis of September 11th hit:

> Nothing kills your company's culture like layoffs. Nobody has ever been furloughed [at Southwest], and that is unprecedented in the airline industry. It's been a huge strength of ours . . . We could have furloughed at various times and been more profitable, but I always thought that was short-sighted. You want to show your people that you value them and you're not going to hurt them just to get a little more money in the short term. (Brooker, 2001, p. 62)

> Not furloughing people breeds loyalty. It breeds a sense of security. It breeds a sense of trust. So in bad times you take care of them, and in good times they're thinking, perhaps, 'We've never lost our jobs. That's a pretty good reason to stick around'.

Whereas the views expressed by Southwest's leaders are consistent with the critical role of relational reserves in fostering organizational resilience, they contradict the prevailing prescriptions for competitive corporate strategy which advocate layoffs to protect the interests of the shareholders (Tichy, 1993). Consistent with a longstanding stream of thought by US economists regarding the dominance of shareholder rights over those of other stakeholders (Friedman, 1970; Jensen, 1989), prevailing prescriptions for competitive corporate strategy often treat the relationship between organizations and their employees as a contingent one. However, as argued by theories of high-performance work systems, job security is essential for sustained innovation and productivity, due to the high levels of trust and commitment such an approach tends to engender (for example, Kochan and Osterman, 1994). Indeed, Cappelli (1999, p. 22) observed that employers who moved toward a more contingent approach to employment 'were shocked by the collapse of employee morale' and often ended up backpedaling to regain the employee commitment without which it was difficult to operate. As even *Business Week* pointed out, there are practical benefits of a no-layoff approach in the face of crisis, namely 'fierce loyalty, higher productivity, and the innovation needed to enable them to snap back once the economy recovers' (Conlin, 2001, p. 42). These arguments are consistent with the negative association between layoffs and stock price recovery discussed earlier.

The Role of Financial Reserves

According to our model and supported by our quantitative analyses, it is not sufficient to want to avoid layoffs. One must be able to sustain a downturn without resorting to layoffs, and financial reserves are one critical factor that enables companies to do so. Avoiding personnel reductions requires that an organization be financially *able* to sustain short-term losses. In particular, two factors – cash on hand and debt load – are important contributors to resilience, or the extent to which a firm can withstand financial crisis. Organizations without sufficient financial reserves may be forced to break their commitments with employees and customers when faced with crisis simply because they cannot meet payroll. On the other hand, relational reserves can be significantly enhanced in the presence of financial slack.

Southwest's ability to resist layoffs can be attributed to its longstanding policy of maintaining low debt levels and an abundant supply of cash on hand. As people throughout the company have pointed out repeatedly over the years: 'At Southwest, we manage in good times as though we were in bad times' (Gittell, 2003, p. 242). Maintaining high levels of financial reserves had not been common business practice in the airline industry, however. For years, Southwest was the only single-A-rated airline company. Prior to September 11th, Kelleher explained Southwest's financial policy and how it enabled Southwest to thrive during past downturns (Brooker, 2001, p. 62):

> Most people think of us as this flamboyant airline, but we're really very conservative from the fiscal standpoint. We have the best balance sheet in the industry. We've always made sure that we never overreached ourselves. We never got dangerously in debt, and never let costs get out of hand. And that gave us a real edge during [the Gulf War crisis of 1990 to 1994].

Over time, Southwest Airlines made a conscious, strategic choice to maintain substantially greater reserves than was the norm in its industry. Southwest protected these reserves by sticking to its policy of gradual steady growth, despite the fact that there was sufficient demand for

Southwest's service to permit a far-faster rate of growth. According to John Denison, Southwest's Executive Vice President of Corporate Services (Gittell, 2003, p. 245):

> We promise the marketplace 10 percent growth, but we are only going to grow as fast as we can manage. Sometimes we have grown faster strategically. We acquired Morris Air in 1994 at the right time to compete. But we try to maintain the balance sheet. It is no accident that we are the only single-A-rated company in the industry.

US Airways, like other US airlines, had taken on high levels of debt over the years, responding to pressures from Wall Street. At the time of the 9/11 attacks, for example, the airline had incurred almost $8 billion in debt obligations and had lost money for eight straight quarters. Wall Street analysts complained publicly about the high costs of labor contracts at US Airways, and the airline's strategy for coping with these financial pressures had been to borrow more money from sources such as J.P. Morgan Chase ($71 million), Wilmington Trust ($50 million) and EDS ($47 million) (*CNN Money*, 2002). High debts levels coupled with high labor costs led to severely restricted financial flexibility when the 9/11 crisis occurred.

These results suggest that it is not merely the desire to invest in relational reserves by avoiding layoffs that accounts for resilience in the face of crisis. *Wanting* to avoid layoffs to preserve relationships is different from being *able* to do it. It is both the desire and ability to avoid damaging relational reserves that account for long-term resilience. Financial reserves, particularly in the form of low debt levels, serve as a supplementary coping resource for organizations by giving them room to maneuver in the face of crisis. Organizations can avoid relying on layoffs as the primary response to crisis if they have the necessary financial reserves.

These results are expected to be relevant beyond the airline industry as well. Indeed, financial theory indicates that interest payments create a form of financial risk that becomes greater when interest payments are higher and when there is variability of operating earnings (Freear, 1980). By implication, financial reserves play a more important role in fostering organizational resilience when the variability of operating earnings in the industry is greater.

If the lack of financial reserves makes an organization vulnerable to crisis and more dependent on using layoffs as a coping strategy – and, therefore, less resilient – why then are high levels of debt a common feature of so many companies? One answer is illustrated by the fact that Southwest has been criticized by Wall Street analysts for its policy of maintaining high levels of financial reserves. The business press reported:

> [Southwest's] conservative approach has been criticized by Wall Street analysts, who have argued that the airline should use its extra cash to make acquisitions or buy back stock. Goldman Sachs airline analyst, Glenn Engel, called the balance sheet 'too strong' [although] Engel allowed, 'this has meant that when times are tough, they have a lot more flexibility'. (Mount, 2002, p. 29)

The Role of a Viable Business Model and of Relationships for Achieving It

As suggested by our model and supported by our quantitative analyses, another reason why Southwest could avoid layoffs in the face of September 11th was its low operating costs. Low operating costs had become the *sine qua non* for a viable business model in the airline industry over the course of the 1990s. Consumer behavior shifted in the early 1990s toward greater price sensitivity, with consumers demanding lower costs and comparing airline travel more carefully with other modes of transportation and communication. After a low point in the early 1990s, passenger willingness to pay rebounded somewhat after 1994. But from 2000 to 2001 (prior to 9/11), revenue per passenger mile dropped by more than 10 percent. The *Wall Street Journal* featured two prominent articles reflecting the growing consensus that these changes in customer willingness to pay, as in the case of the retail industry, reflected the new reality for the airline industry (Brannigan et al., 2001; Trottman and McCartney, 2002). 'Anyone who has a modicum of Internet capability and wants to take what is now a modest amount of time can very rapidly find out and comparison shop', said Leo Mullin, CEO of Delta Airlines. 'There is almost perfect information out there' (Brannigan et al., 2001, p. A1).

While most major airlines had operating costs in the range of 9–11 cents per available seat mile prior to 9/11, Southwest's were just 7.7 cents per seat mile. According to Merrill Lynch analyst Michael Linenberg, 'They tend to have some of the lowest costs in the industry, so in times of depressed business, they can make money while others are losing money' (*USA Today*, 2002). US Airways, by contrast, faced the crisis of September 11th with the highest operating costs of any major airline in the US industry, at 15.4 cents per seat mile.

The key to low unit costs in this industry is productivity of the airline's most costly assets – employees and aircraft. Southwest's strategy for achieving low costs focused over the years on increasing productivity, rather than lowering wages and benefits. In his 1995 Message to the Field, CEO Herb Kelleher said, 'We want to reduce all of our costs, except our wages and benefits and profit-sharing. This is Southwest's way of competing, unlike others who lower their wages and benefits' (Gittell, 2003, p. 11). Southwest achieved the highest levels of productivity for both employees and aircraft, based on its legendary levels of teamwork and 'relational coordination', enabling employees to turn aircraft quickly at the gate, thus maximizing the time that aircraft are in the air, earning revenue. The role of positive working relationships for achieving these outcomes is supported by the analyses conducted above, but was also established in a study that compared Southwest in a more detailed way to some of its major competitors (Gittell, 2003). This study showed that relationships of shared goals, shared knowledge and mutual respect support high levels of coordination among employees, and explain much of the variation in employee and aircraft productivity between Southwest and its competitors.

Starting in the 1990s, US Airways also made concerted efforts to achieve low unit costs. In the case of US Airways, however, there were two distinct strategies – improving productivity, and reducing wages and benefits. After losing most of the intra-California market to Southwest in 1992, US Airways endured a strike with its mechanics, resulting in lower wages than several of its key competitors, including Southwest. After Southwest's invasion of its Baltimore hub in 1993, US Airways tried a second strategy for reducing unit costs – setting up its own quick turnaround operation under the name 'Project High Ground' in 1994. Project High Ground was designed to 'speed the way [USAir] services and turns around planes at airports. Borrowing a page from Southwest, USAir is trying to halve the time its planes spend on the ground' (McCarthy and O'Brian, 1994, p. A1). According to US Airways' Boston station manager, 'On our regular flights, we get 40 to 45 minutes to turn the plane around. Under High Ground, we do it in 20 to 25 minutes for some and 30 to 35 minutes for others, depending on the routing' (Shuckman, 1994).

US Airways announced in spring 1994 that it would expand its new business model from 22 to 100 aircraft. But some industry observers expected US Airways to fail unless pay and benefits were cut further (McCarthy and O'Brian, 1994). Strong pressures from investors for pay cuts and strong stands from union leaders against them put top management in a difficult position. Moody's lowered US Airways' bond ratings, citing 'concern that the company's recent operating changes designed to reduce its cost structure will not be sufficient to offset the anticipated revenue losses from price competition' (Feldman, 1994, p. 24). The pilots' union argued that the airline's costs were too high, not because of labor, but because it was not deploying its assets wisely. A June 1994 proposal to the union to save $175 million through management and staff reductions and by subcontracting mail and freight operations was met by a pilot counterproposal for an employee buyout with board representation. A year later, after stalled negotiations, the airline's fifth crash in five years, a perceived threat of bankruptcy, and threats by management to shrink the airline, the parties reached agreement, but the animosity and ill will continued.

In 1997, still suffering from the animosity of negotiations over pay cuts, US Airways sought again to reduce unit costs by raising employee and aircraft productivity, launching a new low cost business model called MetroJet (Carey, 1998). Two dozen volunteers were recruited who wanted to take part in a 'new kind of airline'. The team consisted of representatives from all the major employee groups and spent six months designing MetroJet from the ground up. At the same time, however, US Airways' leadership continued to extract pay concessions from employees. Labor strife continued throughout the rest of the 1990s, culminating in 2000 with a work action threatened by flight attendants, who were fed up after four years of failing to secure a contract for themselves. The bold experiment to build a new and viable business model for US Airways never fully succeeded, due in large part to poor working relationships, and was terminated by Gangwal in the aftermath of September 11th.

Aftermath

As early as February 2002, Southwest announced plans to hire about 4000 new employees, drawing in part from the employees laid off from other airlines. Its ability to hire new employees was even further strengthened by the fact that it did not conduct layoffs in the downturn after 9/11. US Airways, by contrast, continued to rely on layoffs as a strategy of recovery. In 2002, the company announced that an additional 471 pilots would be let go, adding to a previously announced layoff of 286 pilots after 9/11. Moreover, an additional 915 flight attendants were furloughed, making a total reduction of 3675 flight attendants out of 10,000 that were employed before the 9/11 tragedy (*People's Daily*, 2002). US Airways' management gained deep concessions from its employees through bankruptcy bargaining and thus was able to achieve through threat and fear the cost reductions that it had been unable to achieve otherwise. As a result of bankruptcy bargaining, costs went from a high of 15.4 cents per seat mile just prior to 9/11 (14.4 cents per seat mile for all of 2001), to 13.9 cents per seat mile in 2005, while labor costs dropped from 5.6 to 2.8 cents per seat mile, suggesting that the decrease in overall costs was entirely due to labor cost reduction.

Four years after 9/11, US Airways emerged from bankruptcy as part of a merger with a stronger rival, America West, joining together the two carriers that were seventh and eighth in size with complementary route networks, one centered on the East Coast and the other centered on the West Coast. America West had recovered from 9/11 more slowly than Southwest, having laid off 16 percent of its workforce, but recovered far more quickly than US Airways, having had lower unit costs and lower debt levels going into the crisis. To rid US Airways of its legacy of embattled leadership, Doug Parker, CEO of America West, took over as the CEO of the merged company and moved its headquarters to Tempe, Arizona, the home of America West. The new leadership recognized the image problems related to keeping the US Airways' name, but also signaled its desire to focus on relationships as a way to build a more successful company. According to an America West spokesman: 'We feel if we run a great airline, if we take care of the employees and they take care of the customers, the name will take care of itself' (Isidore, 2005).

Recognizing the challenge of integrating these two airlines with their very distinct cultures, CEO Parker appointed Larry LeSueur, a longtime America West employee, as vice president of culture (Nichols, 2005). According to LeSueur: 'It's not my job just to go around the system and have barbecues. You can't just say, "here's a hamburger, have a nice day". The message is "we do expect you to treat customers well and we, as a leadership team, will do the same for you"'. LeSueur recognized that US Airways' workers had lost trust in management and planned to 'spend time listening to their pent-up frustration and anger'. As for Parker, he visited employees in each of the merged airline's biggest cities, saying 'It's important that they understand what the people that are running the place are like, and get to know us a little better. So far, I think, it's been going well' (Nichols, 2005, p. D1). Considering the other key sources of resilience, the combined unit costs of the merged carrier were still high at 11.1 cents per seat mile in 2006, versus 8.7 cents for Southwest, and the debt/equity ratio was still high, but both indicators were much healthier than they were for US Airways going into September 11th. Customer service remained a serious concern, however, with customer complaints and baggage mishandling nearly the worst in the industry in 2006. Altogether, the jury was out as to whether the new leadership could credibly repair the serious damage done to relationships at US Airways both pre- and post-9/11.

CONCLUSION

The contrasting stories of Southwest Airlines and US Airways illustrate the model of resilience we tested in this chapter. A viable low-cost business model achieved through long-term positive employee relations, along with adequate financial reserves, enabled Southwest to return quickly to business as usual, following the crisis of 9/11. These cases illustrate how the development of relational reserves enables organizations to establish viable business models and build up financial reserves over time, thus minimizing the need for layoffs in times of crisis, and reveal that the failure to build relational reserves leads to very different results.

In the normal pattern of organizational behavior, an organization's leadership responds to financial crises with layoffs and cutbacks. The organization's performance subsequently suffers because of the resulting deterioration in relationships. The relationships that could serve as a collective coping mechanism in the face of adversity are, instead, weakened by layoffs. This scenario represents a dilemma for organizations, in which measures taken for short-term survival appear to undermine the conditions for longer-term success.

However, some organizations respond differently to crisis, accepting the short-term costs of excess staffing levels in order to maintain positive human relationships in the face of adversity. By avoiding layoffs, these organizations maintain or even strengthen human relationships, creating coping resources that enable organizational members to respond cohesively to the crisis in innovative ways. As a result, the deterioration of organizational performance caused by the crisis is ameliorated. Furthermore, once the immediate crisis has passed, organizational performance can return more quickly to pre-crisis performance levels due to the maintenance of relationships during the period of the crisis. To avoid layoffs, however, organizations must be financially able to do so. Financial reserves and viable business models thus play a significant role in minimizing layoffs and in sustaining the relationships that enable organizations to return more quickly to pre-crisis performance levels. Moreover, the achievement of a viable business model over time is itself enabled by positive working relationships, as in the case of Southwest Airlines, and prevented by their absence, as has thus far been the case for US Airways. While viable business models play an important role in minimizing layoffs, the high levels of productivity and low unit costs underlying these models are made possible in part by the careful nurturing of relational reserves over the years. Financial reserves contribute further to the ability to minimize layoffs in the face of crisis. The results of our investigation of US airline companies provide support for this model.

The model developed here combines insights from the literature on high-performance work systems and mutual gains enterprise (for example, Kochan and Osterman, 1994), with insights from the literature on high reliability work organizations and relational models of resilience (for example, Weick et al., 1999), and with insights from positive relationships in organizations (for example, Dutton and Ragins, 2006). The literature on high-performance work systems and mutual gains enterprise has long argued for the role that employee commitment plays in achieving productivity and other outcomes, and for the role that employment security plays in achieving employee commitment. This literature has further highlighted pressures from financial stakeholders as working to undermine employment security. The literature on high reliability work organizations and relational models of resilience has highlighted the importance of relationships and financial reserves for achieving resilience in the face of environmental jolts. The literature on positive social relationships uncovers salutary individual and organizational effects of strong relational reserves.

The most powerful implication of our model is that relational and financial reserves tend to be mutually reinforcing, forming a virtuous cycle that contributes to resiliency. Positive relationships tend to produce lower costs and lower debt levels over time, making it easier to sustain external shocks without breaking commitments, thus further strengthening relationships and performance. Likewise, negative relationships tend to produce high costs and high debt levels over time, making it harder to sustain external shocks without breaking commitments, thus further weakening relationships and performance. This vicious cycle can be replaced by a virtuous cycle of resilience when managers and other key stakeholders commit themselves to building relational and financial reserves over time.

REFERENCES

Aiello, J.R. and Kolb, K.J. (1995), 'Electronic performance monitoring and social context: impact on productivity and stress', *Journal of Applied Psychology*, **80**(3), 339–53.

Barakat, M. (2001), 'US Airways loses $766 million in third quarter, worse than analysts expected', *Arlington Journal*, October 31; http://www.co.arlington.va.us/NewsDigest/Scripts/ViewDetail.asp?Index=593, retrieved September 20, 2002.

Bennett, A. (1991), 'Downsizing doesn't necessarily bring an upswing in corporate profitability', *Wall Street Journal*, June 6, B1, B4.

Brannigan, M., Carey, S. and McCartney, S. (2001), 'First class mutiny: fed up with airlines, business travelers start to fight back', *Wall Street Journal*, August 28, A1.

Brooker, K. (2001), 'The chairman of the board looks back', *Fortune*, May 28, 62.

Bunderson, J.S. and Sutcliffe, K.M. (2002), 'Comparing alternative conceptualizations of functional diversity in management teams: process and performance effects', *Academy of Management Journal*, **45**, 875–93.

Cameron, K.S. (1994), 'Strategies for successful organizational downsizing', *Human Resource Management Journal*, **33**, 189–212.

Cameron, K.S. (1998), 'Strategic organizational downsizing: an extreme case', *Research in Organizational Behavior*, **20**, 185–229.

Cameron, K.S., Freeman, S.J. and Mishra, A.K. (1991), 'Best practices in white collar downsizing: managing contradictions', *Academy of Management Executive*, **5**, 57–73.

Cameron, K.S., Kim, M.U. and Whetten, D.A. (1987), 'Organizational effects of decline and turbulence', *Administrative Science Quarterly*, **32**, 222–40.

Cameron, K.S., Bright, D.S. and Caza, A. (2004), 'Exploring the relationships between virtuousness and organizational performance', *American Behavioral Scientist*, **47**, 766–90.

Cappelli, P. (1999), *The New Deal at Work: Managing the Market-driven Workforce*, Boston, MA: Harvard Business School Press.

Carey, S. (1998), 'US air "Peon" team pilots start-up of company's low-fare airline', *Wall Street Journal*, March 24, B1.

Carver, C.S., Scheier, M.F. and Weintraub, J.K. (1989), 'Assessing coping strategies: a theoretical based approach', *Journal of Personality and Social Psychology*, **56**, 267–83.

Cascio, W.F., Young, C.E. and Morris, J. (1997), 'Financial consequences of employment change decisions in major U.S. corporations', *Academy of Management Journal*, **40**, 1175–89.

CNN Money (2002), 'US Airways files for bankruptcy', August 12, http://money.cnn.com/2002/08/11/news/companies/usair.

Cole, R.E. (1993), 'Learning from learning theory: implications for quality improvement in turnover, use of contingent workers, and job rotation policies', *Quality Management Journal*, **1**, 9–25.

Conlin, M. (2001), 'Where layoffs are a last resort', *Business Week*, October 8, 42.

Denver Post (2000), 'Histories of United, Stephen Wolf intertwined', May 28, p. 1.G.

DeWitt, R.L. (1993), 'The structural consequences of downsizing', *Organization Science*, **4**, 30–40.

Dutton, J.E. and Ragins, B.R. (2006), *Exploring Positive Relationships at Work: Building a Theoretical and Research Foundation*, Mahwah, NJ: Lawrence Erlbaum.

Edmondson, A.C. (1999), 'Psychological safety and learning behavior in work teams', *Administrative Science Quarterly*, **44**, 350–83.

Feldman, J. (1994), 'USAir: where have all the good times gone?', *Air Transport World*, June, 24.

Freear, J. (1980), *The Management of Business Finance*, London: Prentice-Hall.

Friedman, M. (1970), 'The social responsibility of business is to increase its profits', *New York Times Magazine*, September 13.

Gittell, J.H. (2003), *The Southwest Airlines Way: Using the Power of Relationships to Achieve High Performance*, New York: McGraw-Hill.

Gittell, J.H., Cameron, K., Lim, S. and Rivas, V. (2006), 'Relationships, layoffs and organizational resilience', *Journal of Applied Behavioral Science*, **42**(3), 300–329.

Gittell, J.H., von Nordenflycht, A. and Kochan, T.A. (2004), 'Mutual gains or zero sum? Labor relations and firm performance in the airline industry', *Industrial and Labor Relations Review*, **57**(2), 163–79.

Gladwell, M. (2002), *The Tipping Point: How Little Things Can Make a Big Difference*, Boston, MA: Back Bay.

Henkoff, R. (1990), 'Cost cutting: how to do it right', *Fortune*, April 9, 17–19.

Isidore, C. (2005), 'Should the US Air name keep flying?', *CNNMoney*, May 20; http://money.cnn.com/2005/05/20/news/fortune500/usair_americawest_name/in dex.htm, retrieved May 20.

Jensen, M. (1989), 'The eclipse of the public corporation', *Harvard Business Review*, **67**(5), 61–75.

Kochan, T.A. and Osterman, P. (1994), *The Mutual Gains Enterprise: Forging a Winning Partnership Among Labor, Management and Government*, Boston, MA: Harvard Business School Press.

Luthans, F. (2002), 'Positive organizational behavior: developing and managing psychological strengths for performance improvement', *Academy of Management Executive*, **16**, 57–75.

Masten, A.S. and Reed, M. (2002), 'Resilience in development', in C.R. Synder and S.J. Lopez (eds), *Handbook of Positive Psychology*, Oxford: Oxford University Press, pp. 74–88.

McCarthy, M. and O'Brian, B. (1994), 'Lean, nimble airlines head east, targeting region's plump prices', *Wall Street Journal*, February 28, A1.

McKinley, W., Sanchez, C.A. and Schick, A.G. (1995), 'Organizational downsizing: constraining, cloning, learning', working paper, Southern Illinois University Department of Management.

Meyer, A.D. (1982), 'Adapting to environmental jolts', *Administrative Science Quarterly*, **27**, 515–37.

Meyer, G.C. and Meyer, S. (2000), *Dealers, Healers, Brutes, and Saviors*, New York: Wiley.

Mishra, A.K. and Mishra, K.E. (1994), 'The role of mutual trust in effective downsizing strategies', *Human Resource Management*, **33**(2), 261–79.

Mount, I. (2002), 'Southwest's Gary Kelly: a tip of the hat to the CFO at the one airline still making money', *Business 2.0*, March, 29.

Moyle, P. and Parkes, K. (1999), 'The effects of transition stress: a relocation study', *Journal of Organizational Behavior*, **20**(5), 625–46.

Murphy, K. (2001), Presentation at MIT Sloan School of Management, Cambridge, MA, October.

Nichols, J. (2005), 'New US Airways making sure disparate cultures will fly as one', *Arizona Republic*, October 2, D1.

People's Daily (2002), 'US Airways plans to lay off more pilots', October 28, p. 1.

Ryff, C.D. and Singer, B.H. (2001), *Emotion, Social Relationships and Health*, New York: Oxford University Press.

Schaubroeck, J. and Fink, L.S. (1998), 'Facilitating and inhibiting effects of job control and social support on stress outcomes and role behavior: a contingency model', *Journal of Organizational Behavior*, **19**(2), 167–95.

Seligman, M.E.P. (2002), 'Positive psychology, positive prevention and positive therapy', in C. Snyder and S.J. Lopez (eds), *Handbook of Positive Psychology*, New York: Oxford University Press, pp. 3–9.

Sharkey, J. (2004), 'Fewer are flying on short routes', *New York Times*, February 17, C9.

Shuckman, G. (1994), Interview at Logan Airport, Boston, MA.

Sitkin, S.B. (1992), 'Learning through failure: the strategy of small losses', in B.M. Staw and L.L. Cummings (eds), *Research in Organizational Behavior*, vol. 14, Greenwich, CT: JAI Press, pp. 231–66.

Spreitzer, G.S., Sutcliffe, K., Dutton, J., Sonenshein, S. and Grant, A. (2005), 'Enabling human thriving at work', *Organization Science*, **16**, 537–49.

Staw, B., Sandelands, L. and Dutton, J. (1981), 'Threat-rigidity effects in organizational behavior: a multi-level analysis', *Administrative Science Quarterly*, **26**, 501–24.

Sutcliffe, K.M. and Vogus, T. (2003), 'Organizing for resilience', in K.S. Cameron, J.E. Dutton and R.E. Quinn (eds), *Positive Organizational Scholarship: Foundations of a New Discipline*, San Francisco, CA: Berrett-Koehler, pp. 94–110.

Tichy, N.M. (1993), *Control Your Destiny or Someone Else Will*, New York: Doubleday.

Trottman, M. (2001a), 'Southwest Airlines considers cutbacks, halts delivery of planes from Boeing', *Wall Street Journal*, September 21; http://online.wsj.com/article/SB1001038818285461 20.html, retrieved February 9, 2006.

Trottman, M. (2001b), 'Up in the air: amid crippled rivals, Southwest again tries to spread its wings', *Wall Street Journal*, October 11, A1.

Trottman, M. and McCartney, S. (2002), 'The age of Wal-Mart airlines crunches the biggest carriers', *Wall Street Journal*, June 18, B1.

USA Today (2002), 'Southwest to add 4,000 jobs this year', February 18; http://www.usatoday.com/money/biztravel/2002-02-18-southwest-hires.htm, retrieved 13 February, 2006.

Von Nordenflycht, A. and Kochan, T.A. (2003), 'Labor contract negotiations in the airline industry', *Monthly Labor Review*, **126**(7), 18–28.

Weick, K.E. (2003), 'Positive organizing and organizational tragedy', in K.S. Cameron, J.E. Dutton and R.E. Quinn (eds), *Positive Organizational Scholarship: Foundations of a New Discipline*, San Francisco, CA: Berrett-Koehler, pp. 66–80.

Weick, K.E., Sutcliffe, K.M. and Obstfeld, D. (1999), 'Organizing for high reliability: processes of collective mindfulness', in R. Sutton and B.M. Staw (eds), *Research in Organizational Behavior*, vol. 21, Greenwich, CT: JAI Press, pp. 81–124.

Wildavsky, A. (1988), *Searching for Safety*, New Brunswick, NJ: Transaction Books.

Worline, M.C., Dutton, J.E., Frost, P.J., Janov, J., Lilius, J. and Maitlis, S. (2004), 'Creating fertile soil: the organizing dynamics of resilience', working paper, University of Michigan School of Business.

Index